Advances in Interdisciplinary Applied Discrete Mathematics

INTERDISCIPLINARY MATHEMATICAL SCIENCES

Published

Vol. 1: Global Attractors of Nonautonomous Dissipative Dynamical Systems
David N. Cheban

Vol. 2: Stochastic Differential Equations: Theory and Applications
A Volume in Honor of Professor Boris L. Rozovskii
eds. Peter H. Baxendale & Sergey V. Lototsky

Vol. 3: Amplitude Equations for Stochastic Partial Differential Equations
Dirk Blömker

Vol. 4: Mathematical Theory of Adaptive Control
Vladimir G. Sragovich

Vol. 5: The Hilbert–Huang Transform and Its Applications
Norden E. Huang & Samuel S. P. Shen

Vol. 6: Meshfree Approximation Methods with MATLAB
Gregory E. Fasshauer

Vol. 7: Variational Methods for Strongly Indefinite Problems
Yanheng Ding

Vol. 8: Recent Development in Stochastic Dynamics and Stochastic Analysis
eds. Jinqiao Duan, Shunlong Luo & Caishi Wang

Vol. 9: Perspectives in Mathematical Sciences
eds. Yisong Yang, Xinchu Fu & Jinqiao Duan

Vol. 10: Ordinal and Relational Clustering (with CD-ROM)
Melvin F. Janowitz

Vol. 11: Advances in Interdisciplinary Applied Discrete Mathematics
eds. Hemanshu Kaul & Henry Martyn Mulder

Interdisciplinary Mathematical Sciences – Vol. 11

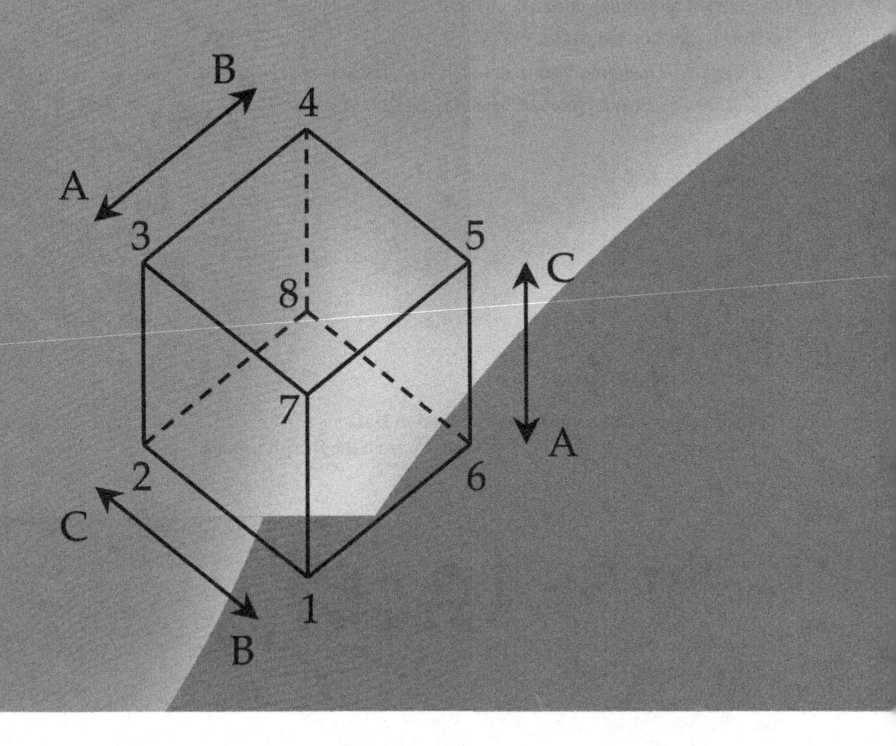

Advances in Interdisciplinary Applied Discrete Mathematics

Editors

Hemanshu Kaul
Illinois Institute of Technology, USA

Henry Martyn Mulder
Erasmus University Rotterdam, The Netherlands

NEW JERSEY • LONDON • SINGAPORE • BEIJING • SHANGHAI • HONG KONG • TAIPEI • CHENNAI

Published by

World Scientific Publishing Co. Pte. Ltd.
5 Toh Tuck Link, Singapore 596224
USA office: 27 Warren Street, Suite 401-402, Hackensack, NJ 07601
UK office: 57 Shelton Street, Covent Garden, London WC2H 9HE

British Library Cataloguing-in-Publication Data
A catalogue record for this book is available from the British Library.

ADVANCES IN INTERDISCIPLINARY APPLIED DISCRETE MATHEMATICS
Interdisciplinary Mathematical Sciences — Vol. 11

ISBN-13 978-981-4299-14-5
ISBN-10 981-4299-14-6

Printed in Singapore by World Scientific Printers.

To Buck McMorris

Preface

End of August 2008 F.R. McMorris, Buck for his friends and colleagues, retired as Dean of the College of Science and Letters at IIT, Illinois Institute of Technology, Chicago. He also celebrated his 65-th birthday. To commemorate these events a two-day conference was held early May 2008 at IIT. In addition this volume is written in honor of his contributions to mathematics and its applications. The focus of the volume is on areas to which he contributed most. The chapters show the broadness of his interests and his influence on many co-authors and other mathematicians. It has been a pleasure for all the contributors to this volume to participate in this project. Not only to honor our esteemed colleague and good friend, but also because of the fascinating mathematics and applications represented in this volume. It is impossible to touch all topics in the areas of focus in one book, let alone to attempt a survey the whole field. So choices had to be made, but we hope that the volume at hand will invoke the interest of the reader for these topics that are so dear to the authors, and that this will incite further interest for the rich literature, of which the various chapters give a small but representative selection.

On the cover of this book one finds a figure with a 3-cube depicting a *voting* situation. It is chosen from Chapter 10. Voting is way to reach *consensus*, which is discussed in Chapter 7. A special case of reaching consensus is finding an optimal location, *location functions* being the topic of Chapter 4. Here the 3-cube also occurs in a figure as a crucial example. Within location theory *centrality* concepts are a main point of focus, see Chapters 6 and 8. A major centrality concept is that of *median*, and median graphs are discussed in depth in Chapter 5. Here again the 3-cube occurs in a figure elucidating an important result. Consensus plays a major role in *mathematical biology*, see Chapter 1, and in various other areas, such as *psychology*, see Chapter 2. Problems in molecular biology involve *intersection graphs*, see Chapter 3. We find applications of location theory to economics in Chapter 5. Another problem of optimization in economics can be found in Chapter 9. All of the above problem areas appear in one way or another in the survey of Buck McMorris's work in Chapter 11. Thus the figure on the cover symbolizes the unity underlying the various themes discussed in this book.

Hemanshu Kaul, Henry Martyn Mulder

Contents

Preface vii

Introduction 1
H. Kaul and H.M. Mulder

1. Contributions of F.R. McMorris to Character Compatibility Analysis 5
G.F. Estabrook

2. Families of Relational Statistics for 2×2 Tables 25
W.J. Heiser and M.J. Warrens

3. Applications of Spanning Subgraphs of Intersection Graphs 53
Terry A. McKee

4. Axiomatic Characterization of Location Functions 71
F.R. McMorris, H.M. Mulder and R.V. Vohra

5. Median Graphs. A Structure Theory 93
H.M. Mulder

6. Generalized Centrality in Trees 127
M.J. Pelsmajer and K.B. Reid

7. Consensus Centered at Majority Rule 149
R.C. Powers

8. Centrality Measures in Trees 167

K.B. Reid

9. The Port Reopening Scheduling Problem 199

F.S. Roberts

10. Reexamining the Complexities in Consensus Theory 211

D.G. Saari

11. The Contributions of F.R. McMorris to Discrete Mathematics and its Applications 225

G.F. Estabrook, T.A. McKee, H.M. Mulder, R.C. Powers and F.S. Roberts

Author Index 243

Subject Index 249

Symbol Index 257

Introduction

Hemanshu Kaul, Henry Martyn Mulder
Department of Applied Mathematics, Illinois Institute of Technology
Chicago, IL, USA
kaul@iit.edu

Econometrisch Instituut, Erasmus Universiteit
P.O. Box 1738, 1066NK Rotterdam, Netherlands
hmmulder@ese.eur.nl

One of the fascinating features of mathematics is that the same ideas, concepts, techniques, models, and structures find applications in diverse disciplines. Or that results developed in one application find unexpected applications elsewhere. Within mathematics itself one finds a similar phenomenon: the same structure occurs with different guises in various mathematical problem areas.

This book focuses on these aspects of some discrete structures. In the past sixty years, these structures have appeared in such diverse areas as consensus theory, voting theory, optimization, location theory, clustering, classification, representation, and other areas of discrete mathematics. The ideas, techniques and concepts discussed here have found applications in different disciplines as Biology, Psychology, Economics, Operations Research, Social Choice, Physics, and Chemistry.

A whole series of books could have been written on the topics in book. The aim of a single volume necessarily has to be quite modest. What we would like to achieve is to raise the interest of the reader for the discrete structures and techniques discussed here and also for the many applications and future possibilities of the ideas presented here. We hope that graduate students and researchers from one area will get acquainted with other areas and be able to use these new ideas towards a interdisciplinary approach to Discrete Applied Mathematics.

The design of the book is rather loose: we have asked twelve authors to contribute a chapter on a topic of their own choice within the areas described above. This has resulted in ten chapters that cover a wide range, from applications in biology, economics, and logistics to discussions of areas in discrete mathematics such as centrality in trees, intersection graphs, and median graphs, from surveys on specific topics such as 2×2 tables, the majority rule, and location functions to discussions of future possibilities, but also new results such as generalized centrality in trees and axiomatic characterization of the median function on cube-free networks. The idea of consensus in various forms reappears in several chapters, sometimes in the guise of voting procedures or location problems. On closer inspection the common features may become more apparent.

Another way by which the connections of the chapters in this book are defined is that all authors have collaborated more or less closely with F.R. 'Buck' McMorris. This may have been as co-author, as member of the same PhD-committee, or as having a major common interest in some problems or theories. When we look at the topics in the chapters of this book, then these are precisely the ones to which McMorris has dedicated his mathematical career. The idea for this book arose at the occasion of his retirement as dean of the College of Science and Letters at the Illinois Institute of Technology, IIT, in Chicago. To highlight this unifying idea, a chapter is added at the end of the book discussing the contributions of McMorris to discrete mathematics and its applications.

In the remaining part of the introduction, we will give a short overview of the individual chapters so as to help orient the reader in the technical landscape of the research topics covered therein.

A phylogenetic tree in evolutionary biology is a tree showing the inferred evolutionary relationships among various biological species based upon similarities and differences in their genetic traits. Evolutionary systematics uses the similarities and differences that can be observed among the species in a group under study to estimate their ancestor relation. Such a basis for comparison is called a character, and the groups that result are called its character states. The character states are arranged into a character state tree to indicate where speciation events associated with the observed changes are hypothesized to have occurred. These hypotheses are expressed as an ancestor relation diagram in which the character states play the role of individual species, and a change from one state to another represents a speciation event on the phylogenetic tree. By mid 20th century, some natural scientists realized that some pairs of such hypotheses based on different bases for comparison could be logically incompatible, i.e., they could not both be true, and they began to develop tests for, and ways to resolve, incompatibilities to estimate the ancestor relation from these hypotheses. In Chapter 1, G.F. Estabrook reviews the basic concepts of character compatibility analysis, gives a survey of related work, especially McMorris' contributions to character compatibility analyses, reviews some of their applications, and presents ideas for future applications.

In biometrics, psychology, ecology, and various other fields of science, data is often summarized in a 2×2 table. In general, such tables arise when two objects/outcomes are compared on the basis of the presence/absence of a set of attributes. For example, in ecology two species could be compared on the basis of genetic encodings, in psychology and biometrics the table could store the dichotomous response of two observers, in epidemiology it could be the results of a clinical trial of two variants of a vaccine, and so on. In many applications, there is a need to summarize such data by a single relational statistic. In Chapter 2, W.J. Heiser and M.J. Warrens give a broad overview of such statistics, and how the different statistics may be interpreted in the contexts of various other statistics. Their discussion should be of interest to practitioners in deciding what statistic to use, and to theoreticians for studying properties of these statistics.

Graphs are ubiquitous in modern applied mathematics since they can model any binary (pairwise) relationship. For example, protein interaction graphs, where the two proteins are related if they interact in some specific biochemical process. In Chapter 3, T.A. McKee discusses intersection graphs, a paradigm for deriving graphs as intersection of a family of mathematical objects: vertices correspond to the objects and edges corresponds to pairwise intersecting objects. In particular, McKee surveys results related to spanning subgraphs of such intersection graphs and their new applications in computational biology and in combinatorial probability. Certain kinds of spanning trees of the protein interaction graph (represented as a particular intersection graph) are applied to show how proteins enter and leave cellular processes. At the other end of the spectrum of applications, construction of spanning subgraphs for intersection graph representations of a family of sets are applied to generalize the Inclusion-Exclusion formula or Bonferroni-type probabilistic inequalities when cardinalities of only certain kinds of intersections of the sets are known.

A common problem in any service-related organization is the decision on where to locate their facilities, such as shipping centers, shopping malls, fire stations, elementary schools, etc., so that they can optimally serve those who benefit from them while minimizing costs. The location of such facilities is discussed in context of certain networks - transportation, communication, etc. - that connect populations centers, manufacturing sites, etc. In Chapter 4, F.R. McMorris, H.M. Mulder and R.V. Vohra discuss the consensus theory, as opposed to optimization, approach to solving this problem. A consensus function takes the potential locations as input and outputs are the locations that satisfy the optimality criterion. Such a "rational" function is built using a list of intuitively natural axioms. McMorris, Mulder and Vohra survey three popular location functions - center, mean, and median, and also include some new results on the median location function.

The graph structure underlying the discrete networks in the location theory above have the property that for any three locations there must be a unique location that lies on a shortest path between any pair of the three locations. Such graphs are called median graphs. These graphs include the often-used graphs in applications - trees, hypercubes, and grids. In Chapter 5, H.M. Mulder surveys the rich structural theory of median graphs and median-type structures, and their applications in Location Theory, Consensus functions, Chemistry, Biology and Psychology, Literary history, Economics and Voting Theory.

The three locations functions discussed in Chapter 4 give a consensus-theoretic generalization of a notion of "center" of a tree based on three distinct norms used for measuring distance. In Chapter 6, M.J. Pelsmajer and K.B. Reid introduce three families of central substructures of trees that generalize these notions of "centrality" - center, centroid, and median - in a tree in a graph-theoretic sense. Their new results give a theoretical framework for each of these new concepts which naturally generalize the previous classical results for center, centroid, and median of a tree as well as their other generalizations, and algorithms for finding these substructures in trees. In the closely related Chapter 8, K.B. Reid gives a thorough survey of various concepts that have been defined and studied as a measure of "central" substructure in a tree. This survey by Reid can be read as a prologue to the discussion in Chapter 6.

Consensus theory, last discussed in Chapter 4 in the context of location theory, is also a natural framework for studying voting systems. In 1952, Kenneth May showed that the simple majority rule is the only two-candidate election procedure in which each candidate is treated equally, each voter is treated equally, and a candidate is never hurt by receiving more votes - three very natural axioms or rules for consensus. In Chapter 7, R.C. Powers discusses various generalizations of the majority rule, as based on various natural axioms that should be satisfied by the consensus function representing the voting procedure, that have been studied in the sixty years since May's seminal result.

In Chapter 9, F.S. Roberts describes and discusses many variants of an optimization model for the problem of scheduling the unloading of waiting ships at a port that takes into account the desired arrival times, quantities, and designated priorities of goods on those ships, when the said port needs to be reopened after closure due to a natural disaster or terrorist event or domestic dispute. Roberts discusses the subtleties involved in defining the objective function, and the algorithmic challenges involved in the solution, and he surveys the related literature.

In Chapter 10, D.G. Saari takes us back to consensus theory as applied to voting systems. Starting with the famous Arrow's theorem that dictatorship is the only voting rule that satisfies certain natural axioms, many important results in consensus theory state that it is impossible to have reasonable consensus functions that satisfy certain natural and innocuous properties (axioms). Saari gives an accessible discussion of what lies at the root of these obstacles to consensus, suggests a way around such difficulties by defining appropriate compatibility conditions, and illustrates his conclusions using simple examples from voting theory.

We conclude with Chapter 11, in which G.F. Estabrook, T.A. McKee, H.M. Mulder, R.C. Powers and F.S. Roberts give a survey of Buck McMorris' work over the past forty years in discrete mathematics, especially Evolutionary Biology, Intersection Graph, Theory, Competition Graphs and related phylogeny graphs, Location Functions on Graphs, and Consensus Theory. This survey helps unify the themes explored in the previous chapters of this book under the aegis of wide-ranging scholarship of Buck McMorris in Discrete Applied Mathematics.

Chapter 1

Contributions of F.R. McMorris to Character Compatibility Analysis

George F. Estabrook

Department of Ecology and Evolutionary Biology, University of Michigan
Ann Arbor MI 4809-1048 USA
gfe@umich.edu

Beginning in the mid 1970s, F.R. McMorris made significant contributions to character compatibility analysis by using his considerable mathematical talent to "prove" algorithms that I had previously conjectured, among them the so-called potential compatibility algorithm for qualitative taxonomic characters. Since then, this algorithm has become the basis for a wide variety of character compatibility analyses. Here I review the basic concepts of character compatibility analysis, describe McMorris' contributions to character compatibility analyses, review some of their applications, and present some ideas for future applications.

Introduction

In the late 19th century, systematic biologists realized that similarities and differences with respect to a basis for comparison among a group of related species under study could be the basis for an hypothesis about the relationships among species and the ancestors from which they evolved, the so-called ancestor relation. Such hypotheses are expressed as characters, which group species together into the same character state if they are considered to be the same with respect to a basis for comparison, and then arrange the character states into a character state tree to indicate where speciation events associated with the observed changes are hypothesized to have occurred. By mid 20th century, some natural scientists also realized that some pairs of such hypotheses based on different bases for comparison could be logically incompatible, i.e., they could not both be true, and they began to develop tests for, and ways to resolve, incompatibilities to estimate the ancestor relation from these hypotheses. Wilson (1964) is among the earliest published works to present an explicit test to the compatibility of (two state) characters.

Estabrook (2008) provides an in-depth discussion of biological concepts of character state change. Here, I will simply assume that relevant character state changes are associated with speciation events. Changes associated with speciation interrupt phyletic continuity over time (Estabrook 1972). This concept gives rise to a historical-biological species concept in which a species evolves at a time in the past, usually in a somewhat restricted geographic area, persists through time possibly dispersing to other geographic areas, and ultimately goes extinct (a few still extant species have not gone extinct yet). During the life of a species, a population may become isolated and independently evolve enough genetic difference so that its members can no longer breed with members of the species from which it was isolated, as described above. In this way, one species becomes the immediate ancestor of another that evolved from it. This process may happen repeatedly, so that one species may come to be the immediate ancestor of several distinct species. In this way, the study of evolutionary relationships among extant species implicates ancestral species that existed over past time, and has for one of it principal objectives an estimate of the ancestor relation among related species, past and present. This view of speciation and systematics has come to be called evolutionary systematics and is represented by Simpson (1961), Mayr (1969) among many others over the past 40 years, including more recently Skelton (1993). To understand and appreciate the context in which McMorris worked with me, I would like to make more explicit some basic concepts.

1.1. Basic Concepts

We rarely know for sure the branching pattern of phyletic lines leading up through time to the extant species under study, but to illustrate the concepts, we consider a hypothetical case in which we do. Suppose that we are studying a group S of six extant species $\{a, b, c, d, e, f\}$ whose phyletic lines branch upward through time as shown on the left in Figure 1.1.

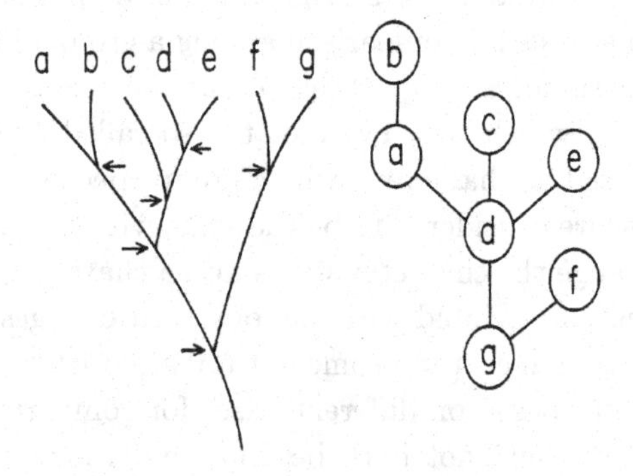

Fig. 1.1.

Whenever a phyletic line branches, one branch represents a new species and the other represents the continuation of the ancestral species. Arrows identify speciation events and point to the branch created by changes that produced a new species. The diagram on the right of Figure 1.1, shows the ancestor relation that results from these speciation events; a line is drawn upward from an ancestral species to any immediate descendant species. Thus, each line in the diagram represents a speciation event on the phylogenetic tree. We say that species x is an ancestor of species y if there is a series of one or more upward lines leading from x to y. The relation "is an ancestor of" is a tree partial order on species; its Hasse diagram is shown on the right of Figure 1.1.

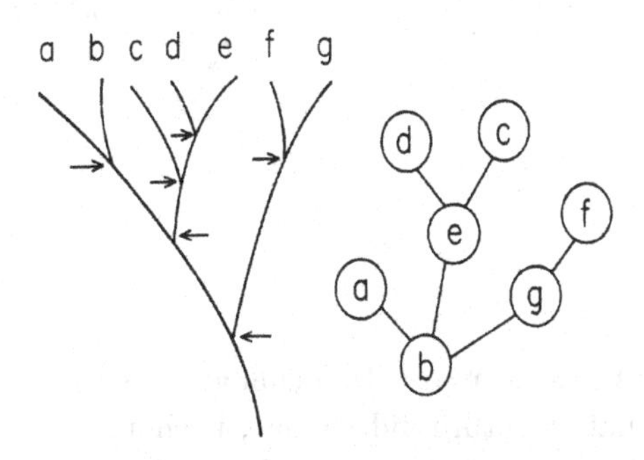

Fig. 1.2.

On the same hypothetically true branching pattern of phyletic lines, the speciation events could have occurred at different times and places. Figure 1.2. shows an example. You can see from the diagram of the resulting ancestor relation that it is quite different from the ancestor relation of Figure 1.1. This example should make it clear that there is not a one-to-one relationship between phyletic branching patterns (phylogenetic trees) and ancestor relations, because the latter are a result of the historical speciation events that created the species we study. Although it is possible to have extinct species represented in S, some ancestral species may not be represented in S because of extinction. Figure 1.3 shows the same speciation events as Figure 1.2 together with an additional speciation, indicated with *, following which the ancestral species, x, goes extinct and is not represented among the species in S; on the right is the diagram of the resulting ancestor relation.

You can imagine the variety of speciations and extinctions that could occur on a phylogenetic tree and the resulting variety of ancestor relations.

One of the tasks of systematics is to use the similarities and differences that can be observed among the species in a group under study to estimate their ancestor relation. If one can observe a given structure for each species in a collection S of species under study, and (using it as a basis for comparison) recognize distinct variations, then the species can be placed into groups so that those in the same group look the same with respect to that basis for comparison, but those in different groups look different. Such a basis for comparison is called a character, and the groups that result are called its character states. For a character to be relevant to the ancestor relation, its states should be based on changes associated with some of the speciation events that created the species in S, as discussed above. Of course, not all

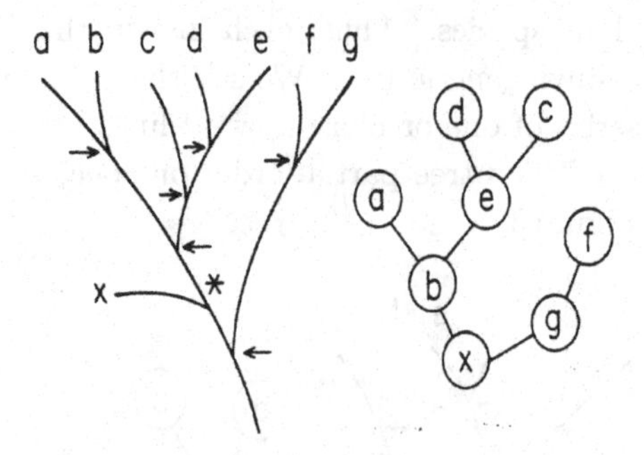

Fig. 1.3.

speciation events need be associated with a change in the structure defining a given character, but when that structure did change, it should have been in association with a speciation event. Such a character can be used as the basis for an hypothesis about the ancestor relation. This hypothesis is expressed as an ancestor relation diagram in which the character states play the role of individual species, and as before, a line leading up from one state to another represents a speciation event on the phylogenetic tree, as shown in Figure 1.4.

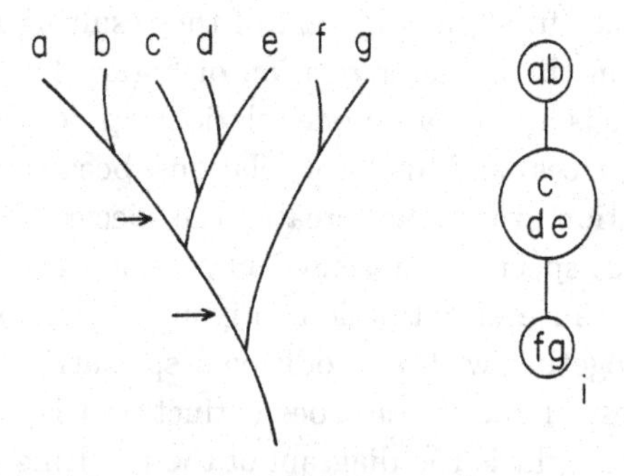

Fig. 1.4.

The *character state tree* (CST) shown on the right represents the two speciation events indicated by arrows near the phylogenetic tree on the left. The speciation indicated by the lower arrow changed a structure from the form exhibited by species *f* and *g* to the form exhibited by species *c*, *d* and *e*; the speciation indicated by the upper arrow changed that form to the one exhibited by species *a* and *b*. If the phylogenetic tree and the speciation events shown on the left of Figure 1.4 are historically correct, then we would say that the hypothesis of the CST is true (or simply that the CST is true) because it corresponds to speciations on the true phylogenetic tree (Estabrook, et al. 1975).

If two CSTs are true, then by considering all the speciation events that correspond to one or the other or both of them, another true CST (called the sum of the first two) is determined, as shown in Figure 1.5.

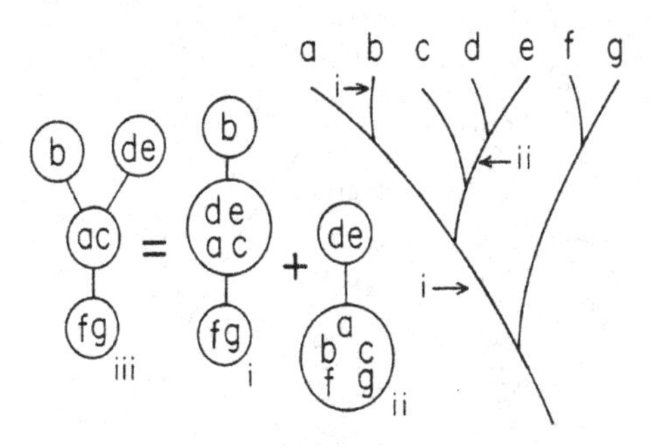

Fig. 1.5.

The "sum" CST is a refinement of either of the two CSTs that were added, because it represents all the speciation events of either. In the same way, another true CST could be added to this sum to create an even more refined CST, as shown in Figure 1.6. The two changes distinguishing state $\{d, e\}$ represented by CSTs iii and iv may have been associated with the same speciation event, or with different speciation events suggesting the possibility of an extinct ancestral species represented by the dotted circle. However, the most ancestral state contains only extinct ancestors because of speciation events on both branches of the phyletic lines leading up to the extant taxa. A moment's consideration should make it clear that if enough true CSTs are added, then the sum CST becomes the diagram of the ancestor relation itself, in which will be shown, in their historical places, ancestors not represented in S. Thus, an ancestor relation is a CST sufficiently refined so that each state has at most one species.

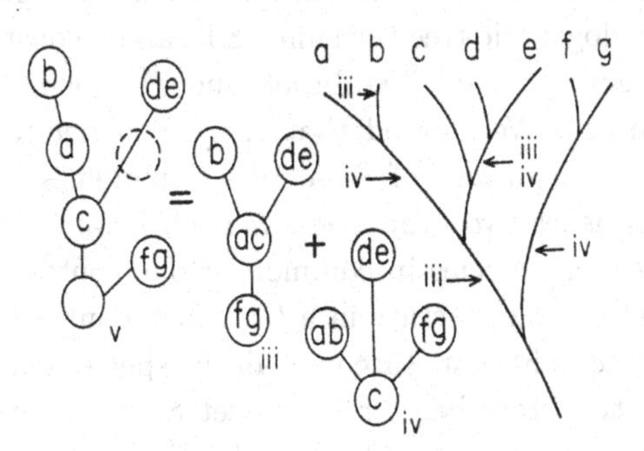

Fig. 1.6. Expansion

Of course, not all CSTs are true; there are three basic ways in which they can be false, as shown in Figure 1.7. On the right of Figure 1.7 we again see our hypothetically true phylogeny, and on the left three false CSTs.

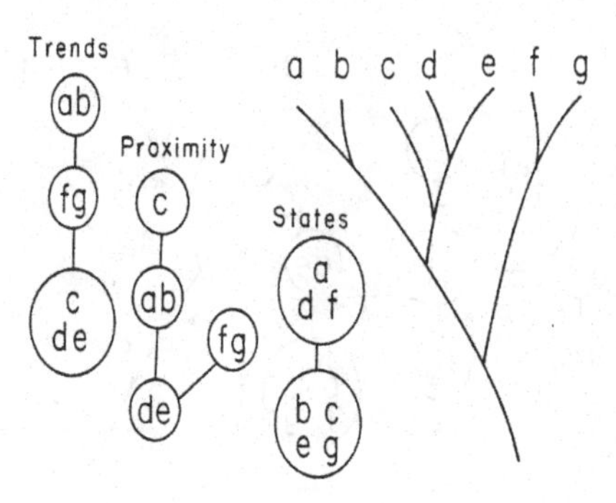

Fig. 1.7.

They are false because there is no possible way that speciations could have occurred on this true phylogenetic tree so that they would correspond to the lines in the CSTs. The leftmost misrepresents the direction (trend) of the changes, because changing the direction of change so that state $\{f, g\}$ is primitive does make it possible to put speciation events on the phylogenetic tree so that this CST would be true. The middle CST connects pairs of states that cannot all be next to each other and still represent speciation events on the phylogenetic tree; redirecting these proximity relations cannot make a CST whose speciation events can be placed on the true phylogenetic tree. However, attaching state $\{c\}$ to state $\{d, e\}$ instead of to state $\{a, b\}$ does make this possible. For the rightmost CST, there is no way to place a speciation on the phylogenetic tree so that even its states would result.

Two CSTs do not have to be both true for it to be possible to add them; they can be added so long as there is some phylogenetic tree (true or not) on which all their speciations can be simultaneously placed. Then from the placement of these speciations on this phylogenetic tree the sum CST can be constructed. But how can we find such a phylogenetic tree? Estabrook and McMorris (1980) demonstrated that we do not need to. We showed that there is a one-to-one correspondence between CSTs and trees of subsets of S. A collection of subsets of S is called a tree of subsets of S if it satisfies two properties: S itself is one of the subsets, and any two subsets either have no species in common or one contains all the species that are in the other. Each character state in a CST has a subset in its tree of subsets consisting of all the species in that state plus all the species in any descendant state. Thus, the most primitive state has for its subset S, the entire study collection of species. States with no descendant states have for their subset only the species that they contain themselves. The correspondence is shown in Figure 1.8, where below each CST is shown its tree of subsets, arranged so that derived states are above their ancestors. The sum is determined as the union of the subsets from both trees of subsets; if the result is itself a tree of subsets, then the sum is the corresponding CST. The process is illustrated in Figure 1.8, where the trees of subsets of the two CSTs to be added are combined to make the tree of subsets in the lower right; finally the CST above is constructed using the principles described above. This CST is the sum of the first two.

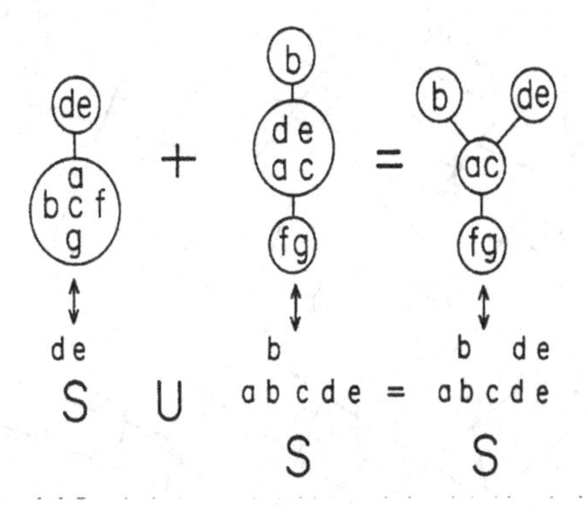

Fig. 1.8.

Two CSTs do not have to be true for one to be a refinement of the other. We can readily see from the sum of two CSTs that if the tree of subsets of one CST contains all the subsets in the tree of subsets of another CST, then the first is a refinement of the second. The diagram of the relation "is a refinement of" among CSTs makes a semi-lattice. Meacham (1989) presents a figure of this semi-lattice for the simple case in which S contains only 3 species. It is reproduced in Figure 1.9. This refinement relation has been studied theoretically by Estabrook and McMorris (1980), McMorris and Zaslavsky (1981), and Janowitz (1984).

Not every pair of CSTs can be added. When the union of their two trees of subsets is NOT a tree of subsets, as shown in Figure 1.10, then there is no phylogenetic tree on which to place speciations that correspond to the lines in both CSTs. Typically we do not know the true phylogenetic tree so we cannot test a CST to discover whether it is true or not. However, if two CSTs cannot be added, then they cannot both be true; they are incompatible as hypotheses about the ancestor relation among the species and their ancestors under study (Estabrook, 1984). Two CSTs that can be added are compatible as hypotheses about the ancestor relation. This concept of character compatibility forms the basis of a variety of compatibility methods developed and used over the past 50 years.

Estabrook (2008) discusses the early history of compatibility concepts in some detail. Here it will suffice to recall the reasons, illustrated in Figure 1.7, why a character state tree might be false. These same reasons may explain the incompatibility of two characters: redirecting the characters state tree of one of the characters

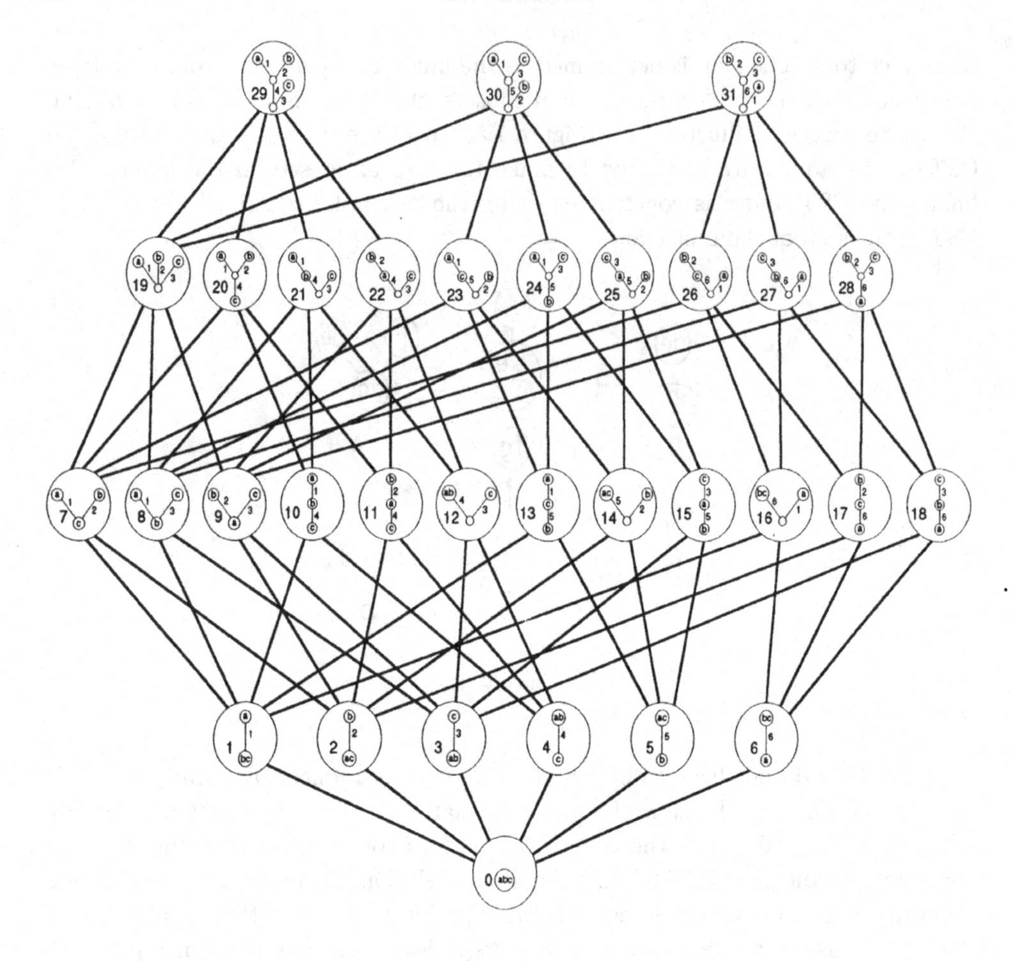

Fig. 1.9.

(trends); re-arranging the way in which the character states are connected (proximities); or recognizing a different concept of "similar" and "different" (states) may resolve the conflict. Estabrook and Meacham (1979) presented a test for undirected multi-state CSTs, i.e., is the incompatibility a "trends" problem? They proved that in a CST there is always a state that can be designated as the most primitive so that the number of species in any state x immediately derived from it, plus the sum of the number of species in all the states derived from state x, is never more than half the number of species in S. A CST directed with such a state most primitive is said to be directed "common equal primitive". They then proved that if two common equal primitive CSTs are incompatible, then they will remain so directed in any other way. Thus if CSTs are directed "common equal primitive", a compatibility test by trees of subsets (Estabrook and McMorris 1980) will also test them as undi-

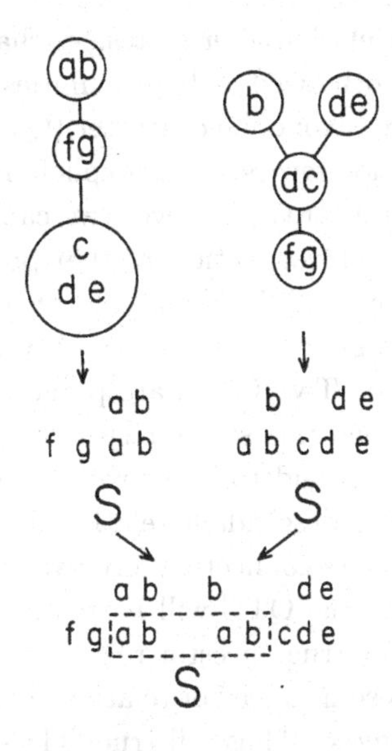

Fig. 1.10.

rected hypotheses. Estabrook (1977) suggested that systematists might be tempted to believe that common equal primitive because directing change that way eliminates conflicts due to direction alone. Meacham (1984) made the then controversial suggestion that the role of hypothesizing direction of evolutionary change before estimating the ancestor relation among species under study could be reduced, or even eliminated, by reasoning with undirected CSTs, especially in cases with little or no a priori evidence to identify a primitive condition. Donoghue and Maddison (1986) objected on philosophical grounds that were later shown to be irrelevant to modern methods.

Proximity tests were independently suggested by Fitch (1975), Sneath et al. (1975) and Estabrook and Landrum (1975). These determine whether incompatibilities can be resolved by hypothesizing different proximities among character states. With two-state CSTs, there are no incompatibilities that can be resolved by changing proximities of states in CSTs because there is only one proximity. However, for two incompatible multi-state CSTs we can ask whether there is some other compatible pair of CSTs with the same respective states, but different proximities. This is especially relevant when the data source provides predominantly undirected, multi-state characters. In the 1970s protein sequencing became more common. Species could be compared based on which amino acid appeared in any given position of a sequenced protein; Boulter et al. (1979) present an early example. The resulting characters often had more than two states, and could have as many as 20 states. Multi-state characters continued to be relevant through the 1980s and up until present times, as DNA sequencing became more available. In many current data sets, nucleotide bases represent character states. A character consisting of just its character states, with no hypothesized state proximities or direction of evolutionary change, is called a qualitative taxonomic character (QTC). Without a tree ordering for its states, a QTC does not represent an explicit hypothesis about speciation events on the true phylogenetic tree. However, we can hypothesize that there are speciation events on the true phylogenetic tree that produce a CST with the same states as a QTC. In this way, a QTC represents an hypothesis about speciation events; it is weaker than the hypothesis associated with a CST and asserts nothing about direction of change. Two QTCs are potentially compatible if there are two compatible CSTs with the same states respectively. Those CSTs realize that potential. A character state is said to be convex (in the mathematical sense) in the undirected tree of ancestor-descendant relations, if the unique path of (ancestor, immediate descendant) pairs connecting any two species in that state contains only species in that state. A true QTC will have states that are "convex" on the undirected tree diagram of the true ancestor relation. If two QTCs are potentially compatible, then there exists some ancestor relation (not necessarily true) for which all the states of both are convex. Thus, all true QTCs are necessarily potentially compatible with each other. Often, when working within a context of only QTCs, potentially compatible QTCs are called simply compatible. Estabrook and McMorris (1977) mathematically proved the validity of the test algorithm of Estabrook and Landrum (1975). The mathematical form of this proof is entirely the contribution of McMorris. This has become a work-horse algorithm for many applications to follow. What follows will explicate this algorithm, sketch McMorris' proof, present several published examples of its application, and suggest ideas for future applications.

1.2. Potential Compatibility Algorithm

To test the potential compatibility of two QTCs, EUs (evolutionary units) are entered in the cells of a matrix: the states of the first character label the rows, and the states of the second character label the columns. Each EU is placed in the cell whose row and column labels are the states to which it belongs, as shown in Figure 1.11. Moving only from one occupied cell to another in a straight line horizontally or vertically but never retracing a path already taken, if you can return to an occupied cell you have already visited then the two characters are incompatible, as for I and III, otherwise the two characters are potentially compatible, as for I and II. In Figure 1.11, possible moves are shown with dashed lines. To discover two CSTs that realize potential compatibility, first connect pairs of cells in the same row or column to make a one-component graph with all the occupied cells included among the vertices, in any way that does not close a loop, and then designate as primitive any vertex (cell) in this graph. Direct the edges away from this primitive cell. The CSTs for each character are inherited from this directed graph. Thus, for QTCs I and II in Figure 1.11, in addition to the connections already shown, you could connect cell (IG IIT) containing $\{d, f\}$ to the empty cell (IC IIT) on the path between cell (IC IIA) containing $\{b\}$ and cell (IC IIC) containing $\{e, g\}$; then designate cell (IC IIT) primitive. The resulting CST for I has state C primitive with states A and G separately derived from it. The resulting CST for II has state T primitive with states C and A separately derived from it and state G derived from state A.

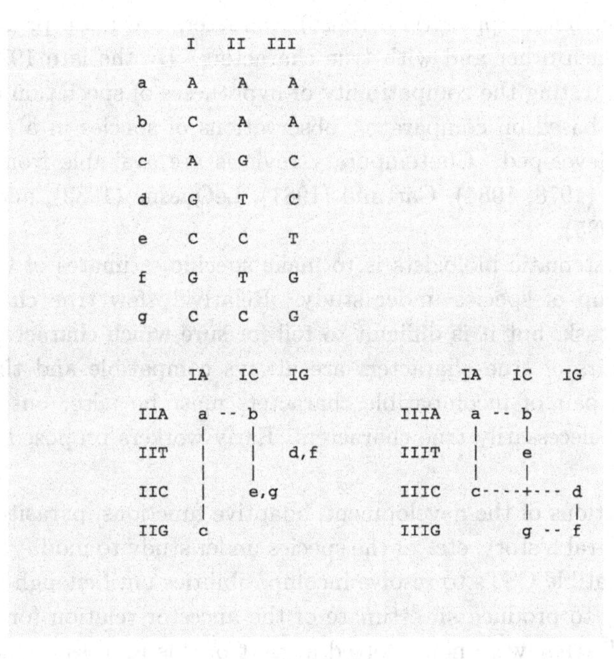

Fig. 1.11.

To implement this algorithm computationally, I conjectured (and McMorris proved) that the matrix of a pair of potentially compatible QTCs would always contain at least one row or column with only one cell occupied; if the EUs in this cell were removed, in the context of the remaining EUs the two QTCs would continue to be compatible, thus presenting a matrix with at least one row or column with only one cell occupied, whose EUs could be removed, etc; two QTCs are potentially compatible iff all EUs can be eventually removed from S in this way (Estabrook and McMorris 1977).

1.3. Applications of Compatibility using McMorris' Results

Some characters are inappropriate no matter what their compatibility relations are with other characters because their character state changes are not associated with speciation events, such as "random" within-population changes, or because changes to a character state are associated with several speciation events (parallel or reversed evolution), or because they are based on mistaken homology (A structure in one species is homologous to a structure in another species if each structure evolved from the same structure in the most recent common ancestor of those two species. This is particularly problematic with DNA sequence data, in which alignment estimates homology). I have discussed in some detail the nature of inappropriate characters (Estabrook 1998, 2008); the basic argument is that true characters are always compatible, but inappropriate (false) characters are more prone to be incompatible with each other and with true characters. By the late 1970's, all the basic concepts for testing the compatibility of hypotheses of speciation events on a phylogenetic tree, based on comparative observations of species in a group under study, had been developed. Contemporary reviews are available from McMorris (1975), Estabrook (1978, 1984), Cartmill (1981), LeQuesne (1982), and Meacham and Estabrook (1985).

One goal of systematic biologists is to make specific estimates of the ancestor relation for a group of species under study. Relatively few true characters are sufficient for this task, but it is difficult to tell for sure which characters are true. We know that pairs of true characters are always compatible and that at least one member of a pair of incompatible characters must be false, but compatible characters are not necessarily true characters. Early workers proposed three basic approaches.

1) Make considerations of the development, adaptive functions, parasites, diseases, biogeography, natural history, etc. of the species under study to modify one or both of pairs of incompatible CSTs to resolve incompatibilities until enough CSTs could be added together to produce an estimate of the ancestor relation for the species under study. CSTs that were not resolved as part of this process are set aside, as less reliable or problematic.

2) Leave characters as originally constructed, but apply some operational (often quantitative) criterion to choose one character (or a compatible group) to make an initial partial estimate. Then, within the context of subsets of S that are convex on this partial estimate, apply the criterion again, until the ancestor relation is resolved to the satisfaction of the investigator.

3) Use an automatic incompatibility resolving procedure. One such procedure is to sub-divide character states without consideration of the development, form, adaptive functions, parasites, diseases, biogeography, natural history, etc. to create new characters with more, smaller states, that are all mutually compatible. For any two incompatible CSTs or QTCs, it is always possible to sub-divide states enough to create two new compatible characters. In general, there are very many ways to subdivide the states of characters to resolve all incompatibilities in a data set, especially if S is large. For this reason, this approach imposes the additional criterion (parsimony) that as few as possible new character states should be created to resolve all the incompatibilities among the characters. Especially if S is large, there may still be several, often disparate, ways to satisfy this criterion. Another

procedure assumes that all state changes are actually "random" and chooses (from a large family) the probability model of evolution that makes the evolution of the states of all characters most likely (maximum likelihood). There are many other automatic procedures.

Among practitioners of the first approach, one of the most influential in his time was Hennig (1966). This book is a translation from German into English by R. Zangerl of an unpublished MS composed by Hennig shortly before that date as a major revision of a less well known book he had published 15 years earlier, also in German. In his preface, Zangerl himself warns us of the linguistic difficulties of making such a translation. Indeed, many of the terms in Hennig (1966) are taken from evolutionary biologists writing in English, where their meanings have been well understood for decades, and given different meanings, either by Hennig (or by his translator). Unfortunately, this resulted in some serious misunderstandings during the 1970's and 1980's, which are only now beginning to be resolved. Mayr (1974) and Sokal (1975) provide a contemporary discussion of some of these issues. These misunderstandings continued to confuse Hennig's followers (of which there were many) over the next three decades, and ramifications of this confusion are still with us today.

Many others who did not consider themselves followers of Hennig also considered natural, biological criteria to restructure CSTs to provide more consistent estimates. Good examples of natural criteria for estimating CSTs are given by Voss and Voss (1983). DeMarco et al (1985), which publication is in Italian, but their results are presented in Table 2 of Estabrook (2001), Stein et al. (1984), Gardner and LaDuke (1979), LaDuke and Crawford (1979), and more recently Strasser and Delson (1987). Few investigators use this first approach today, in part because molecular data have come to dominate as the basis for estimating relationships, and it is not yet clear how to apply considerations from the natural world explicitly to restructure QTCs arising from molecular data. This approach may become more useful again as macromolecular data, e.g., chromosomal rearrangements or other large scale genetic changes, become more widely implicated in the estimation of ancestor relations (Qui et al 2006).

The second approach uses an operational criterion to select some of the CSTs or QTCs to use compatibly to make a (possibly only partial) estimate of the ancestor relation. LeQuesne (1969. 1974) was an early proponent of this approach. In the case of CSTs, directed or undirected, I conjectured (and McMorris proved) that if all pairs of CSTs in a collection of CSTs are compatible, then the entire collection is compatible, i.e. there are ancestor relations (namely all refinements of their sum) that are refinements of every CST in the collection (Estabrook et al. 1976). Thus, a large collection of pairwise compatible CSTs could be chosen as the basis for a first (perhaps only partial) estimate of the ancestor relation. Discovering such a collection is equivalent to discovering the maximal cliques in an undirected graph, a computationally difficult (NP complete) problem as S becomes large. Bron and Lerbosch (1973) published an algorithm to discover the largest collections (cliques) of pairwise compatible CSTs, which has been used in computer programs implementing this appraoch.

Many investigators were not comfortable hypothesizing CSTs. The observable states of a QTC seemed to have more objective reality than a CST, which includes a (possibly questionable) hypothesis about how those states evolved from one another. A collection of QTCs can also be tested for potential compatibility and maximal groups of pairwise potentially compatible QTCs discovered, but there may be no possible estimate of the ancestor relation on which all the states of the QTCs are convex, i.e., these QTCs may not be group-wise potentially compatible (Fitch, 1975). This poses an interesting problem whose mathematical analog has been studied by Gavril (1974), who was not aware of the analog, and by McMorris et al (1994), who were fully aware.

Of course, for a group of pairwise compatible QTCs there may be an ancestor relation on which all or most of them have convex states. Boulter et al (1979) using amino acid sequences of a protein from flowering plants were among the first to apply compatibility of QTCs to a major study. Estabrook (1991) and Camacho et al (1997) provide later examples, and recently Grupta and Sneath (2007) use a maximal clique of compatible 2-state characters to estimate the ancestor relation in a large molecular study.

With the advent of molecular data sets with multi-state characters, it became unclear how to hypothesize CSTs, or how to take the first approach to resolve incompatibilities. Especially in data sets with a large number of more distantly related EUs, the second approach often resulted in typically only a small fraction of the data participating in estimates of the ancestor relation. In molecular data sets, high levels of incompatibility may result from a larger fraction of molecular data evolved as random changes not associated with particular speciation events. The third approach (use an automatic incompatibility resolving procedure such as parsimony or maximum likelihood) has been commonly taken in these cases. When levels of compatibility are very low, the vast majority of characters are false. Automatically resolving incompatibility among so many false characters may result in weaker estimates of the ancestor relation. Recognition and elimination of characters whose level of compatibility can not be distinguished from "random" may leave a reduced data set with a higher fraction of true characters, from which a more reliable estimate could be made. To do this we need to define a reasonable concept of "random".

LeQuesne (1972) presented a formula to calculate the probability that two 2-state characters would be potentially compatible; and used it to select characters for further consideration. Meacham (1981) defined clearly the concept of "at random" implicit in LeQuesne (1972) and generalized it to two (or a group) of CSTs (with any number of states), and showed that CSTs with many large advanced states were less likely to be compatible at random than those with fewer, smaller advanced states.

This suggested an alternative to choosing the largest clique, which might have many CSTs more likely to be compatible at random; instead choose the clique of CSTs least likely to be compatible at random. The explicit approach of Meacham (1981) to calculate exact probabilities, becomes computationally intractable with more than just a few EUs, and was not applicable to QTCs. To address these problems, Meacham (1994), using the same concept of "random" (any character with the same numbers of EUs in its respective sates is equally likely) estimated probabilities of compatibility by simulation. The algorithm of Estabrook and McMorris (1977) is at the center of these simulations.

Especially in data sets with a large number of EUs, the number of characters in the largest (or least likely) clique was often a very small fraction of the total number of characters, which could result in an unsatisfactory estimate of the ancestor relation. Flagrantly false characters, whose states would have to be subdivided many times to become convex on the true ancestor relation, might be as likely to be compatible with other characters as a character to whose states EUs were assigned at random. For a given character, the probabilities with which it would be compatible at random with each other character could be summed to give the number of other characters with which it would be expected to be compatible at random. This could be compared with the number with which it was actually compatible; If this number were not more than would be expected at random, then the character could be set aside as indistinguishable from random. Remaining characters could be dealt with in any of the three approaches described above.

Meacham (1994) construed the number of other characters with which a given character is compatible as a random variable, and undertook to estimate its distribution under the hypothesis that the given character was random. Because Meacham (1994) used simulation to make close approximations to compatibility related probabilities, he could estimate the probability that a given character (QTC or CST) would be compatible with at least as many other characters as it actually was, under the hypothesis that it was a random character. Meacham (1994) called this probability *Cf*, the Frequency of Compatibility Attainment. Camacho et al (1997), Qui and Estabrook (2008), Pisani (2004) used this approach to eliminate characters with levels of compatibility whose random probability exceeded a given threshold, and then took one of the three basic approaches above to estimate an ancestor relation. Pisani (2004), using a large data set (866 DNA QTC characters and 47 species), suggested that removing characters with $Cf > 0.5$, i.e., a random character would be expected to be compatible with more other characters than observed, before using those remaining with a maximum likelihood method, may reduce the effects of long branch attraction. Qiu and Estabrook (2008) observed increased clarity of parsimony estimates of relationships among key groups of angiosperms when all characters with $Cf > 0.2$ were removed. Day et al. (1998) used the number of compatible pairs of characters in a whole data set as a random variable under the hypothesis that all the characters in the data set were random in the sense of

Meacham (1994). They analyzed 102 published data sets, of which 12 had fewer compatible pairs than would be expected at random. In general, they observed that inclusion of out groups increased the probability that compatibility levels are random, sometimes substantially so.

Estabrook and McMorris (2006) was the last compatibility application on which we collaborated, so called stratigraphic compatibility. Rock strata containing fossils have been used to estimate the interval of time from the evolution to the extinction of species (or other higher taxon). These estimates could place so-called stratigraphic constraints on estimates of the ancestor relation: a species whose first appearance in rock strata is more recent than the first appearance of another species cannot be one of its ancestor in the true ancestor relation; and a species whose last appearance is before the first appearance of another species cannot be its immediate ancestor.

These constraints can be expressed as a graph whose directed edges are the stratigraphically possible ancestor - immediate descendant pairs. Estimates of the ancestor relation must be a spanning directed tree subgraph. In the context of stratigraphic constraints, a CST or QTC, or even a single character state, can be incompatible with the stratigraphy, and a pair of otherwise compatible characters (CST or QTC) can become incompatible. Sometimes stratigraphic constraints can be quite severe, which make them unpopular with some investigators, especially when they conflict with comparative data.

1.4. Possible Future Applications

Compatibility concepts have been applied to several other areas, beyond the scope of this paper. Many of them are discussed in Estabrook (2008). But I want to mention two applications, using variations of the test for the compatibility of two QTCs (Estabrook and McMorris 1977), that are currently under development.

Sometimes one or a few EUs in a data set are anomalous: perhaps they do not really belong to the study group; perhaps mistakes have been made in assigning them to characters states (alignment problems); or perhaps they have evolved substantially more rapidly than the other EUs in S. Their inclusion in the data set could confound estimates of the ancestor relation by generating incompatibilities that would disappear if they were removed. After a good estimate of the ancestor relation among the remaining EUs has been made, perhaps these EUs could be added. To identify such anomalous EUs (if such there be) each EU (or small group) is removed and compatibilities of all pairs of characters are tested again (using, of course the algorithm of Estabrook and McMorris 1977); if removal of any EUs (or small groups) results in a sharp increase in the number of compatible pairs of QTCs, the EUs removed are candidates for consideration as anomalous EUs.

The second application under development uses a variation on the algorithm of Estabrook and McMorris (1977). In DNA data sets with large numbers of EUs, there are often many small errors that arise from the use of automatic sequencing technology. These might result in the misplacement of one or a few EUs into the wrong character state for some characters. In the contingency matrix of Estabrook and McMorris (1977) this might result in a box occupied by a single EU (the misplaced one). So-called "almost compatible" pairs of QTCs are those that pass the potential compatibility test when boxes in their contingency matrix with only one EU are considered empty. This concept could be simulated as easily as strict compatibility to calculate the probability of almost compatibility, or the realized significance of the number of other characters in a data set with which a given character is almost compatible. Eliminating characters not sufficiently almost compatible with the rest of the other characters might result in the retention of more characters useful in parsimony, maximum likelihood, or other third approach automatic methods.

References

1. D. Boulter, D. Peacock, A. Guise, T.J. Gleaves, G.F. Estabrook, Relationships between the partial amino acid sequences of plastocyanin from members of ten families of flowering plants. *Phytochenistry* **18** (1979) 603–608.
2. C. Bron, J. Lerbosch, Algorithm 457: Finding all cliques of an undirected graph, *Communications of the Association for Computing Machines* **16** (1973) 575–577.
3. A.I. Camacho, E. Bello, G.F. Estabrook, A statistical approach to the evaluation of characters to estimate evolutionary history among the species of the aquatic subterranean genus, Iberobathynella (Crustacea, Snycardia). *Biological Journal of the Linnean Society 60* (1997) 221–241.
4. M. Cartmill, Hypotheses testing and phylogenetic reconstruction, *Zeitschrift für Zoologische Systematik Evolutionsforschung* **19** (1981) 93–96.
5. W.H.E. Day, G.F. Estabrook, F.R. McMorris, Measuring the phylogentic randomness of biological data sets. *Systematic Biology* **47** (1998) 604–616.
6. G.A. DeMarco, A. Altieri, G.F. Estabrook, Relazioni evolutivi e biogeografiche dei popolamenti ad areale disguinte di Genista ephedroides DC, *Biogeographia* **11** (1985) 115–131.
7. M.J. Donoghue, W.P. Maddison, Polarity Assessment in Phylogenetic Systematics: A Response to Meacham, *TAXON* **35** (1986) 534–545.
8. G.F. Estabrook, Cladistic methodology, *Annual Reviews of Ecology and Systematics* **3** (1972) 427–456.
9. G.F. Estabrook, L.R. Landrum. A simple test for the possible simultaneous evolutionary divergence of two amino acid positions. *TAXON* **24** (1975) 609–613.
10. G.F. Estabrook, C.S. Johnson, F.R. McMorris, An idealized concept of the true cladistic character, *Mathematical BioSciences* **23** (1975) 263–272.
11. G.F. Estabrook, C.S. Johnson, F.R. McMorris, A mathematical foundation for the analysis of cladistic character compatibility, *Mathematical BioSciences* **29** (1976) 181–187.
12. G.F. Estabrook, F.R. McMorris, When are two qualitative taxonomic characters compatible? *Journal of Mathematical Biology* **4** (1977) 195–200.
13. G.F. Estabrook, Does common equal primitive? *Systematic Botany* **2** (1977) 36–42.
14. G.F. Estabrook, Some concepts for the estimation of evolutionary relationships in systematic botany, *Systematic Botany* **3** (1978) 146–158.
15. G.F. Estabrook, C.A. Meacham, How to determine the compatibility of undirected character state trees, *Mathematical BioSciences* **46** (1979) 251–256.

16. G.F. Estabrook, F.R. McMorris, When is one estimate of evolutionary relationships a refinement of another? *Journal of Mathematical Biology* **10** (1980) 367–373.
17. G.F. Estabrook, Phylogenetic trees and character state trees, in: Duncan T, Stuessy TF. eds. *Cladistics. perspectives on the reconstruction of evolutionary history*, Columbia University Press, New York, 1984, pp. 135–151.
18. G.F. Estabrook, The use of distance measures, such as from DNA-DNA hybridization, in phylogenetic reconstruction and in convex phenetics, *Bollettino di Zoologia* **58** (1991) 299–305.
19. G.F. Estabrook, Ancestor-descendant relations and incompatible data: Motivation for research in discrete mathematics, in: B. Mirkin et al., *Hierarchical Structures in Biology*, DIMACS Applied Mathematics Series Vol 37. American Mathematical Society, Providence RI USA, 1998.
20. G.F. Estabrook, Vicariance or Dispersal: the use of natural history data to test competing hypotheses of disjunction on the Tyrrhenian coast, *Journal of Biogeography* **28** (2001) 95–103.
21. G.F. Estabrook, F.R. McMorris, The compatibility of stratigraphic and comparative constraints on estimates of ancestor-descendant relations, *Systematics and Biodiversity* **4** (2006) 9–17.
22. G.F. Estabrook, Fifty years of character compatibility concepts at work, *Journal of Systematics and Evolution* **46** (2008) 109–129.
23. W.M. Fitch, Towards finding the tree of maximum parsimony. in: G.F. Estabrook, ed. *Proceedings of the Eighth International Conference on Numerical Taxonomy*, Freeman. San Francisco USA, 1975., pp. 189–230.
24. R.C. Gardener, J.C. LaDuke, Phyletic and Cladistic Relationships in Lipochaeta (Compositae), *Systematic Botany* **3** (1979) 197–207.
25. F. Gavril, The intersection graphs of subtrees in trees are exactly the chordal graphs, *J. Combin. Theory, Series B* **16** (1974) 47–56.
26. B.S. Gupta, P.H.A. Sneath, Application of the Character Compatibility Approach to Generalized Molecular Sequence Data: Branching Order of the Proteobacterial Subdivisions, *Journal of Molecular Evolution* **64** (2007) 90–100.
27. W. Hennig, *Phylogenetic Systematics*, (Translated from German manuscript by D. D. Davis and R. Zangerl) University of Illinois Press, 1966.
28. M.F. Janowitz, On the semi-lattice of the weak orders of a set, *Mathematical Social Sciences* **8** (1984) 229–239.
29. J. LaDuke, D.J. Crawford, Character compatibility and phyletic relationships in several closely related species of Chenopodium of the western United States, *TAXON* **28** (1979) 307–314.
30. W.J. LeQuesne, A method of selection of characters in numerical taxonomy, *Systematic Zoology* **18** (1969) 201–205.
31. W.J. LeQuesne, Further studies based on the uniquely derived character concept, *Systematic Zoology* **21** (1972) 281–288.
32. W.J. LeQuesne, Uniquely derived character concept and its cladistic application, *Systematic Zoology* **23** (1974) 513–517.
33. W.J. LeQuesne, Compatibility analysis and its applications, *Zoological Journal of the Linnaean Society* **74** (1982) 267–275
34. E. Mayr, *Principles of Systematic Zoology*, McGraw Hill New York USA, 1969.
35. E. Mayr, Cladistic analysis or cladistic classification? *Zeitschrift fur Zoologische Systematik und Evolutionsforchung* **12** (1974) 94–128.
36. F.R. McMorris, Compatibility criteria for cladistic and qualitative taxonomic characters, in: G.F. Estabrook, ed. *Proceedings of the Eighth International Conference on Numerical Taxonomy*, Freeman, San Francisco USA, 1975, pp. 189–230.

37. F.R. McMorris, On the compatibility of binary qualitative taxonomic characters, *Bulletin of Mathematical Biology* **39** (1977) 133–138.
38. F.R. McMorris, T. Zaslavsky, The number of cladistic characters, *Mathematical BioSciences* **54** (1981) 3–10.
39. F.R. McMorris, T.J. Warnow, T. Wimmer, Triangulating vertex colored graphs., *SIAM J. of Discrete Math.* **7** (1994) 296–306.
40. C.A. Meacham, A probability measure for character compatibility *Mathematical BioSciences* **57** (1981) 1–18.
41. C.A. Meacham, The role of hypothesized direction of characters in the estimation of evolutionary history, *TAXON* **33** (1984) 26–38.
42. C.A. Meacham, G.F. Estabrook, Compatibility methods in Systematics, *Annual Review of Ecology and Evolution* **16** (1985) 431–446.
43. C.A. Meacham, *Hasse diagrams of the "is a refinement" semi- lattice*, Class Notes, University of California, Berkeley, 1989.
44. C.A. Meacham, Phylogenetic relationships at the basal radiation of Angiosperms: further study by probability of compatibility, *Systematic Botany* **19** (1994) 506–522.
45. D. Pisani, Identifying and removing fast-evolving sites using compatibility analysis: An example from the Arthropoda, *Systematic Biology* **53** (2004) 978–989.
46. Y.-L. Qiu, G.F. Estabrook, Phylogenetic relationships among key angiosperm lineages using a compatibility method on a molecular data set, *Journal of Systematics and Evolution* **46** (2008) 130–141.
47. Y.-L. Qiu, L. Li, B. Wang, Z. Chen, V. Knoop, M. Groth-Malonek, O. Dombrovska, J.L. Kent, J. Rest, G.F. Estabrook, T.A. Hendry, D.W. Taylor, C.M. Testa, M. Ambros, B. Crandall-Stotler, R.J. Duff, M. Stech, W. Frey, D. Quandt, C.C. Davis, The deepest divergence in land plants inferred from phylogenetic evidence, *Proceedings of the National Academy of Science* **103** (2006) 15511–15516, and Supporting Text.
48. G.G. Simpson, *Principles of Animal Taxonomy*, Columbia University Press, New York USA, 1961.
49. P.W. Skelton, Adaptive radiation: definition and diagnostic test, in: R.D. Lees, D. Edwards, eds. *Evolutionary Patterns and Processes*, Linnaean Society Symposium Series **14** (1993) 46–58.
50. P.H.A. Sneath, M.J. Sackin, R.P. Ambler, Detecting evolutionary incompatibilities from protein sequences, *Systematic Zoology* **24** (1975) 311–332.
51. R.R. Sokal, Mayr on Cladism - and his critics, *Systematic Zoology* **24** (1975) 257–262.
52. W.E. Stein Jr, D.C. Wight, C.B. Beck, Possible alternatives for the origin of Sphenopsida, *Systematic Botany* **9** (1984) 102–118.
53. E. Strasser, E. Delson, Cladistic analysis of Cercopithecid relationships, *Journal of Human Evolution* **16** (1987) 81–99.
54. N.A. Voss, R.S. Voss, Phylogenetic relationships in the cephalopod family Cranchiidae (Oegopsida), *Malacologia* **23** (1983) 397–426.
55. E.O. Wilson, A consistency test for phylogenies based on contemporary species, *Systematic Zoology* **14** (1965) 214–220.

Chapter 2

Families of Relational Statistics for 2 × 2 Tables

Willem J. Heiser and Matthijs J. Warrens
Psychologisch Instituut, Leiden University
Wassenaarseweg 52, 2333 AK Leiden, Netherlands
heiser@fsw.leidenuniv.nl, warrens@fsw.leidenuniv.nl

2 × 2 Tables are frequently encountered in various fields of science, including biometrics, psychology and ecology. To summarize the data in a 2 × 2 table it is convenient to calculate a statistic that quantifies in some way the degree of association or agreement between the rows and columns. The literature contains a vast amount of such relational statistics for 2 × 2 tables, many of which are just functions of the four cells a, b, c and d of the 2 × 2 table. In this paper we review the important ones. Furthermore, we show how many of the functions based on the four quantities a, b, c and d are related to one another by reviewing families of relational statistics from the literature. A statistic must be considered in the context of the data-analytic study of which it is a part. Studying families of statistics provides insight into how one statistic can be interpreted and is related to other statistics. The overview provides insights that may be helpful to researchers from various fields of science in deciding what statistic to use in applications or for studying theoretical properties.

Introduction

Data that can be summarized in a 2 × 2 table are encountered in many fields of science. For example, in psychology and biometrics the data may be the result of a reliability study where two observers classify a sample of objects using a dichotomous response (Fleiss, Fleiss 1975; Martín Andrés and Femia-Marzo, 2008). In epidemiology, a 2 × 2 table can be the result of a randomized clinical trial with a binary outcome of success (Kraemer, 2004). Furthermore, in ecology a 2 × 2 table may be the cross-classification of the presence/absence codings of two species types in a number of locations (Janson and Vegelius, 1981; Warrens, 2008a,b). Finally, in cluster analysis a 2×2 table may be the result of comparing partitions obtained from two different clustering methods (Steinley, 2004; Albatineh, Niewiadomska-Bugaj and Mihalko, 2006).

In many applications the researcher wants to further summarize the 2 × 2 table by a single relational statistic. Sometimes reporting a single statistic concludes the data-analytic part of a study. In other cases, multiple statistics or matrices of coefficients are used as input in techniques in data mining, cluster analysis, or information retrieval. The frequent occurrence of binary data has led to the fact that the literature contains a vast amount of relational statistics for 2×2 tables (Cheetham and Hazel, 1969; Janson and Vegelius, 1981; Hubálek, 1982; Gower and Legendre, 1986; Krippendorff, 1987; Baulieu, 1989, 1997; Albatineh et al., 2006; Deza and Deza, 2006; Martín Andrés and Femia-Marzo, 2008; Warrens, 2008a,b,c,d,e,f, 2009; Lesot, Rifgi and Benhadda, 2009). Well-known examples are the phi coefficient, Cohen's (1960) kappa and the simple matching coefficient. Relational statistics have different names depending on the field of science or the analytic context. Example are: similarity measures, association coefficients, agreement indices, reliability statistics or presence/absence coefficients. All relational statistics can basically be used to express in one number (quantify) the strength of the relationship between two binary variables or the amount of agreement (similarity, resemblance) between two objects.

In choosing a particular statistic to summarize the data in a 2×2 table, a function has to be considered in the context of the data analysis of which it is a part. Relational statistics can be distinguished according to the type of data they apply to or how they may be interpreted. In this paper, we only review the most important statistics. In Section 2.2, we distinguish three general types of statistics, called type A, type B and type C. Furthermore, we show how many of the functions for 2×2 tables are related to one another by distinguishing a variety of parameter families of statistics in Section 2.3. For some parameter families all special cases are of the same type (A, B, C), while other families contain special cases of different types.

The aim of this broad overview is to show how the different statistics may be interpreted in the contexts of various other statistics. The overview provides insights that may be beneficial to both practitioners in deciding what statistic to use, and theorists for studying properties of these statistics. The paper is organized as follows. In Section 2.1, we present definitions and we discuss three types of relational statistics for 2×2 tables. In Section 2.2, we present various examples of relational statistics and fields of science where they are used. Parameter families of statistics for 2×2 tables are considered in Section 2.3. Some applications of these families are discussed in Section 2.4.

2.1. Definitions

Relational statistics for 2×2 tables have been classified in a number of different ways (Sokal and Sneath, 1963; Krippendorff, 1987; Baulieu, 1989; Lesot et al., 2009). Here we distinguish three general types of statistics, called type A, type B and type C, although many other classifications are possible. Before considering the three types, we discuss some preliminaries.

2.1.1. *Preliminaries*

In general, a 2×2 table is obtained if two objects are compared on the presence/absence of a set of attributes, or if a set of objects is cross-classified by two binary variables. To simplify the presentation, we presuppose that the 2×2 table is a cross-classification of two binary (1/0) variables X and Y. The table below is an example of a 2×2 table. The four proportions a, b, c, and d characterize the joint distribution of the variables X and Y. Quantities a and d are often called, respectively, the *positive and negative matches*, whereas b and c are the *mismatches.* The row and column totals are the marginal distributions that result from summing the joint proportions. We denote these by p_1 and q_1 for variable X and by p_2 and q_2 for variable Y. Instead of proportions, the 2×2 may also be defined on counts or frequencies; proportions are used here for notational convenience.

Proportions	$Y = 1$	$Y = 0$	Totals
$X = 1$	a	b	p_1
$X = 0$	c	d	q_1
Totals	p_2	q_2	1

Relational statistics for 2×2 tables are functions that quantify the extent to which two binary variables are associated or the extent to which two objects resemble one another. These functions take as arguments pairs of variables and return numerical values that are higher if the variables are more associated. We will use S as a general symbol for a relational statistic. Following Sokal and Sneath (1963, p. 128), Albatineh et al. (2006), Warrens (2008a,b,c,d,f, 2009) and Lesot et al. (2009), the convention is adopted of calling a measure by its originator or the first we know to propose it. Moreover, we will study the formulas in this paper as sample statistics and not as population parameters.

The term symmetry is usually associated with the mismatches b and c. A statistic is called *symmetric* if the values of b and c can be interchanged without changing the value of the statistic. Although the majority of statistics discussed in this paper are symmetric in b and c, relational statistics are not required to be symmetric (see, for example, Lesot et al., 2009). Asymmetric statistics have natural interpretations if, for example, the variable X is a criterion against which variable Y is evaluated (see Section 2.2.2).

Consider the statistic

$$S_{\text{SM}} = \frac{a+d}{a+b+c+d} \qquad \text{(Sokal and Michener, 1958; Rand, 1971).}$$

Function S_{SM} is also known as the *simple matching coefficient* or the proportion of observed agreement. Statistic S_{SM} is symmetric in b and c since we can interchange the two quantities without effecting the value of S_{SM}. The complement of the simple matching coefficient is a special case of the symmetric set difference (see, for example, Boorman and Arabie, 1972; Margush and McMorris, 1981; Barthélémy and McMorris, 1986). Function $1 - S_{\text{SM}}$ can also been seen as a special case of the Hamming distance between two profiles (see, for example, Day and McMorris, 1991). The *Hamming distance* is a count of the number of positions where two profiles of the same length differ.

Consider the statistics $a/(a+b) = a/p_1$ and $a/(a+c) = a/p_2$. Statistic a/p_2 can be interpreted as measuring the extent to which X is included in Y, whereas a/p_1 reciprocally measures the extent to which Y is included in X (Lesot et al., 2009). The quantities $a/(a+b) = a/p_1$ and a/p_2 are asymmetric statistics.

Sokal and Sneath (1963) made a classical distinction between functions that include the positive matches a only and functions that include both the positive and negative matches a and d (see also, Gower and Legendre, 1986; Baulieu, 1989; Warrens, 2008a,b; Lesot et al., 2009). A binary variable can be an ordinal or a nominal variable. If X is an ordinal variable, then $X = 1$ is more in some sense than $X = 0$. For example, if a binary variable is a coding of the presence or absence of a list of attributes or features, then d reflects the number of negative matches. In the field of numerical taxonomy quantity d is generally felt not to contribute to similarity, and hence should not be included in the definition of the relational statistic.

We are now ready to discuss the three different types of relational statistics.

2.1.2. *Type A statistics*

Type A statistics satisfy the two requirements

$$(A1) \qquad S = 1 \;\text{ if }\; b = c = 0,$$
$$(A2) \qquad S = 0 \Leftrightarrow a = 0.$$

Property $(A1)$ states that $S = 1$ if there are no mismatches (two species types always occur together), whereas $(A2)$ states that $S = 0$ if and only if the proportion $a = 0$ (two species types do not coexist). Type A statistics are typically functions that are increasing in a, decreasing in b and c, and have a range $[1, 0]$. Type A statistics are suitable for ordinal variables and are similar to what are called type 1 statistics in Lesot et al. (2009) (see also Janson and Vegelius, 1981).

Symmetric examples of type A statistics are

$$S_{\text{J}} = \frac{a}{a+b+c} \qquad \text{(Jaccard, 1912)},$$

$$S_{\text{DK}} = \frac{a}{\sqrt{p_1 p_2}} \qquad \text{(Driver and Kroeber, 1932; Ochiai, 1957)},$$

$$S_{\text{K}} = \frac{1}{2}\left(\frac{a}{p_1} + \frac{a}{p_2}\right) \qquad \text{(Kulczyński, 1927)}.$$

Asymmetric examples are a/p_1 and a/p_2 (Dice, 1945; Wallace, 1983). Statistics

S_{DK} and S_{K} are the geometric and arithmetic means (see Section 2.3.3) of the quantities a/p_1 and a/p_2.

Type A statistics can be functions that are increasing in d. An example is the statistic

$$S_{\text{BB}} = \frac{a + \sqrt{ad}}{a + \sqrt{ad} + b + c} \qquad \text{(Baroni-Urbani and Buser, 1976).}$$

The statistic $S_{\text{RR}} = a$ (Russel and Rao, 1940) is a hybrid type A statistic (Sokal and Sneath, 1963), since it does not satisfy $(A1)$. The statistic S_{RR} does satisfy $(A2)$, one of the requirements of a type A statistic. Furthermore, statistic S_{RR} satisfies the requirement $S = 1$ if $b = c = d = 0 \Leftrightarrow a = 1$. The complement of S_{RR} is a special case of a distance used in Goddard, Kubicka, Kubicki and McMorris (1994) that counts the number of leaves which have to be pruned from two trees to obtain a common substructure.

Type A statistic

$$S_{\text{Si}} = \frac{a}{\min(p_1, p_2)} \qquad \text{(Simpson, 1943)}$$

satisfies the stronger condition

$$(A1)^* \qquad S = 1 \text{ if } b = 0 \vee c = 0.$$

Statistic $S_{\text{Si}} = 1$ if one species type only occurs in locations where a second species type exists.

2.1.3. *Type B statistics*

Type B statistics satisfy the two requirements

$$(A1) \qquad S = 1 \text{ if } b = c = 0,$$
$$(B2) \qquad S = 0 \text{ if } a = d = 0.$$

Condition $(B2)$ states that $S = 0$ if there are no positive and negative matches. Type B statistics are typically functions that are increasing in a and d, decreasing in b and c, and have a range $[1, 0]$. Type B statistics are suitable for nominal variables and are similar to what are called type 2 statistics in Lesot et al. (2009).

Examples of type B statistics are S_{SM} (Section 2.1.1) and

$$S_{\text{SS1}} = \frac{1}{4}\left(\frac{a}{p_1} + \frac{a}{p_2} + \frac{d}{q_1} + \frac{d}{q_2}\right) \qquad \text{(Sokal and Sneath, 1963),}$$

$$S_{\text{SS2}} = \frac{ad}{\sqrt{p_1 p_2 q_1 q_2}} \qquad \text{(Sokal and Sneath, 1963).}$$

Sokal and Sneath (1963) proposed coefficients S_{SS1} and S_{SS2} as alternatives to type A statistics S_{DK} and S_{K} for nominal variables. Statistic S_{K} is the arithmetic mean of conditional probabilities a/p_1 and a/p_2, whereas S_{SS1} is the arithmetic mean of conditional probabilities a/p_1, a/p_2, d/q_1 and d/q_2. Statistic S_{DK} is the geometric mean of a/p_1 and a/p_2, whereas S_{SS2} is the square root of the geometric mean of a/p_1, a/p_2, d/q_1 and d/q_2 (Janson and Vegelius, 1981). Note that S_{SS2} actually satisfies the stronger requirement $S = 0$ if $a = 0 \vee d = 0$.

2.1.4. *Type C statistics*

Type C statistics satisfy the three conditions

$$\begin{aligned} &(A1) \qquad S = 1 \text{ if } b = c = 0, \\ &(C2) \qquad S = 0 \text{ under statistical independence}, \\ &(C3) \qquad S = -1 \text{ if } a = d = 0. \end{aligned}$$

Requirement $(C3)$ states that $S = -1$ if there are no matches. It expresses perfect negative association, in which the categories of one variable must be reversed to match the categories of the other variable. Furthermore, $(C2)$ specifies that $S = 0$ if the variables X and Y are statistically independent, that is, $ad = bc \Leftrightarrow a = p_1p_2$ (see Section 2.2.1). Type C statistics are functions that are increasing in a and d, decreasing in b and c, and have a range $[1, -1]$. Type C statistics are suitable for nominal variables.

Examples of type C statistics are

$$S_{Y1} = \frac{ad - bc}{\sqrt{p_1p_2q_1q_2}} \qquad \text{(Yule, 1912)},$$

$$S_{Coh} = \frac{2(ad - bc)}{p_1q_2 + p_2q_1} \qquad \text{(Cohen, 1960)}.$$

Function S_{Y1} is also known as the phi coefficient (cf. Zysno, 1997). Function S_{Coh} is Cohen's kappa for the case of two categories (Kraemer, 1979; Bloch and Kraemer, 1989; Guggenmoos-Holzmann, 1996).

Some type C statistics satisfy the stronger conditions

$$\begin{aligned} &(A1)^* \qquad S = 1 \text{ if } b = 0 \vee c = 0, \\ &(C3)^* \qquad S = -1 \text{ if } a = 0 \vee d = 0. \end{aligned}$$

An example is the relational statistic

$$S_{Y2} = \frac{ad - bc}{ad + bc} \qquad \text{(Yule, 1900)}$$

which is also known as Yule's Q.

2.2. Examples of relational statistics for 2 × 2 tables

In this section we discuss some important relational statistics for 2×2 tables, namely, the tetrachoric correlation and the odds ratio, statistics used in epidemiological studies, measures of ecological association, measures for comparing two partitions, and a measure for test homogeneity.

2.2.1. *Tetrachoric correlation and odds ratio*

A traditional measure for the 2×2 table is the tetrachoric correlation (Pearson, 1900; Divgi, 1979). It is an important statistic because the tetrachoric correlation is an estimate of the Pearson product-moment correlation coefficient between hypothetical row and column variables with normal distributions, that would reproduce the observed contingency table if they were divided into two categories in the appropriate proportions. Because an approximate estimate of the Pearson correlation may well be as adequate in many applications, particularly in small samples, various authors have introduced approximations to the tetrachoric correlation (Digby, 1983; Castellan, 1966; Pearson, 1900). The tetrachoric correlation cannot be expressed in the proportions a, b, c and d.

Another classic statistic is the odds ratio or cross-product ad/bc. The odds ratio is defined as the ratio of the odds of an event occurring in one group (a/b) to the odds of it occurring in another group (c/d). These groups might be any other dichotomous classification. An odds ratio of 1 indicates that the condition or event under study is equally likely in both groups. An odds ratio greater than 1 indicates that the event is more likely in the first group. Probability theory tells us that two binary variables are statistically independent if the odds ratio is equal to unity, that is, $ad/bc = 1$. Due to the simple formula of its standard error, the logarithm of the odds ratio is sometimes preferred over the ordinary odds ratio. The value of the odds ratio lies between zero and infinity.

Edwards (1963) discussed several relational statistics as a function of the cross-product ad/bc. Functions that transform the odds ratio to a correlation-like range $[1, -1]$ are

$$S_{\mathrm{Y2}} = \frac{\frac{ad}{bc} - 1}{\frac{ad}{bc} + 1} = \frac{ad - bc}{ad + bc} \qquad \text{(Yule, 1900)}$$

$$S_{\mathrm{Dy}} = \frac{(ad)^{3/4} - (bc)^{3/4}}{(ad)^{3/4} + (bc)^{3/4}} \qquad \text{(Digby, 1983)}$$

$$S_{\mathrm{Y3}} = \frac{(ad)^{1/2} - (bc)^{1/2}}{(ad)^{1/2} + (bc)^{1/2}} \qquad \text{(Yule, 1912).}$$

Statistics S_{Y2} and S_{Y3} are also known as Yule's Q and Yule's Y. Statistics S_{Y2}, S_{Dy}, and S_{Y3} are nonlinear transformations of ad/bc. All three statistics have been proposed as alternatives (approximations) to the tetrachoric correlation and all three are type C statistics (Section 2.1.4) that satisfy $(A1)^*$ and $(C3)^*$. Some properties of S_{Y2} and S_{Y3} are discussed in Castellan (1966). Digby (1983) uses the symbol H for S_{Dy} and shows that the statistic performs better than S_{Y2} and S_{Y3} as an approximation to the tetrachoric correlation.

For 2×2 tables, Yule (1912) also proposed the relational statistic

$$S_{\mathrm{Y1}} = \frac{ad - bc}{\sqrt{p_1 p_2 q_1 q_2}} \qquad \text{(Yule, 1912).}$$

Function S_{Y1} is also known as the phi coefficient (cf. Zysno, 1997). Statistic S_{Y1} is what Pearson's product-moment correlation becomes when it is applied to binary variables. The statistic

$$S_{\text{Do}} = \frac{(ad-bc)^2}{p_1p_2q_1q_2} \qquad \text{(Doolittle, 1885)}$$

is better known as the mean square contingency (Pearson, 1926; Goodman and Kruskal, 1959, p. 126). Function S_{Y1} is the square root of S_{Do}. Statistics S_{Y1} and S_{Do} are type C statistics (Section 2.1.4).

2.2.2. *Epidemiological studies*

The odds ratio is probably the most widely used relational statistic in epidemiology (Kraemer, 2004). In addition, a variety of other statistics can be used. In general, two cases can be distinguished in epidemiology. In the first case the variable X is a criterion against which variable Y is evaluated (Kraemer, 2004). Examples are the evaluation of a new medical test against a gold standard diagnosis, or a risk factor against a disorder, or assessing the validity of a binary measure against a binary criterion. In these cases a and d are the proportions of true positives and true negatives, whereas b and c are the proportions of false positives and false negatives. A researcher is interested in statistics like sensitivity (a/p_1), specificity (d/q_1), the predictive value of a positive Y (a/p_2) and the predictive value of a negative Y (d/q_2). These four functions are type A statistics (Section 2.1.2). Other examples of asymmetric statistics are

$$S_{\text{P1}} = \frac{ad-bc}{p_1q_2} \quad \text{and}$$

$$S_{\text{P0}} = \frac{ad-bc}{p_2q_1} \qquad \text{(Peirce, 1884).}$$

Statistics S_{P1} and S_{P0} were also proposed in Light (1971) and are generally known as weighted kappas (Kraemer, 2004). Functions S_{P1} and S_{P0} are type C statistics (Section 2.1.4).

In the second case the variables X and Y are equally important, for example, in studies of inter-rater reliability or test-retest reliability. Suppose the variables are observers and that the 2×2 table is the cross classification of the judgments by the raters on the presence or absence of a trait. An obvious measure of agreement that has been proposed independently for this situation by various authors (Fleiss, 1975; Goodman and Kruskal, 1954) is the proportion of all subjects on whom the two raters agree, that is $a+d$. The observed proportion of agreement

$$S_{\text{SM}} = \frac{a+d}{a+b+c+d} \qquad \text{(Sokal and Michener, 1958)}$$

is also known as the simple matching coefficient. Since in the present notation $a+b+c+d=1$, $S_{\text{SM}} = a+d$. Statistic S_{SM} is thus the sum of the positive and negative matches and can be interpreted as the number of 1s and 0s shared by the variables in the same positions, divided by the total number of positions.

In reliability studies it is considered a necessity that a relational statistic measures agreement over and above chance agreement. Examples of statistics that control for chance agreement are the phi coefficient S_{Y1} and

$$S_{\text{Coh}} = \frac{2(ad-bc)}{p_1q_2+p_2q_1} \qquad \text{(Cohen, 1960)},$$

$$S_{\text{MP}} = \frac{2(ad-bc)}{p_1q_1+p_2q_2} \qquad \text{(Maxwell \& Pilliner, 1968)}.$$

Function S_{Coh} is Cohen's kappa for the case of two categories (Kraemer, 1979; Bloch and Kraemer, 1989; Feinstein and Cicchetti, 1990). Warrens (2008e) proves that the 2×2 kappa S_{Coh} is equivalent to the Hubert-Arabie (1985) adjusted Rand index for cluster validation (cf. Steinley, 2004; Albatineh et al., 2006). Although statistics S_{Y1}, S_{Coh}, and S_{MP} have a correlation-like range $[1,-1]$ (type C statistics), the measures are commonly used to distinguish between positive and zero association.

For recommendations and guidelines on what statistics to use under what circumstances in epidemiological studies, we refer to Kraemer (2004).

2.2.3. *Ecological association*

In ecological biology, one may distinguish several contexts where association coefficients can be used (Sokal and Sneath, 1963; Janson and Vegelius, 1981). One such case deals with measuring the degree of coexistence between two species types over different locations. A second situation is measuring association between two locations over different species types. In the first situation a binary variable is a coding of the presence or absence of a species type in a number of locations. The joint proportion a then equals the proportion of locations where both species types are found.

Dice (1945) discussed the asymmetric quantities a/p_1 and a/p_2 (see also Wallace, 1983; Post and Snijders, 1992). Statistic a/p_1 is equal to the number of locations where both species types are found divided by the number of locations where only the first species type is found. Quantity a/p_2 is then the number of locations where both species types exist divided by the number of locations of the second species type.

Popular statistics for ecological association are

$$S_{\text{J}} = \frac{a}{a+b+c} = \frac{a}{p_1+p_2-a} \qquad \text{(Jaccard, 1912)},$$

$$S_{\text{Di}} = \frac{2a}{2a+b+c} = \frac{2a}{p_1+p_2} \qquad \text{(Dice, 1945)},$$

$$S_{\text{DK}} = \frac{a}{\sqrt{p_1p_2}} \qquad \text{(Driver and Kroeber, 1932)}.$$

Coefficient S_J can be interpreted as the number of 1s shared by X and Y in the same positions, divided by the total number of positions were 1s occur. Statistic S_{Di} seems to be independently proposed by both Dice (1945) and Sørenson (1948), but is often contributed to the former. The statistic is a special case of statistics by Czekanowski (1913, 1932) and Gleason (1920). With respect to S_J coefficient S_{Di} gives twice as much weight to a. The latter function is regularly used with presence/absence data in the case that there are only a few positive matches relatively to the number of mismatches. Statistic S_{Di} is also derived in Nei and Li (1979). Coefficient S_{DK} by Driver and Kroeber (1932) is often attributed to Ochiai (1957). Coefficient S_{DK} is also proposed by Fowlkes and Mallows (1983) for the comparison of two clustering algorithms. Functions a/p_1 and a/p_2, and S_J, S_{Di} and S_{DK} are type A statistics.

Statistics S_J, S_{Di} and S_{DK} are popular measures of ecological association, and they have been empirically compared to other functions for 2×2 tables in numerous studies. For example, Duarte, Santos and Melo (1999) evaluated association measures in clustering and ordination of common bean cultivars analyzed by RAPD type molecular markers. The genetic distance measures obtained by taking the complement of statistic S_{Di} were considered the most adequate. Boyce and Ellison (2001) studied similarity coefficients for 2×2 tables in the context of fuzzy set ordination, and concluded that the statistics S_J, S_{Di} and S_{DK}, are the preferred association measures.

2.2.4. *Comparing two partitions*

In many applications of cluster analysis, for example cluster validity, one is interested in comparing the partitions from two different clustering methods (Rand, 1971; Popping, 1983; Fowlkes and Mallows, 1983; Wallace, 1983; Hubert and Arabie, 1985; Steinley, 2004; Albatineh et al., 2006). An equivalent problem in psychology is that of measuring agreement among judges in classifying answers to open-ended questions, or psychologists rating people on categories not defined in advance (Brennan and Light, 1974; Janson and Vegelius, 1982; Popping, 1983, 1984).

Suppose we have two partitions of the same objects. The two clustering partitions can be summarized by a 2×2 table with quantities a, b, c, and d, by counting the number of pairs of objects that were placed in the same cluster in both partitions (a), in the same cluster in one partition but in different clusters in the other partition (b and c), and in different clusters in both (d). A statistic for a 2×2 table can then be used to quantify the amount of agreement between the two partitions (Steinley, 2004; Albatineh et al., 2006; Warrens, 2008e).

For some time, the Rand index

$$S_R = \frac{a+d}{a+b+c+d} \qquad \text{(Rand, 1971)}$$

was a popular measure for comparing two partitions. Coefficient S_R is equivalent to the simple matching coefficient S_{SM}. Statistic S_R was also proposed in Brennan and Light (1974) for measuring agreement among psychologists rating people on categories not defined in advance. Nowadays, there seems to be considerable agreement in the cluster community that the Hubert-Arabie (1985) adjusted Rand index (cf. Steinley, 2004) is the preferred measure for comparing two partitions. Warrens (2008e) showed that the Hubert-Arabie adjusted Rand index can be written as

$$S_{\text{HA}} = \frac{2(ad - bc)}{p_1 q_2 + p_2 q_1} \qquad \text{(Hubert and Arabie, 1985)}.$$

The adjusted Rand index S_{HA} is thus equivalent to Cohen's kappa for two categories, where the categories are "same cluster" and "different cluster".

Fowlkes and Mallows (1983) used the measure $a/\sqrt{p_1 p_2}$ (S_{DK}, Driver and Kroeber, 1932) and Wallace (1983) discussed the quantities a/p_1 and a/p_2 for comparing two partitions.

2.2.5. *Test homogeneity*

The statistic

$$S_{\text{Be}} = \frac{ad - bc}{\min(p_1 q_2, p_2 q_1)} \qquad \text{(Benini, 1901)}$$

is a central quantity in Mokken scale analysis (Sijtsma and Molenaar, 2002), a methodology that may be used to select a subset of binary test items that are sensitive to the same underlying dimension. Coefficient S_{Be} was attributed to Loevinger (1947, 1948) by Mokken (1971) and Sijtsma and Molenaar (2002). Goodman and Kruskal (1959, p. 134; 1979) and Krippendorff (1987) reported that statistic S_{Be} was first proposed by Benini (1901). Goodman and Kruskal (1959, p. 134; 1979) reported that the statistic is also proposed in Jordan (1941). Relational statistic S_{Be} was also considered in Johnson (1945).

Although coefficient S_{Be} has a correlation-like range $[1, -1]$ (type C statistic), it is usual to assume that two items are at least positively dependent. The function S_{Be} is a type C statistic that satisfies the condition $(A1)^*$ (Section 2.1.4). We have $S_{\text{Be}} = 1$ if two items form a so-called Guttman pair. In this case, all subjects that pass the first item also pass the second item, or vice versa. Using S_{Be} we may have perfect association with different marginal distributions, that is, the item popularities or difficulties p_1 and p_2 may be different.

The three statistics of ecological association S_{J}, S_{Di} and S_{DK} (Section 2.2.3) measure the degree to which two species types occur jointly in a number of locations. Several authors proposed coefficients of ecological association that measure the degree to which the observed proportion of joint occurrences of two species types exceeds or falls short of the proportion of joint occurrences expected on the basis of chance alone (cf. Cole, 1949). A measure introduced in Cole (1949), can be written as

$$S_{\mathrm{Col}} = \begin{cases} (ad - bc)/\min(p_1 q_2, p_2 q_1) & \text{if } ad > bc \\ 0 & \text{if } ad = bc \\ (ad - bc)/\min(p_1 p_2, q_1 q_2) & \text{if } ad < bc. \end{cases}$$

The formula S_{Col} can be found in Ratliff (1982) and Warrens (2008d). Statistic S_{Col} is equivalent to S_{Be} if $ad \geq bc$, that is, if the two binary variables are positively dependent. Although S_{Col} is less popular than measures S_{J}, S_{Di} and S_{DK}, the coefficient has been used in various applications by animal and plant ecologists (Hurlbert, 1969; Ratliff, 1982). A variant of S_{Col} proposed in Hurlbert (1969), is less influenced by the species' frequencies. Hurlbert (1969) examined both S_{Col} and the variant as approximations to the tetrachoric correlation. Function S_{Col} is a type C statistic that satisfies the conditions $(A1)^*$ and $(C3)^*$.

2.3. Families of relational statistics for 2 × 2 tables

In this section we consider various families that have been proposed in the literature. Many relational statistics are special cases of a one-parameter family. The families specify how various statistics are related to one another and provide ways for interpreting them.

2.3.1. *Rational functions*

Gower and Legendre (1986, p. 13) defined two parameter families of which all members are rational functions, linear in both numerator and denominator. The families are given by

$$S_{\mathrm{FG}}(\theta) = \frac{a}{a + \theta(b + c)} \qquad \text{(Fichet, 1986; Gower, 1986)},$$

$$S_{\mathrm{GL}}(\theta) = \frac{a + d}{a + \theta(b + c) + d} \qquad \text{(Gower and Legendre, 1986)},$$

where θ is a positive real number to avoid negative values. According to Heiser and Bennani (1997) function $S_{\mathrm{FG}}(\theta)$ was first studied by Fichet (1986) and Gower (1986). Special cases of $S_{\mathrm{FG}}(\theta)$ are type A statistics (Section 2.1.2). Using $\theta = 2, 1, \frac{1}{2}$ in $S_{\mathrm{FG}}(\theta)$ we obtain the special cases

$$S_{\mathrm{SS3}} = \frac{a}{a + 2(b + c)} \qquad \text{(Sokal and Sneath, 1963)},$$

$$S_{\mathrm{J}} = \frac{a}{a + b + c} \qquad \text{(Jaccard, 1912)},$$

$$S_{\mathrm{Di}} = \frac{2a}{2a + b + c} \qquad \text{(Dice, 1945; Sørenson, 1948)}.$$

Function S_{Di} gives twice as much weight to proportion a compared to S_{J}, whereas S_{SS3} gives twice as much weight to $b + c$ compared to S_{J}. Function S_{Di} is regularly used if there are only a few positive matches relatively to the number of mismatches. Lesot et al. (2009, p. 67) report the special cases of $S_{\rm FG}(\theta)$ for $\theta = \frac{1}{4}, \frac{1}{8}$.

Janson and Vegelius (1981) show an interesting relationship between the special cases of $S_{\rm FG}(\theta)$ that can sometimes be useful when comparing two of them (see also, Snijders et al., 1992; Lesot et al., 2009). Statistics $S_{\rm Di}$ and $S_{\rm J}$ are related by $S_{\rm Jac} = S_{\rm Di}/(2 - S_{\rm Di})$. In general it holds that

$$S_{\rm FG}(2\theta) = \frac{S_{\rm FG}(\theta)}{2 - S_{\rm FG}(\theta)}.$$

Let δ and ω be positive real numbers. Function $S_{\rm FG}(\theta)$ is a special case of the contrast model

$$S_{\rm T}(\delta, \omega) = \frac{a}{a + \delta b + \omega c}$$

proposed by Tversky (1977). In contrast to $S_{\rm FG}(\theta)$ the function $S_{\rm T}(\delta, \omega)$ does not impose the symmetry property. Using $\delta = \omega$ in $S_{\rm T}(\delta, \omega)$ we obtain $S_{\rm FG}(\theta)$. Using $\delta = 0, 1$ and $\omega = 1 - \delta$ in $S_{\rm T}(\delta, \omega)$ we obtain the conditional probabilities a/p_1 and a/p_2.

The special cases of $S_{\rm GL}(\theta)$ include the negative matches d in the numerators and denominators. Special cases of $S_{\rm GL}(\theta)$ are type B statistics (Section 2.1.3). Using $\theta = 2, 1, \frac{1}{2}$ in $S_{\rm GL}(\theta)$ we obtain

$$S_{\rm RT} = \frac{a + d}{a + 2(b + c) + d} \qquad \text{(Rogers and Tanimoto, 1960)},$$

$$S_{\rm SM} = a + d \qquad \text{(Sokal and Michener, 1958)},$$

$$S_{\rm SS4} = \frac{2(a + d)}{2a + b + c + 2d} \qquad \text{(Sokal and Sneath, 1963)}.$$

Statistic $S_{\rm SM}$ is the simple matching coefficient. It holds that

$$S_{\rm GL}(2\theta) = \frac{S_{\rm GL}(\theta)}{2 - S_{\rm GL}(\theta)}.$$

Function $S_{\rm GL}(\theta)$ is a special case of the complement of the dissimilarity function

$$D_{\rm B}(\delta, \omega) = \frac{b + c}{\delta a + b + c + \omega d},$$

derived in Baulieu (1989). Function $D_{\rm B}(\delta, \omega)$ satisfies a variety of desiderata in a formal framework considered in Baulieu (1989). Using $1/\delta = 1/\omega$ in $1 - D_{\rm B}(\delta, \omega)$ we obtain $S_{\rm GL}(\theta)$.

Let ε be a real number. Warrens (2009) considered another type of family of rational functions, linear in both numerators and denominators. This family is given by

$$S_{\rm W}(\varepsilon) = \frac{\varepsilon a + (2 - \varepsilon d)}{\varepsilon a + b + c + (2 - \varepsilon)d}.$$

Using $\varepsilon = 0, 1, 2$ in function $S_{\text{W}}(\varepsilon)$ we obtain the special cases

$$S_{\text{CF}} = \frac{2d}{b+c+2d} \quad \text{(Cicchetti and Feinstein, 1990)}$$

$$S_{\text{SM}} = a + d \quad \text{(Sokal and Michener, 1958)}$$

$$S_{\text{Di}} = \frac{2a}{2a+b+c} \quad \text{(Dice, 1945; Sørenson, 1948).}$$

2.3.2. *Chance-corrected statistics*

Functions S_{Y1}, S_{Y2}, S_{Y3} and S_{Coh} from Sections 2.2.1 and 2.2.2 are examples of statistics that have zero value if binary variables X and Y are statistically independent. Warrens (2008a, 2009) considered a family of statistics that correct for chance and that can be expressed in the form

$$G[E(a+d)] = \frac{a+d-E(a+d)}{1-E(a+d)},$$

where $0 < E(a+d) \leq 1$. In $G[E(a+d)]$, $a+d = S_{\text{SM}}$ is the observed proportion of agreement. The quantity $E(a+d)$ is called the expected proportion of agreement, and is conditional on fixed marginal proportions of the 2×2 table. The value 1 is the maximum value of $a+d$.

Four definitions of $E(a+d)$ from the literature are

$$E(a+d)_{\text{BAG}} = 1/2,$$

$$E(a+d)_{\text{GK}} = \frac{\max(p_1+p_2, q_1+q_2)}{2},$$

$$E(a+d)_{\text{Sc}} = \left(\frac{p_1+p_2}{2}\right)^2 + \left(\frac{q_1+q_2}{2}\right)^2,$$

$$E(a+d)_{\text{Coh}} = p_1p_2 + q_1q_2.$$

Using $E(a+d)_{\text{BAG}}$, $E(a+d)_{\text{GK}}$, $E(a+d)_{\text{S}}$ and $E(a+d)_{\text{C}}$ in $G[E(a+d)]$, we obtain, respectively, Bennett, Alpert and Goldstein's S, Goodman and Kruskal's lambda, Scott's pi and Cohen's kappa, given by

$$S_{\text{BAG}} = 2(a+d) - 1 \quad \text{(Bennett et al., 1954),}$$

$$S_{\text{GK}} = \frac{2\min(a,d) - b - c}{2\min(a,d) + b + c} \quad \text{(Goodman and Kruskal, 1954),}$$

$$S_{\text{Sc}} = \frac{4ad - (b+c)^2}{(p_1+p_2)(q_1+q_2)} \quad \text{(Scott, 1955),}$$

$$S_{\text{Coh}} = \frac{2(ad-bc)}{p_1q_2 + p_2q_1} \quad \text{(Cohen, 1960).}$$

Statistic $S_{\text{BAG}} = 2(a+d) - 1$ is actually a special case of the function proposed in Bennett et al. (1954). The statistic for square agreement tables with two or more categories is equivalent to the measure C proposed in Janson and Vegelius

(1979, p. 260), the measure κ_n discussed in Brennan and Prediger (1981, p. 693) and RE proposed in Janes (1979). The statistic S_{BAG} for 2×2 tables is equivalent to statistics discussed or derived in Hamann (1961), Holley and Guilford (1964), Maxwell (1977) and Byrt, Bishop and Carlin (1993).

Statistics S_{BAG}, S_{GK}, S_{Sc} and S_{Coh} are based on different assumptions and may therefore not be appropriate in all contexts. The assumptions are hidden in the different definitions of $E(a+d)$. Reviews of the rationales behind S_{BAG}, S_{Sc} and S_{Coh} can be found in Zwick (1988), Hsu and Field (2003) and De Mast (2007). Following Krippendorff (1987) and Warrens (2008c, 2009), we distinguish three ways in which chance factors may operate: two, one or none underlying continua.

Suppose the data are a product of chance concerning two different frequency distributions (Cohen, 1960; Krippendorf, 1987), one for each variable. $E(a+d)_{\text{Coh}}$ is the value of $S_{\text{SM}} = a + d$ under statistical independence. $E(a+d)_{\text{Coh}}$ can be obtained by considering all permutations of the observations of one of the binary variables, while preserving the order of the observations of the other variable. For each permutation the value of S_{SM} can be determined. The arithmetic mean of these values is $p_1p_2 + q_1q_2$.

A second possibility is that there are no relevant underlying continua. $E(a+d)_{\text{GK}}$ simply focuses on the most abundant category. Alternatively, if, for example, two raters randomly allocate objects to categories, then, for each rater, the expected marginal probability for each category is 1/2. The probability that two raters assign, by chance, any object to the same category is $(1/2)(1/2) = 1/4$. Summing these probabilities over the two categories, we obtain $2/4 = 1/2 = E(P)_{\text{BAG}}$. Furthermore, in the case of two distributions of which one is the uniform distribution, Brennan and Prediger (1981, p. 693) showed that the probability of chance agreement is also given by $E(a+d)_{\text{BAG}} = 1/2$.

Finally, suppose it is assumed that the frequency distribution underlying the two binary variables is the same for both variables (Scott, 1955; Krippendorf, 1987). The expectation of proportion a must be estimated from the marginals p_1 and p_2. Different functions may be used. Scott (1955) proposed the arithmetic mean $(p_1 + p_2)/2$.

2.3.3. *Power mean*

There are several functions that may reflect the mean value of two real non-negative (or two non-positive) numbers u and v. The harmonic, geometric and arithmetic means, also known as the Pythagorean means, are given by respectively $2/(u^{-1} + v^{-1})$, $\sqrt{uv}$ and $(u+v)/2$ (Janson and Vegelius, 1981). Several coefficients can be expressed in terms of these Pythagorean means.

Consider for example the quantities a/p_1 and a/p_2 (Dice, 1945; Post and Snijders, 1993). Both are special cases of function $S_{\text{T}}(\delta, \omega)$. The quantity a/p_2 can be interpreted as the extent to which variable X is included in variable Y, whereas a/p_1 reciprocally measures the extent to which Y is included in variable X. The harmonic, geometric and arithmetic means of a/p_1 and a/p_2 are respectively

$$S_{\mathrm{Di}} = \frac{2a}{p_1 + p_2} \qquad \text{(Dice, 1945; Sørenson, 1948)},$$

$$S_{\mathrm{DK}} = \frac{a}{\sqrt{p_1 p_2}} \qquad \text{(Driver and Kroeber, 1932; Ochiai, 1957)},$$

$$S_{\mathrm{K}} = \frac{1}{2}\left(\frac{a}{p_1} + \frac{a}{p_2}\right) \qquad \text{(Kulczyński, 1927)}.$$

Different types of statistics can be obtained by considering abstractions of the Pythagorean means. One type of so-called generalized means is the power mean, sometimes referred to as the Hölder mean (see, for example, Bullen, 2003). Let ε be a real number. The power mean $M_\varepsilon(u, v)$ of u and v is given by

$$M_\varepsilon(u, v) = \left(\frac{u^\varepsilon + v^\varepsilon}{2}\right)^{1/\varepsilon}.$$

Special cases of $M_\varepsilon(u, v)$ are the minimum ($\min(u, v)$) and maximum ($\max(u, v)$) and the Pythagorean means. A variety of statistics turn out to be special cases of a power mean. The harmonic, geometric and arithmetic means of a/p_1 and a/p_2 are S_{Di}, S_{DK} and S_{K}. The minimum and maximum of a/p_1 and a/p_2 are given by

$$S_{\mathrm{Br}} = \frac{a}{\max(p_1, p_2)} \qquad \text{(Braun-Blanquet, 1932)},$$

$$S_{\mathrm{Si}} = \frac{a}{\min(p_1, p_2)} \qquad \text{(Simpson, 1943)}.$$

As a second example of a power mean, consider the weighted kappas (Section 2.2.2)

$$S_{\mathrm{P1}} = \frac{ad - bc}{p_1 q_2} \quad \text{and}$$

$$S_{\mathrm{P0}} = \frac{ad - bc}{p_2 q_1} \qquad \text{(Peirce, 1884; Cole, 1949)}.$$

The quantity $ad - bc$ is known as the covariance for two binary variables. If $p_1 \leq p_2$ then $p_1 q_2$ is the maximum value of the covariance $ad - bc$ given the marginal proportions. The harmonic and geometric means and the maximum of S_{P1} and S_{P0} are

$$S_{\mathrm{Coh}} = \frac{2(ad - bc)}{p_1 q_2 + p_2 q_1} \qquad \text{(Cohen, 1960)},$$

$$S_{\mathrm{Y1}} = \frac{ad - bc}{\sqrt{p_1 p_2 q_1 q_2}} \qquad \text{(Yule, 1912)},$$

$$S_{\mathrm{Be}} = \frac{ad - bc}{\min(p_1 q_2, p_2 q_1)} \qquad \text{(Benini, 1901)}.$$

Statistics S_{Coh} and S_{Y1} are Cohen's (1960) kappa and the phi coefficient (cf. Zysno, 1997). Statistic S_{Be} is discussed in Section 2.2.5.

2.3.4. *Linear transformations of S_{SM}*

Warrens (2008c,d, 2009) studied a family of statistics that are linear transformations of $S_{\text{SM}} = a+d$. Members in this family are of the form $\lambda + \mu(a+d)$ where λ and μ, unique for each statistic, depend on fixed marginal proportions of the 2×2 table. Since $a = p_2 - q_1 + d$, proportions a and d are also linear in $(a+d)$. Linear in S_{SM} is therefore equivalent to linear in a and linear in d.

Many relational statistics for 2×2 tables can be expressed in the form $\lambda + \mu(a+d)$. We consider some examples. Statistic S_{Di} can be expressed in the form $\lambda_{\text{Di}} + \mu_{\text{Di}}(a+d)$ where

$$\lambda_{\text{Di}} = -\frac{1}{p_1+p_2} + 1 \quad \text{and} \quad \mu_{\text{Di}} = \frac{1}{p_1+p_2}.$$

In fact, Warrens (2009) showed that all special cases of the parameter family

$$S_{\text{W}}(\varepsilon) = \frac{\varepsilon a + (2 - \varepsilon d)}{\varepsilon a + b + c + (2-\varepsilon)d}$$

are linear transformations of S_{SM} given the marginal proportions. The quantities λ and μ for family $S_{\text{W}}(\varepsilon)$ (Section 2.3.1) are given by

$$\lambda_{\text{W}} = \frac{(\varepsilon-1)(p_2-q_1)}{1+(\varepsilon-1)(p_2-q_1)},$$
$$\mu_{\text{W}} = \frac{1}{1+(\varepsilon-1)(p_2-q_1)}.$$

Statistic S_{Coh} can be expressed in the form $\lambda_{\text{Coh}} + \mu_{\text{Coh}}(a+d)$ where

$$\lambda_{\text{Coh}} = -\frac{p_1p_2+q_1q_2}{p_1q_2+p_2q_1} \quad \text{and} \quad \mu_{\text{Coh}} = \frac{1}{p_1q_2+p_2q_1}.$$

In fact, all functions of the form $G[E(a+d)]$ (Section 2.3.2) can be expressed in the form $\lambda_{\text{G}} + \mu_{\text{G}}(a+d)$ where

$$\lambda_{\text{G}} = -\frac{E(a+d)}{1-E(a+d)} \quad \text{and} \quad \mu_{\text{G}} = \frac{1}{1-E(a+d)}.$$

The power mean of a/p_1 and a/p_2 and the power mean of the weighted kappas S_{P1} and S_{P0} (Section 2.3.3) are also linear transformations of statistic S_{SM} given the marginal proportions. The power mean of a/p_1 and a/p_2 equals

$$M_\varepsilon\left(\frac{a}{p_1}, \frac{a}{p_2}\right) = \frac{a}{p_1p_2}\left(\frac{p_1^\varepsilon + p_2^\varepsilon}{2}\right)^{1/\varepsilon},$$

and is thus linear in the quantity a. The power mean of the weighted kappas S_{P1} and S_{P0}

$$M_\varepsilon\left(\frac{ad-bc}{p_1q_2}, \frac{ad-bc}{p_2q_1}\right)$$

can be written as

$$\frac{a - p_1p_2}{p_1p_2q_1q_2}\left[\frac{(p_1q_2)^\varepsilon + (p_2q_1)^\varepsilon}{2}\right]^{1/\varepsilon}$$

and is thus also linear in the quantity a.

A statistic that cannot be expressed in the form $\lambda + \mu(a + d)$ is

$$S_{\text{J}} = \frac{a}{a+b+c} \qquad \text{(Jaccard, 1912).}$$

Furthermore, the statistics S_{Y2}, S_{Dy} and S_{Y3} (Section 2.2.1) are special cases of the family

$$S_{\text{Dy}}(\beta) = \frac{(ad)^\beta - (bc)^\beta}{(ad)^\beta + (bc)^\beta} \qquad \text{(Digby, 1983)}$$

where $0 < \beta \leq 1$. No special case of function $S_{\text{Dy}}(\beta)$ can be expressed as a linear transformations of the statistic S_{SM} given the marginal proportions.

2.4. Discussion

In this section we consider properties of relational statistics for 2×2 tables that follow from their membership of a particular family.

2.4.1. *Order equivalence*

The formulation of $S_{\text{FG}}(\theta)$ and that of $S_{\text{GL}}(\theta)$ (Section 2.3.1) is closely related to the concept of order equivalence (Sibson, 1972; Gower, 1986; Baulieu, 1989; Batagelj and Bren, 1995; Lesot et al., 2009). If two statistics are order equivalent, they are interchangeable with respect to an analysis method that is invariant under ordinal transformations. The relevant information for these analysis methods is in the ranking induced by the relational statistics, not in the values themselves. Examples are in image retrieval (Lesot et al., 2009) and monotone equivariant cluster analysis (Janowitz, 1979). Any two special cases of $S_{\text{FG}}(\theta)$ are order equivalent, and any two special cases of $S_{\text{GL}}(\theta)$ are order equivalent. Omhover, Rifqi and Detyniecki (2006) showed that two special cases of $S_{\text{T}}(\delta,\omega)$ with parameters (δ,ω) and (δ',ω') are order equivalent if $\delta\omega' = \delta'\omega$. Similarly, Baulieu (1989) showed that two special cases of $D_{\text{B}}(\delta,\omega)$ with parameters (δ,ω) and (δ',ω') are order equivalent if $\delta\omega' = \delta'\omega$.

2.4.2. *Inequalities*

Warrens (2008b) presented inequalities between a variety of statistics for 2×2 tables. Several insights can be obtained from studying inequalities between statistics. For example, if several functions defined on the same quantities have unconditional inequalities between them it is likely that these statistics reflect the association or agreement of the binary variables X and Y in a similar way, but to a different extent (some have lower/higher values than others).

Consider the function

$$S_{\mathrm{FG}}(\theta) = \frac{a}{a+\theta(b+c)} \quad \text{(Fichet, 1986; Gower, 1986)}$$

(Section 2.3.1), with special cases

$$S_{\mathrm{SS3}} = \frac{a}{a+2(b+c)} \quad \text{(Sokal and Sneath, 1963)},$$

$$S_{\mathrm{J}} = \frac{a}{a+b+c} \quad \text{(Jaccard, 1912)},$$

$$S_{\mathrm{Di}} = \frac{2a}{2a+b+c} \quad \text{(Dice, 1945; Sørenson, 1948)}.$$

Since $S_{\mathrm{FG}}(\theta)$ is decreasing in θ, the double inequality $S_{\mathrm{SS3}} \leq S_{\mathrm{J}} \leq S_{\mathrm{Di}}$ is valid. Furthermore, since $S_{\mathrm{GL}}(\theta)$ (Section 2.3.1) is decreasing in θ, the double inequality $S_{\mathrm{RT}} \leq S_{\mathrm{SM}} \leq S_{\mathrm{SS4}}$ is valid.

Consider the function

$$G[E(a+d)] = \frac{a+d-E(a+d)}{1-E(a+d)}$$

(Section 2.3.2), with special cases

$$S_{\mathrm{BAG}} = 2(a+d)-1 \quad \text{(Bennett et al., 1954)},$$

$$S_{\mathrm{GK}} = \frac{2\min(a,d)-b-c}{2\min(a,d)+b+c} \quad \text{(Goodman and Kruskal, 1954)},$$

$$S_{\mathrm{Sc}} = \frac{4ad-(b+c)^2}{(p_1+p_2)(q_1+q_2)} \quad \text{(Scott, 1955)},$$

$$S_{\mathrm{Coh}} = \frac{2(ad-bc)}{p_1q_2+p_2q_1} \quad \text{(Cohen, 1960)}.$$

Using the fact that $G[E(a+d)]$ is decreasing in $E(a+d)$ it can be shown (Warrens, 2008b,c) that $S_{\mathrm{GK}} \leq S_{\mathrm{Sc}} \leq S_{\mathrm{BAG}}, S_{\mathrm{Coh}}$. Furthermore $S_{\mathrm{BAG}} > S_{\mathrm{Coh}}$ if and only if $p_1, p_2 > 0.5$ or $q_1, q_2 > 0.5$. Inequality $S_{\mathrm{BAG}} \leq S_{\mathrm{Coh}}$ otherwise.

Consider the power mean $M_\varepsilon\,(a/p_1, a/p_2)$ (Section 2.3.3) with special cases S_{Di} and

$$S_{\mathrm{Br}} = \frac{a}{\max(p_1,p_2)} \quad \text{(Braun-Blanquet, 1932)},$$

$$S_{\mathrm{DK}} = \frac{a}{\sqrt{p_1p_2}} \quad \text{(Driver and Kroeber, 1932; Ochiai, 1957)},$$

$$S_{\mathrm{K}} = \frac{1}{2}\left(\frac{a}{p_1}+\frac{a}{p_2}\right) \quad \text{(Kulczyński, 1927)},$$

$$S_{\mathrm{Si}} = \frac{a}{\min(p_1,p_2)} \quad \text{(Simpson, 1943)}.$$

Since $M_\varepsilon\,(a/p_1, a/p_2)$ is increasing in ε the inequality $S_{\mathrm{J}} \leq S_{\mathrm{Br}} \leq S_{\mathrm{Di}} \leq S_{\mathrm{DK}} \leq S_{\mathrm{K}} \leq S_{\mathrm{Si}}$ is valid. Using similar arguments, it holds that $|S_{\mathrm{Coh}}| \leq |S_{\mathrm{Y1}}| \leq |S_{\mathrm{Be}}|$.

Finally, consider the function

$$S_{\mathrm{Dy}}(\beta) = \frac{(ad)^{\beta} - (bc)^{\beta}}{(ad)^{\beta} + (bc)^{\beta}} \qquad \text{(Digby, 1983)}$$

where $0 < \beta \leq 1$, with special cases

$$S_{\mathrm{Y2}} = \frac{ad - bc}{ad + bc} \qquad \text{(Yule, 1900),}$$

$$S_{\mathrm{Dy}} = \frac{(ad)^{3/4} - (bc)^{3/4}}{(ad)^{3/4} + (bc)^{3/4}} \qquad \text{(Digby, 1983),}$$

$$S_{\mathrm{Y3}} = \frac{(ad)^{1/2} - (bc)^{1/2}}{(ad)^{1/2} + (bc)^{1/2}} \qquad \text{(Yule, 1912).}$$

Since $S_{\mathrm{Dy}}(\beta)$ is increasing in β, the double inequality $|S_{\mathrm{Y3}}| \leq |S_{\mathrm{Di}}| \leq |S_{\mathrm{Y2}}|$ is valid.

2.4.3. *Correction for chance*

It may be desirable that the theoretical value of a statistic is zero if the two binary variables are statistically independent. In general we have the following requirement.

$$(D1) \qquad S = 0 \ \text{ if } \ a + d = E(a+d) \ \text{ for some } \ E(a+d).$$

Four definitions of $E(a+d)$ from the literature are discussed in Section 2.3.2. If $E(a+d)_{\mathrm{Coh}} = p_1p_2 + q_1q_2$ is the appropriate expectation of $S_{\mathrm{SM}} = a + d$, then $(D1)$ requires that $S = 0$ under statistical independence (condition $(C2)$ in Section 2.1.4).

In several domains of data analysis $(D1)$ is a natural desideratum. In reliability studies and when comparing partitions in cluster analysis, property $(D1)$ is considered a necessity. For example, statistics S_{Y1}, S_{Y2}, S_{Y3} and S_{Coh} each have zero value under statistical independence. Property $(D1)$ is less important for statistics of ecological association (Section 2.2.3), although some authors have argued to look at agreement beyond chance (see statistic S_{Col} in Section 2.2.5). For example, S_{J}, S_{D}, S_{K}, S_{DK}, S_{SM}, and S_{RT} do not have zero value under statistical independence.

If a statistic does not satisfy desideratum $(D1)$, it may be corrected for agreement due to chance (Fleiss, 1975; Krippendorff, 1987; Albatineh et al., 2006; Warrens, 2008c,d, 2009). After correction for chance a similarity coefficient S has a form

$$\frac{S - E(S)}{1 - E(S)},$$

where expectation $E(S)$ is conditional upon fixed marginal proportions of the 2×2 table. Warrens (2009) showed that all special cases of the family

$$S_{\mathrm{W}}(\varepsilon) = \frac{\varepsilon a + (2 - \varepsilon d)}{\varepsilon a + b + c + (2 - \varepsilon)d}$$

have a form

$$G[E(a+d)] = \frac{a+d-E(a+d)}{1-E(a+d)},$$

after correction $[S-E(S)[/[1-E(S)]$. Note that $E(a+d)$ is unspecified in $G[E(a+d)]$. Hence, all special cases of the function $S_{\mathrm{W}}(\varepsilon)$ coincide after correction $[S-E(S)[/[1-E(S)]$, irrespective of which $E(a+d)$ is used. Using $E(a+d)_{\mathrm{BAG}}$, $E(a+d)_{\mathrm{GK}}$, $E(a+d)_{\mathrm{Sc}}$ or $E(a+d)_{\mathrm{Coh}}$ (Section 2.3.2) in $G[E(a+d)]$, we obtain, respectively,

$$S_{\mathrm{BAG}} = 2(a+d) - 1 \qquad \text{(Bennett et al., 1954)},$$

$$S_{\mathrm{GK}} = \frac{2\min(a,d) - b - c}{2\min(a,d) + b + c} \qquad \text{(Goodman and Kruskal, 1954)},$$

$$S_{\mathrm{Sc}} = \frac{4ad - (b+c)^2}{(p_1+p_2)(q_1+q_2)} \qquad \text{(Scott, 1955)},$$

$$S_{\mathrm{Coh}} = \frac{2(ad-bc)}{p_1q_2 + p_2q_1} \qquad \text{(Cohen, 1960)}.$$

Thus, all special cases of the function $S_{\mathrm{W}}(\varepsilon)$ become S_{Coh} using linear transformation $[S-E(S)[/[1-E(S)]$ and $E(a+d)_{\mathrm{Coh}}$.

2.4.4. *Correction for maximum value*

In general we speak of positive association or positive agreement between two variables X and Y if the value of a relational statistic $S \geq 0$. Furthermore, we have perfect association or perfect agreement between the variables if $S = 1$. However, we may require the stronger property

$$(A1)^* \qquad S = 1 \text{ if } b = 0 \vee c = 0.$$

Functions that satisfy requirement $(A1)^*$ are statistics S_{Y1}, S_{Dy}, and S_{Y2} (Section 2.2.1), S_{Si} (Section 2.2.3) and S_{Be} and S_{Col} (Section 2.2.5).

For various statistics for 2 × 2 tables the maximal attainable value depends on the marginal distributions. For example, proportion a in Table 1 cannot exceed its marginal probabilities p_1 and p_2. Statistics S_{J} or S_{Di} (Section 2.2.3) for example, can therefore only attain the maximum value of unity if $p_1 = p_2$, that is, in the case of marginal symmetry. The maximum value of a, denoted by $a_{\max}$, equals $a_{\max} = \min(p_1, p_2)$. The maximum value of S_{Di} given the marginal distributions equals $2\min(p_1,p_2)/(p_1+p_2)$.

The maximum value of the covariance $(ad-bc)$ between two binary variables, given the marginal distributions, is equal to $(ad-bc)_{\max} = \min(p_1q_2, p_2q_1)$. The maximum value of the phi coefficient S_{Y1} and other statistics with the covariance $ad-bc$ in the numerator is thus also restricted by the marginal distributions (Cureton, 1959; Guilford, 1965; Zysno, 1997). In the literature on this phenomenon, it was suggested to use the ratio S_{Y3} divided by the maximum value of S_{Y1} given the marginal probabilities (cf. Davenport and El-Sanhurry, 1991). In general, for coefficients of which the maximum value depends on the marginal probabilities, authors from the phi/phimax literature suggest the linear transformation

$$\frac{S}{S_{\text{max}}},$$

where S_{max} is the maximum value of S given the marginal distributions.

Consider the power mean $M_\varepsilon(a/p_1, a/p_2)$ with special cases

$$S_{\text{Br}} = \frac{a}{\max(p_1, p_2)} \qquad \text{(Braun-Blanquet, 1932)},$$

$$S_{\text{Di}} = \frac{2a}{p_1 + p_2} \qquad \text{(Dice, 1945; Sørenson, 1948)},$$

$$S_{\text{DK}} = \frac{a}{\sqrt{p_1 p_2}} \qquad \text{(Driver and Kroeber, 1932)},$$

$$S_{\text{K}} = \frac{1}{2}\left(\frac{a}{p_1} + \frac{a}{p_2}\right) \qquad \text{(Kulczyński, 1927)}.$$

Warrens (2008f) showed that all special cases of $M_\varepsilon(a/p_1, a/p_2)$ become

$$S_{\text{Si}} = \frac{a}{\min(p_1, p_2)} \qquad \text{(Simpson, 1943)}$$

after the linear transformation S/S_{max}.

Furthermore, consider the power mean $M_\theta(S_{\text{P1}}, S_{\text{P0}})$ with special cases

$$S_{\text{Coh}} = \frac{2(ad - bc)}{p_1 q_2 + p_2 q_1} \qquad \text{(Cohen, 1960)},$$

$$S_{\text{Y1}} = \frac{ad - bc}{\sqrt{p_1 p_2 q_1 q_2}} \qquad \text{(Yule, 1912)},$$

where S_{Coh} and S_{Y1} are Cohen's kappa and the phi coefficient, respectively. Various authors (for example, Fleiss, 1975) have observed that phi/phimax is equal to kappa/kappamax. Warrens (2008f) showed that all special cases of $M_\varepsilon(S_{\text{P1}}, S_{\text{P0}})$ become

$$S_{\text{Be}} = \frac{ad - bc}{\min(p_1 q_2, p_2 q_1)} \qquad \text{(Benini, 1901)}.$$

after the linear transformation S/S_{max}.

We end this paper with some words on statistics that satisfy $(A1)^*$ and $(C2)$. Examples are S_{Y2}, S_{Y3} and S_{Dy} (Section 2.2.1) and statistics S_{Be} and S_{Col} (Section 2.2.5). In Section 2.3.4 it is shown that the family of statistics of a form $\lambda + \mu(a+d)$ has been given a lot of attention in the literature. Warrens (2008d) showed that there is only one statistic of the form $\lambda + \mu(a + d)$ that has a maximum value of unity independent of the marginals and zero value under statistical independence. This statistic happens to be S_{Be}.

References

1. Albatineh, A. N., Niewiadomska-Bugaj, M., & Mihalko, D. (2006). On similarity indices and correction for chance agreement. *Journal of Classification, 23*, 301-313.
2. Baroni-Urbani, C. and Buser, M. W. (1976). Similarity of binary data. *Systematic Zoology, 25,* 251-259.
3. Barthélémy, J.-P., & McMorris, F. R. (1986). The median procedure for n-Trees. *Journal of Classification, 3,* 329-334.
4. Batagelj, V. & Bren, M. (1995). Comparing resemblance measures. *Journal of Classification, 12,* 73-90.
5. Baulieu, F. B. (1989). A classification of presence/absence based dissimilarity coefficients. *Journal of Classification, 6,* 233246.
6. Baulieu, F. B. (1997). Two variant axiom systems for presence/absence based dissimilarity coefficients. *Journal of Classification, 14,* 159170.
7. Benini, R. (1901). *Principii di Demografia.* G. Barbèra, Firenzi. No. 29 of *Manuali Barbèra di Scienza Giuridiche Sociale e Politiche.*
8. Bennett, E. M., Alpert, R., & Goldstein, A. C. (1954). Communications through limited response questioning. *Public Opinion Quarterly, 18,* 303-308.
9. Bloch, D. A., & Kraemer, H. C. (1989). 2 × 2 Kappa coefficients: measures of agreement or association. *Biometrics, 45,* 269-287.
10. Boorman, S. A., & Arabie, P. (1972). Structural measures and the method of sorting. In R.N. Shepard, A.K. Romney, and S.B. Nerlove (Eds.), *Multidimensional Scaling: Theory and Applications in the Behavioral Sciences, vol. 1: Theory,* pp. 225-249. New York: Seminar Press.
11. Boyce, R. L., & Ellison, P. C. (2001). Choosing the best similarity index when performing fuzzy set ordination on binary data. *Journal of Vegetational Science, 12,* 711720.
12. Braun-Blanquet, J. (1932). *Plant Sociology: The Study of Plant Communities.* Authorized English translation of Pflanzensoziologie. New York: McGraw-Hill.
13. Brennan, R. L., & Light, R. J. (1974). Measuring agreement when two observers classify people into categories not defined in advance. *British Journal of Mathematical and Statistical Psychology, 27,* 154163.
14. Brennan, R. L., & Prediger, D. J. (1981). Coefficient kappa: some uses, misuses, and alternatives. *Educational and Psychological Measurement, 41,* 687-699.
15. Bullen, P. S. (2003). *Handbook of Means and Their Inequalities.* Dordrecht: Kluwer.
16. Byrt, T., Bishop, J., & Carlin, J. B. (1993). Bias, prevalence and kappa. *Journal of Clinical Epidemiology, 46,* 423-429.
17. Castellan, N. J. (1966). On the estimation of the tetrachoric correlation coefficient. *Psychometrika, 31,* 6773.
18. Cheetham, A. H., & Hazel, J. E. (1969). Binary (presence-absence) similarity coefficients. *Journal of Paleontology, 43,* 1130-1136.
19. Cicchetti, D. V. & Feinstein, A. R. (1990). High Agreement but low kappa: II. Resolving the paradoxes. *Journal of Clinical Epidemiology, 43,* 551-558.
20. Cohen, J. A. (1960). A coefficient of agreement for nominal scales. *Educational and Psychological Measurement, 20,* 213-220.
21. Cole, L. C. (1949). The measurement of interspecific association. *Ecology, 30,* 411-424.
22. Cureton, E. E. (1959). Note on $\phi/\phi_{\max}$. *Psychometrika, 24,* 8991.
23. Czekanowski, J. (1913). *Zarys Metod Statystycnck.* Warsaw: E. Wendego.
24. Czekanowski, J. (1932). "Coefficient of racial likeness" und "Durchschnittliche Dif-

ferenz." *Anthropologischer Anzeiger, 9,* 227-249.
25. Davenport, E. C., & El-Sanhurry, N. A. (1991). Phi/phimax: review and synthesis. *Educational and Psychological Measurement, 51,* 821-828.
26. Day, W. H. E., & McMorris, F. R. (1991). Critical comparison of consensus methods for molecular sequences. *Nucleic Acids Research, 20,* 1093-1099.
27. De Mast, J. (2007). Agreement and kappa-type indices. *The American Statistician, 61,* 148-153.
28. Deza, E., & Deza, M. M. (2006). *Dictionary of Distances.* Amsterdam: Elsevier.
29. Dice, L. R. (1945). Measures of the amount of ecologic association between species. *Ecology, 26,* 297-302.
30. Digby, P. G. N. (1983). Approximating the tetrachoric correlation coefficient. *Biometrics, 39,* 753757.
31. Divgi, D. R. (1979). Calculation of the tetrachoric correlation coefficient. *Psychometrika, 44,* 169172.
32. Driver, H. E., & Kroeber, A. L. (1932). Quantitative expression of cultural relationship. *The University of California Publications in American Archaeology and Ethnology, 31,* 211-256.
33. Duarte, J. M., Santos, J. B., & Melo, L. C. (1999). Comparison of similarity coefficients based on RAPD markers in the common bean. *Genetics and Molecular Biology, 22,* 427432.
34. Edwards, A. W. F. (1963). The measure of association in a 2 $\times$ 2 table. *Journal of the Royal Statistical Society, Series A, 126,* 109114.
35. Feinstein, A. R., & Cicchetti, D. V. (1990). High agreement but low kappa: I. The problems of two paradoxes. *Journal of Clinical Epidemiology, 43,* 543-549.
36. Fichet, B. (1986). Distances and Euclidean distances for presence-absence characters and their application to factor analysis. In J. de Leeuw, W.J. Heiser, J.J. Meulman and F. Critchley (Eds.), *Multidimensional Data Analysis,* 23-46. Leiden: DSWO Press.
37. Fleiss, J. L. (1975). Measuring agreement between two judges on the presence or absence of a trait. *Biometrics, 31,* 651-659.
38. Fowlkes, E. B., & Mallows, C. L. (1983). A method for comparing two hierarchical clusterings. *Journal of the American Statistical Association, 78,* 553-569.
39. Gleason, H. A. (1920). Some applications of the quadrat method. *Bulletin of the Torrey Botanical Club, 47,* 21-33.
40. Goddard, W., Kubicka, E. Kubicki, G., & McMorris, F. R. (1994). The agreement metric for labeled binary trees. *Mathematical Biosciences, 123,* 215-226.
41. Goodman, G. D., & Kruskal, W. H. (1954). Measures of association for cross classifications. *Journal of the American Statistical Association, 49,* 732-764.
42. Goodman, G. D., & Kruskal, W. H. (1959). Measures of association for cross classifications II: further discussion and references. *Journal of the American Statistical Association, 54,* 123-163.
43. Goodman, L. A., & Kruskal, W. H. (1979). *Measures of Association for Cross Classifications.* New York: Springer-Verlag.
44. Gower, J. C. (1986). Euclidean distance matrices. In J. de Leeuw, W.J. Heiser, J.J. Meulman and F. Critchley (Eds.), *Multidimensional Data Analysis,* 11-22. Leiden: DSWO Press.
45. Gower, J.C., & Legendre, P. (1986). Metric and Euclidean properties of dissimilarity coefficients. *Journal of Classification, 3,* 5-48.
46. Guggenmoos-Holzmann, I. (1996). The meaning of kappa: probabilistic concepts of reliability and validity revisited," *Journal of Clinical Epidemiology, 49,* 775-783.

47. Guilford, J. P. (1965). The minimal phi coefficient and the maximal phi. *Educational and Psychological Measurement, 25,* 38.
48. Hamann, U. (1961). Merkmalsbestand und Verwandtschaftsbeziehungen der Farinose. Ein Betrag zum System der Monokotyledonen. *Willdenowia, 2,* 639-768.
49. Heiser, W. J., & Bennani, M. (1997). Triadic distance models: axiomatization and least squares representation. *Journal of Mathematical Psychology, 41,* 189-206.
50. Holley, J. W., & Guilford, J. P. (1964). A note on the G index of agreement. *Educational and Psychological Measurement, 24,* 749-753.
51. Hsu, L. M., & Field, R. (2003). Interrater agreement measures: Comments on kappa$_n$, Cohen's kappa, Scott's π and Aickin's α. *Understanding Statistics, 2,* 205-219.
52. Hubálek, Z. (1982). Coefficients of association and similarity based on binary (presence-absence) data: an evaluation. *Biological Reviews, 57,* 669-689.
53. Hubert, L. J., & Arabie, P. (1985). Comparing partitions. *Journal of Classification, 2,* 193-218.
54. Hurlbert, S. H. (1969). A coefficient of interspecific association. *Ecology, 50,* 19.
55. Jaccard, P. (1912). The distribution of the flora in the Alpine zone. *The New Phytologist, 11,* 37-50.
56. Janes, C. L. (1979). An extension of the random error coefficient of agreement to $N \times N$ tables. *British Journal of Psychiatry, 134,* 617-619.
57. Janson, S., & Vegelius, J. (1979). On generalizations of the G index and the Phi coefficient to nominal scales. *Multivariate Behavioral Research, 14,* 255-269.
58. Janson, S., & Vegelius, J. (1981). Measures of ecological association. *Oecologia, 49,* 371-376.
59. Janson, S., & Vegelius, J. (1982). The J-index as a measure of nominal scale response agreement. *Applied Psychological Measurement, 6,* 111-121.
60. Janowitz. M. F. (1979). Monotone equivariant cluster analysis. *Journal of Mathematical Psychology, 37,* 148-165.
61. Johnson, H. M. (1945). Maximal selectivity, correctivity and correlation obtainable in a 2 × 2 contingency table. *American Journal of Psychology, 58,* 65-68.
62. Jordan, K. (1941). A Korreláció számitása I. *Magyar Statisztikai Szemle Kiadványai, 1.* Szám.
63. Kraemer, H. C. (1979). Ramifications of a population model for κ as a coefficient of reliability. *Psychometrika, 44,* 461-472.
64. Kraemer, H. C. (2004). Reconsidering the odds ratio as a measure of 2×2 association in a population. *Statistics in Medicine, 23,* 257-270.
65. Krippendorff, K. (1987). Association, agreement, and equity. *Quality and Quantity, 21,* 109-123.
66. Kulczyński, S. (1927). Die Pflanzenassociationen der Pienenen. *Bulletin International de LAcademie Polonaise des Sciences et des Letters, classe des sciences mathematiques et naturelles, Serie B, Supplement II, 2,* 57-203.
67. Lesot, M.-J., Rifgi, M., & Benhadda, H. (2009). Similarity measures for binary and numerical data: a survey. *International Journal of Knowledge Engineering and Soft Data Paradigms, 1,* 63-84.
68. Light, R. J. (1971). Measures of response agreement for qualitative data: Some generalizations and alternatives. *Psychological Bulletin, 76,* 365-377.
69. Loevinger, J. A. (1947). A systematic approach to the construction and evaluation of tests of ability. *Psychometrika Monograph No. 4.*
70. Loevinger, J. A. (1948). The technique of homogeneous tests compared with some aspects of "scale analysis" and factor analysis. *Psychological Bulletin, 45,* 507530.
71. Margush, T., & McMorris, F. R. (1981). Consensus n-Trees. *Bulletin of Mathematical*

Biology, 43, 239-244.

72. Martín Andrés, A., & Femia-Marzo, P. (2008). Chance-corrected measures of reliability and validity in 2×2 tables. *Communications in Statistics, Theory and Methods, 37,* 760-772.
73. Maxwell, A. E. (1977). Coefficients of agreement between observers and their interpretation. *British Journal of Psychiatry, 130,* 79-83.
74. Maxwell, A. E., & Pilliner, A. E. G. (1968). Deriving coefficients of reliability and agreement for ratings. *British Journal of Mathematical and Statistical Psychology, 21,* 105116.
75. Mokken, R. J. (1971). *A theory and procedure of scale analysis.* Hague: Mouton.
76. Nei, M., & Li, W.-H. (1979). Mathematical model for studying genetic variation in terms of restriction endonucleases. *Proceedings of the National Academy of Sciences, 76,* 5269-5273.
77. Omhover, J.F., Rifqi, M., & Detyniecki, M. (2006). Ranking invariance based on similarity measures in document retrieval. In M. Detyniecki, J.M. Jose, A. Nürnberger and C.J.K. Rijsbergen (Eds.), *Adaptive Multimedia Retrieval: User, Context and Feedback. Third International Workshop, AMR 2005, Revised Selected Papers,* 55-64. LNCS, Springer.
78. Ochiai, A. (1957). Zoogeographic studies on the soleoid fishes found in Japan and its neighboring regions. *Bulletin of the Japanese Society for Fish Science, 22,* 526530.
79. Pearson, K. (1900). Mathematical contributions to the theory of evolution. VII. On the correlation of characters not quantitatively measurable. *Philosophical Transactions of the Royal Society of London, Series A, 195,* 147.
80. Pearson, K. (1926). On the coefficient of racial likeness. *Biometrika, 9,* 105-117.
81. Peirce, C. S. (1884). The numerical measure of the success of prediction. *Science, 4,* 453-454.
82. Popping, R. (1983). *Overeenstemmingsmaten Voor Nominale Data.* Doctoral dissertation. Groningen: Rijksuniversiteit Groningen.
83. Popping, R. (1984). Traces of agreement. On some agreement indices for open-ended questions. *Quality and Quantity, 18,* 147158.
84. Post, W. J., & Snijders, T. A. B. (1993). Nonparametric unfolding models for dichotomous data. *Methodika, 7,* 130-156.
85. Rand, W. M. (1971). Objective criteria for the evaluation of clustering methods. *Journal of the American Statistical Association, 66,* 846-850.
86. Ratliff, R.D. (1982). A correction of Coles C_7 and Hurlberts C_8 coefficients of interspecific association. *Ecology, 50,* 19.
87. Rogers, D. J., & Tanimoto, T. T. (1960). A computer program for classifying plants. *Science, 132,* 1115-1118.
88. Russel, P. F., & Rao, T. R. (1940). On habitat and association of species of Anopheline larvae in South-Eastern Madras. *Journal of Malaria Institute India, 3,* 153-178.
89. Scott, W. A. (1955). Reliability of content analysis: The case of nominal scale coding. *Public Opinion Quarterly, 19,* 321-325.
90. Sibson, R. (1972). Order invariant methods for data analysis. *Journal of the Royal Statistical Society, Series B, 34,* 311-349.
91. Sijtsma, K., & Molenaar, I. W. (2002). *Introduction to nonparametric item response theory.* Thousand Oaks: Sage.
92. Simpson, G. G. (1943). Mammals and the nature of continents. *American Journal of Science, 24,* 1131.
93. Snijders, T. A. B., Dormaar, M., Van Schuur, W. H., Dijkman-Caes, C., & Driessen, G. (1990). Distribution of some similarity coefficients for dyadic binary data in the

case of associated attributes. *Journal of Classification, 7,* 5-31.
94. Sokal, R. R., & Michener, C. D. (1958). A statistical method for evaluating systematic relationships. *University of Kansas Science Bulletin, 38,* 1409-1438.
95. Sokal, R. R., & Sneath, P. H. A. (1963). *Principles of numerical taxonomy.* San Francisco: Freeman.
96. Sørenson, T. (1948). A method of stabilizing groups of equivalent amplitude in plant sociology based on the similarity of species content and its application to analyses of the vegetation on Danish commons. *Kongelige Danske Videnskabernes Selskab Biologiske Skrifter, 5,* 1-34.
97. Steinley, D. (2004). Properties of the Hubert-Arabie adjusted Rand index. *Psychological Methods, 9,* 386-396.
98. Tversky, A. (1977). Features of similarity. *Psychological Review, 84,* 327-352.
99. Wallace, D. L. (1983). A method for comparing two hierarchical clusterings: Comment. *Journal of the American Statistical Association, 78,* 569-576.
100. Warrens, M. J. (2008a). On the indeterminacy of resemblance measures for (presence/absence) data. *Journal of Classification, 25,* 125-136.
101. Warrens, M. J. (2008b). Bounds of resemblance measures for binary (presence/absence) variables. *Journal of Classification, 25,* 195-208.
Warrens, M. J. (2008c). On similarity coefficients for 2 × 2 tables and correction for chance. *Psychometrika, 73,* 487-502.
102. Warrens, M. J. (2008d). On association coefficients for 2 × 2 tables and properties that do not depend on the marginal distributions. *Psychometrika, 73,* 777-789.
103. Warrens, M. J. (2008e). On the equivalence of Cohen's kappa and the Hubert-Arabie adjusted Rand index. *Journal of Classification, 25,* 177-183.
104. Warrens, M. J. (2008f). On resemblance measures for binary data and correction for maximum value. In K. Shigemasu, A. Okada, T. Imaizumi, and T. Hoshino (Eds.), *New Trends in Psychometrics*, 543-548. Tokyo: University Academic Press.
105. Warrens, M. J. (2009). On a Family of Indices for 2 × 2 Tables and Correction for Chance. *Unpublished paper.*
106. Yule, G. U. (1900). On the association of attributes in statistics. *Philosophical Transactions of the Royal Society A, 75,* 257-319.
107. Yule, G. U. (1912). On the methods of measuring the association between two attributes. *Journal of the Royal Statistical Society, 75,* 579-652.
108. Zwick, R. (1988). Another look at interrater agreement. *Psychological Bulletin, 103,* 374-378.
109. Zysno, P. V. (1997). The modification of the phi-coefficient reducing its dependence on the marginal distributions. *Methods of Psychological Research Online, 2,* 41-52.

Chapter 3

Applications of Spanning Subgraphs of Intersection Graphs

Terry A. McKee
Department of Mathematics & Statistics, Wright State University
Dayton, Ohio 45435 USA
terry.mckee@wright.edu

Two applications of intersection graphs are presented—to protein interaction in cellular processes, and to Bonferroni-type inequalities in probability theory. Each application exploits appropriate spanning subgraphs of an intersection graph.

Introduction

A graph G with vertex set $\{v_1, \ldots, v_n\}$ is the *intersection graph* of a family $\mathcal{F} = \{S_1, \ldots, S_n\}$ of subsets of some underlying set if vertices $v_i \neq v_j$ are adjacent exactly when $S_i \cap S_j \neq \emptyset$; in this case $\mathcal{F}$ is called an *intersection representation* of G. (A *family* $\mathcal{F}$ will always mean a *multiset* $\mathcal{F}$—in other words, $S_i = S_j$ will be allowed in $\mathcal{F}$ when $i \neq j$.)

Intersection graphs have long been motivated by and studied alongside their applications. For instance, the monograph [19] contains brief descriptions of applications to biology (compatibility analysis in numerical taxonomy and physical mapping of DNA), to computing (database schemes and consecutive retrieval), to matrix analysis (elimination schemes and determinantal formulas), and to statistics (decomposable log linear models). Sections 3.1 and Sec. 3.2 will describe two additional application-oriented topics.

In Section 3.1, the members of $\mathcal{F}$ will be all the induced subgraphs of a given graph G that possess some selected inherent structure. Particular spanning trees of the intersection graph of $\mathcal{F}$ can illuminate the organization of G in terms of those subgraphs selected for $\mathcal{F}$. Such spanning trees have been employed in computational molecular biology.

In Section 3.2, the members of $\mathcal{F}$ will be arbitrary subsets covering an underlying set $\mathcal{X}$. Particular spanning subgraphs of the intersection graph of $\mathcal{F}$ can give bounds on the cardinality of $\mathcal{X}$ using only limited knowledge of the cardinalities of intersections of sets in $\mathcal{F}$. Such spanning subgraphs correspond to probabilistic approximation formulas.

3.1. A Family-Tree Approach to Graph Classes

One of the central themes in the monograph *Topics in Intersection Theory* [19] is representing a graph class as the intersection graph G of a family of subgraphs of a "host" graph H, where the nodes of H correspond to specific sorts of subgraphs of G. (It is convenient to refer to the *nodes* of the host H to avoid confusing them with the *vertices* of G.) The leading example for this is the representation of the class of *chordal graphs*—the graphs in which triangles are the only induced cycles—as the intersection graphs G of subtrees of a host tree T, where the nodes of T correspond to the *maxcliques*—the inclusion-maximal complete subgraphs—of G.

This "family-tree" approach can start with families of subgraphs other than the family of all maxcliques to produce representations for other graph classes. A 2006 application of graph theory to computational biochemistry by Zotenko, Guimarães, Jothi, and Przytycka [23] contains a perceptive description of this approach:

> "Given a graph, it is usually very useful to be able to represent it using some kind of a tree. Such tree representation exposes a hierarchical organization that a graph may have, allowing for a structured analysis."

Section 3.1.5 will return to this biological application of trees.

3.1.1. *Basic results on $\mathcal{F}$-tree representations*

Suppose $\mathcal{F}$ is any family of induced subgraphs of a graph G such that $\mathcal{F}$ covers $V(G)$ (for convenience, routinely identifying the subgraphs in $\mathcal{F}$ with their vertex sets from $V(G)$). Let $\Omega^w(\mathcal{F})$ denote the complete graph on the node set $\mathcal{F} = \{S_1, \ldots, S_n\}$, where each edge S_iS_j has weight $|S_i \cap S_j| \geq 0$; thus, the positive-weight edges of $\Omega^w(\mathcal{F})$ form the intersection graph of $\mathcal{F}$ (although $\Omega^w(\mathcal{F})$ itself is sometimes also referred to as the intersection graph when the weight-0 edges not drawn). Figure 3.1 shows an artificial example in which $\mathcal{F}$ consists of the four induced subgraphs $\langle\{a, b, c\}\rangle$, $\langle\{c, d, f\}\rangle$, $\langle\{a, c, d, f\}\rangle$, and $\langle\{d, e, f\}\rangle$ of G.

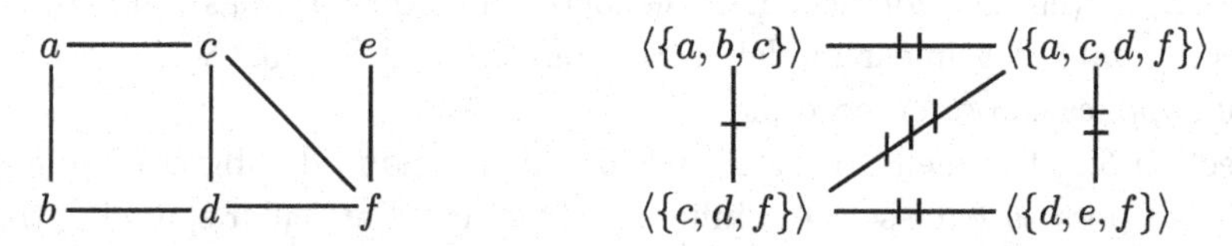

Fig. 3.1. An example of a graph G and a weighted intersection graph based on G (where edge weights are shown by cross-hatchings and the weight-0 edge of $\Omega^w(\mathcal{F})$ is not drawn).

Suppose T is any spanning subtree of $\Omega^w(\mathcal{F})$. For every $v \in V(G)$, let T_v denote the subgraph of T induced by those nodes that contain v. Call T an *$\mathcal{F}$-tree* for G if every T_v is connected (in other words, if every T_v is a tree).

For instance, Fig. 3.2 shows two spanning trees T of $\Omega^w(\mathcal{F})$ from Fig. 3.1. In each, T_a is a single edge of T; T_b and T_e are single nodes of T; and T_c, T_d, and

T_f are length-2 subpaths of T. Both trees T are $\mathcal{F}$-trees for G. Notice that if $\mathcal{F}' = \mathcal{F} \cup \{\langle\{b,d,f\}\rangle\} - \{\langle\{c,d,f\}\rangle\}$, then G would have no $\mathcal{F}'$-tree—not all of T_a, T_b, and T_d can be connected subgraphs of a spanning tree T of $\Omega^w(\mathcal{F}')$.

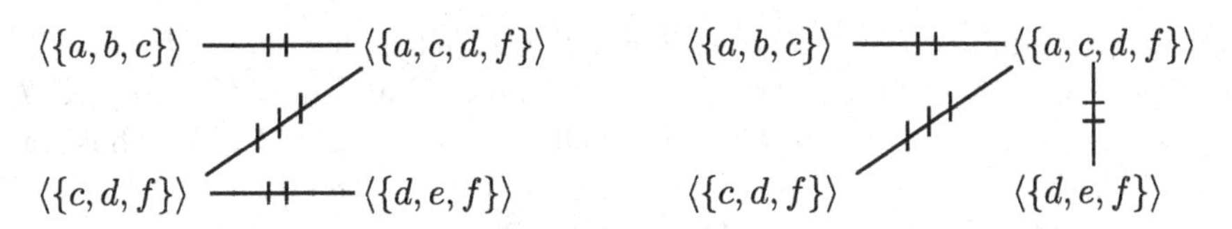

Fig. 3.2. Two $\mathcal{F}$-trees T for the graph G in Fig. 3.1.

Theorems 3.1, 3.2, and 3.3 appear in [14].

Theorem 3.1. *If T is a spanning tree of $\Omega^w(\mathcal{F})$, then*

$$|V(G)| \;\le\; \sum_{S \in V(T)} |S| \;-\; \sum_{SS' \in E(T)} |S \cap S'|, \tag{3.1}$$

with T an $\mathcal{F}$-tree for G if and only if equality holds in (3.1).

Proof. Suppose T is a spanning tree of $\Omega^w(\mathcal{F})$. For each $v \in V(G)$, the subgraph T_v of T is a forest, and so $1 \le |V(T_v)| - |E(T_v)|$ holds, with equality if and only if T_v is connected. Summing these inequalities over all $v \in V(G)$ shows (3.1), with equality if and only if every T_v is connected. □

The right side of (3.1) is the total number of vertices of G that occur in T, counting repetitions, minus the number of cross-hatchings on edges in T. For both of the trees T in Fig 3.2, inequality (3.1) says that $6 \le (3+3+4+3)-(2+3+2) = 13-7 = 6$. Theorem 3.1 shows that both the spanning trees T in Fig. 3.2 are $\mathcal{F}$-trees for G without needing to check each of the six subgraphs T_v.

Theorem 3.2. *If G has at least one $\mathcal{F}$-tree, then the $\mathcal{F}$-trees for G are precisely the maximum spanning trees of $\Omega^w(\mathcal{F})$.*

Proof. Since $|V(G)|$ and $\sum_{S \in V(T)} |S| = \sum_{i=1}^{n} |S_i|$ are both fixed for given G and $\mathcal{F}$, equality can only be achieved in (3.1) when $\sum_{SS' \in E(T)} |S \cap S'|$—which is the sum of the edge weights of T in $\Omega^w(\mathcal{F})$—is maximized. □

Kruskal's well-known greedy algorithm—*repeatedly choose an edge of greatest weight so long as no cycle is formed with the previously-chosen edges*— will find all the maximum spanning trees of $\Omega^w(\mathcal{F})$ and so, using the numerical check in (3.1), will find all the $\mathcal{F}$-trees for a graph. (The two spanning trees in Fig. 3.2 are the only maximum spanning trees of $\Omega(\mathcal{F})$ from Fig. 3.1, and so they are the only $\mathcal{F}$-trees for that G.)

In spite of the possibility that a graph G has more than one $\mathcal{F}$-tree, Theorem 3.3 will show that all the $\mathcal{F}$-trees T for G will have the same edge multiset $\mathcal{E}(T) =$

$\{S_i \cap S_j : S_iS_j \in E(T)\}$. For instance, both the $\mathcal{F}$-trees T in Fig. 3.2 have edge multisets $\mathcal{E}(T) = \{\{a,c\},\{c,d,f\},\{d,f\}\}$.

Theorem 3.3. *If T_1 and T_2 are $\mathcal{F}$-trees for the same graph, then $\mathcal{E}(T_1) = \mathcal{E}(T_2)$.*

Proof. First suppose G has $\mathcal{F}$-trees T and $\widehat{T}$ with edge multisets $\mathcal{E}(T)$ and $\mathcal{E}(\widehat{T})$ that differ on exactly one edge: say $S \cap S' \in \mathcal{E}(T) - \mathcal{E}(\widehat{T})$ and $\widehat{R} \cap \widehat{R}' \in \mathcal{E}(\widehat{T}) - \mathcal{E}(T)$. Edges SS' and $\widehat{R}\widehat{R}'$ will be in the unique cycle in the subgraph $T \cup \widehat{T}$. Thus, each $v \in S \cap S'$ will also be in $\widehat{R} \cap \widehat{R}'$ (since $\widehat{T}_v$ is connected), and so $S \cap S' \subseteq \widehat{R} \cap \widehat{R}'$. Similarly $\widehat{R} \cap \widehat{R}' \subseteq S \cap S'$, and so $S \cap S' = \widehat{R} \cap \widehat{R}'$.

The theorem then follows from—as is true for any graph—spanning trees T and $\widehat{T}$ always being linked by a sequence $T = T_1, \ldots, T_k = \widehat{T}$ of spanning trees such that consecutive trees in the sequence differ by exactly one edge. [Specifically, pick any edge $\widehat{R}\widehat{R}'$ with $\widehat{R} \cap \widehat{R}' \in \mathcal{E}(\widehat{T}) - \mathcal{E}(T)$, then pick an edge SS' with $S \cap S' \in \mathcal{E}(T) - \mathcal{E}(\widehat{T})$ from the unique cycle formed by $E(T) \cup \{\widehat{R}\widehat{R}'\}$, and then define T_1 to have edge set $E(T) - \{SS'\} \cup \{\widehat{R}\widehat{R}'\}$; continue in this way to define T_2, and so on, until reaching T_k with $E(T_k) = E(\widehat{T})$.] □

3.1.2. *$\mathcal{F}$-tree representations for chordal graphs*

Recall that a graph is *chordal* if the only induced cycles are triangles; Figure 3.3 shows an example. A *clique tree* for a graph G is an $\mathcal{F}$-tree for G where $\mathcal{F}$ is the set of all the maxcliques of G. The positive-weight edges of $\Omega^w(\mathcal{F})$ constitute the *clique graph* of G. Theorem 3.4 contains two of the oldest of the many characterizations of chordal graphs in [1; 19]. It is proved in [19], and it is worth noting that, in proving (3.4.2) ⇒ (3.4.3), the family of subtrees of a tree can always be taken to be the family $\{T_v : v \in V(G)\}$ from any clique tree T for G.

Theorem 3.4. *The following are equivalent for any graph G:*

(3.4.1) *G is chordal.*
(3.4.2) *G has a clique tree.*
(3.4.3) *G is the intersection graph of a family of subtrees of a tree.*

Figure 3.4 shows a maximum spanning tree T (which, in this case, is unique) of the clique graph of the chordal graph G in Fig. 3.3. By Theorem 3.1 (with the numerical check $15 = 29 - 14$) and Theorem 3.2, T is the (unique) clique tree for G.

A clique tree T for a chordal graph G displays considerable information about G. For instance, the members of the edge multiset $\mathcal{E}(T)$ are precisely the *minimal vertex separators* of G—the inclusion-minimal sets $S \subset V(G)$ for which there exist vertices $v, w \in V(G)$ that are in a common component of G, but different components of $G - \{v, w\}$; see [1; 19]. (The sets of vertices of a chordal graph G that correspond to positive-weight edges of $\Omega^w(\mathcal{F})$ that are edges of no clique tree of G are precisely the *minimal vertex weak separators* of G as defined in Exercise 2.9 of [19].) Reference [18] studies those chordal graphs in which minimal separators of,

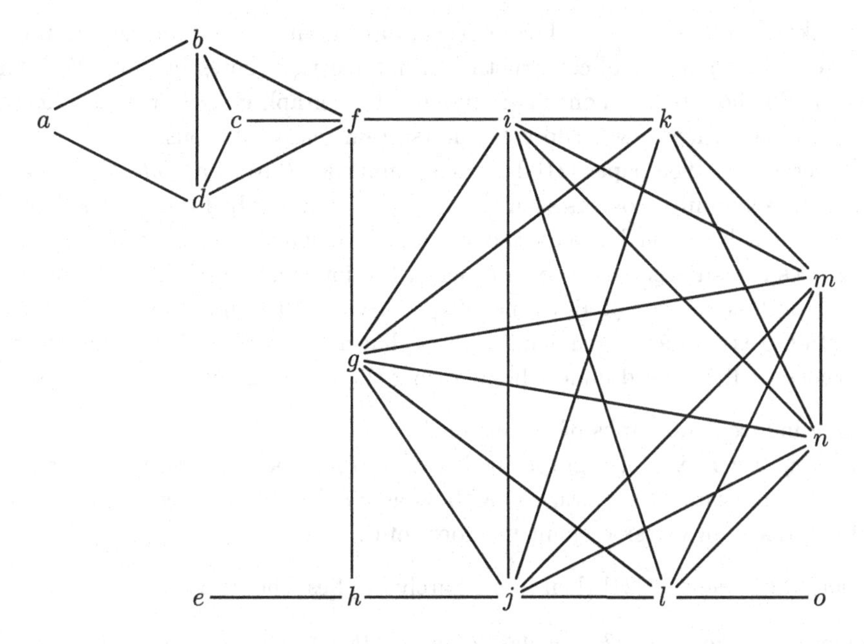

Fig. 3.3. A chordal graph.

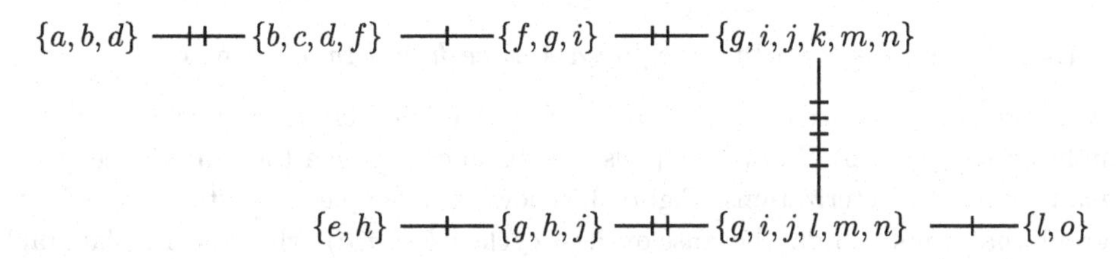

Fig. 3.4. A clique tree for the chordal graph in Fig. 3.3.

simultaneously, three or more vertices correspond to certain substars of their clique trees (where a *star* is a subgraph that is isomorphic to some $K_{1,k}$).

3.1.3. $\mathcal{F}$-tree representations for other graph classes

Many other graph classes can be characterized by having $\mathcal{F}$-trees for particular families $\mathcal{F}$. As one example among those in [14], if $\mathcal{F}$ is the family of all closed neighborhoods of vertices in a graph G, then G has an $\mathcal{F}$-tree if and only if G is a *dually chordal graph* (the clique graph of a chordal graph). Another example will be discussed in Sec. 3.1.4.

It is important to emphasize that, as soon as $\mathcal{F}$ is taken to be something different from the family of all maxcliques, a graph G having an $\mathcal{F}$-tree T might no longer imply that G is the intersection graph of the T_v subtrees—v and w being in a common member of $\mathcal{F}$ might no longer imply that v is adjacent to w in G. Therefore, unlike the special case in Theorem 3.4, having an $\mathcal{F}$-tree representation does not in general give a way of constructing an intersection representation. Yet, Theorem 3.5 will show show—nonconstructively—that graph classes that have $\mathcal{F}$-tree representations in fact always do have intersection representations.

Scheinerman's 1985 paper [21] defines a graph class $\mathcal{C}$ to be an *intersection class* if there exists a family Σ of sets such that G is in $\mathcal{C}$ if and only if there is a subfamily $\mathcal{F} \subset \Sigma$ and a bijection between the vertices of G and the sets in $\mathcal{F}$ such that two vertices are adjacent in G exactly when the intersection of the corresponding members of $\mathcal{F}$ is nonempty. Reference [21]—also see [19, Thm. 1.5]—characterizes intersection classes and, in particular, notes that a graph class $\mathcal{C}$ is an intersection class whenever it is closed under the following three operations:

- Taking induced subgraphs of graphs in $\mathcal{C}$.
- Replacing a vertex v of a graph in $\mathcal{C}$ with two new adjacent vertices v' and v'' that have the same pre-existing neighbors as v did (and deleting v).
- Taking unions of vertex-disjoint members of $\mathcal{C}$.

Scheinerman's result in [21] then immediately implies Theorem 3.5.

Theorem 3.5. *For every $\mathcal{F}$, the class of graphs that have $\mathcal{F}$-trees is an intersection class.*

3.1.4. *$\mathcal{F}$-tree representations for distance-hereditary graphs*

A graph G is *distance-hereditary* if the distance between vertices in a connected induced subgraph of G always equals the distance between them in G; see [1] for many other characterizations. Figure 3.5 shows an example of a distance-hereditary graph that is not chordal (because of the 4-cycle a, b, c, d, a), while the chordal graph in Fig. 3.3 is not distance-hereditary (because the distance between vertices f and h changes from two to three when vertex g is deleted).

A graph G is a *cograph*—short for *complement-reducible graph*—if G has no induced path of length three; see [1; 19] for many other characterizations. A *$\mathcal{CC}$-tree* for G is a $\mathcal{F}$-tree for G where $\mathcal{F}$ is the family of all induced subgraphs that are inclusion-maximal connected cographs of G.

Theorem 3.6 is one of the characterizations of distance-hereditary graphs in [1], based on Nicolai's 1996 hypergraph characterization [1, Thm. 8.4.1].

Theorem 3.6. *A graph is distance-hereditary if and only if it has a $\mathcal{CC}$*-tree.

Figure 3.6 shows a maximum spanning tree (which, in this case, is unique) of $\Omega^w(\mathcal{F})$ where $\mathcal{F}$ is the set of maximal connected cographs in the distance-hereditary graph G shown in Fig. 3.5. By Theorem 3.1 (with the numerical check $15 = 33 - 18$) and Theorem 3.6, this tree is the (unique) $\mathcal{CC}$-tree for G.

Theorem 3.5 shows that the class of distance-hereditary graphs is an intersection class. Echoing Theorem 3.4, the class of distance-hereditary graphs has also been

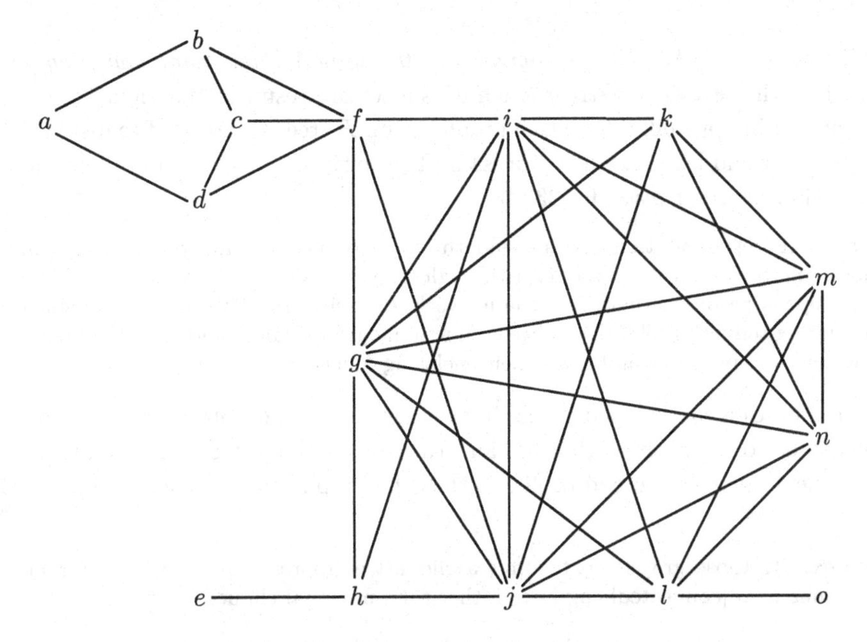

Fig. 3.5. A distance-hereditary graph.

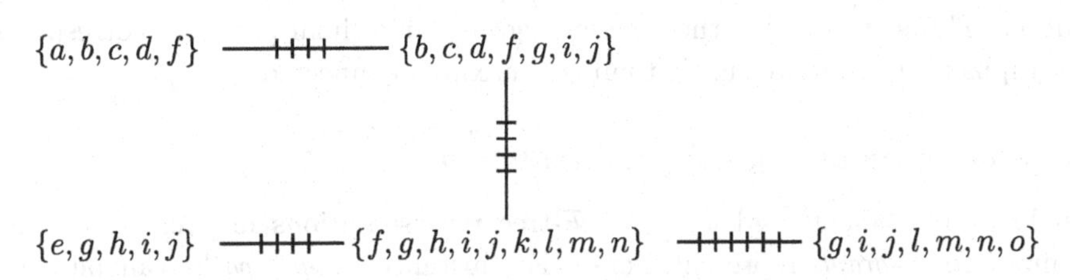

Fig. 3.6. A $\mathcal{CC}$-tree for the distance-hereditary graph in Fig. 3.5.

characterized explicitly by intersection representations in Gioan and Paul's 2007 paper [6]. Although the intersection representations in [6] are based on trees, they are not closely related to $\mathcal{CC}$-tree representations. (They are, instead, closely related to the "one-vertex extensions" characterization of distance-hereditary graphs in [2].)

3.1.5. *Returning to the molecular biology example*

The biological application in [23]—also see the more general survey in [20]—involves *protein interaction graphs*, where the vertices are proteins with two vertices adjacent if and only if the corresponding proteins interact in some specific biochemical process (for instance, a cell signaling pathway). The goal is to represent protein interaction graphs using $\mathcal{F}$-trees where, ideally, $\mathcal{F}$ is the family of "functional groups" of proteins. Two of the figures in [23]—labeled *Complex overlap decomposition* and *A hypothetical protein interaction network*—show the resulting $\mathcal{F}$-trees when $\mathcal{F}$ is the set of all maxcliques (representing "protein complexes"). Two other figures in [23]—labeled *TNFα/NF-χB signaling pathway* and *Pheromone signaling pathway*—show the resulting $\mathcal{F}$-trees when $\mathcal{F}$ is a set of presumed functional groups.

Representing protein interaction graphs using $\mathcal{F}$-trees is intended to display how proteins enter and leave cellular processes. The spirit of the application is contained in the following description in [23]:

> "This [tree] representation shows a smooth transition between functional groups and allows for tracking a protein's path through a cascade of functional groups. Therefore, depending on the nature of the network, the representation may be capable of elucidating temporal relations between functional groups [and capturing] the manner in which proteins enter and leave their enclosing functional groups."

The procedures developed in [23] involve inserting edges into protein interaction graphs so as to make them chordal, and then using clique trees. A tentative conjecture mentioned at the end of [23] is that the graphs that are susceptible to the procedures

> "are exactly those graphs that admit a clique tree representation, with the nodes being maximal [connected] cographs rather than maximal cliques."

In other words, these are exactly the graphs that have $\mathcal{CC}$-trees. The authors of [23] were unaware of Theorem 3.6—none of the graphs in their four examples is distance-hereditary. This suggests that the functional groups in cellular processes correspond to subgraphs that are more subtle than the maximal connected cographs.

3.1.6. *More general $\mathcal{F}$-graph representations*

It should also be mentioned that the $\mathcal{F}$-tree representations in Sec. 3.1 can be generalized to *$\mathcal{F}$-graph* representations. For instance, *clique paths* can be used instead of clique trees, yielding interval graphs as in [19, Chap. 2]. Reference [15] shows one way to use *clique cycles* instead of clique trees. Reference [16] does this even more generally, using, for instance, weighted cycles and other subgraphs of $\Omega^w(\mathcal{F})$ in addition to edges. (Section 3.2.4 below will involve something similar.) Studying $\mathcal{F}$-graph representations with more general families $\mathcal{F}$ of subgraphs seems to be virtually unexplored.

3.2. Hunter–Worsley/Bonferroni-Type Set Bounds

Suppose $\mathcal{F} = \{S_1, \ldots, S_n\}$ is a family of finite subsets of a given underlying set. Let $\Omega = \Omega^w(\mathcal{F})$ denote the complete graph on the node set $\mathcal{F}$, where each edge S_iS_j has weight $|S_i \cap S_j| \geq 0$. Let $\mathcal{Q}(\Omega)$ denote the set of all the complete subgraphs of Ω (equivalently, $\mathcal{Q}(\Omega)$ consists of all the nonempty subsets of $\mathcal{F}$).

Figure 3.7 shows an example with $|\mathcal{F}| = 6$ where the cardinalities of the six sets are shown by the numerals in the circles and the weight of each edge S_iS_j in $\Omega^w(\mathcal{F})$ is shown by the number of cross-hatchings. (The two weight-0 edges are not drawn.)

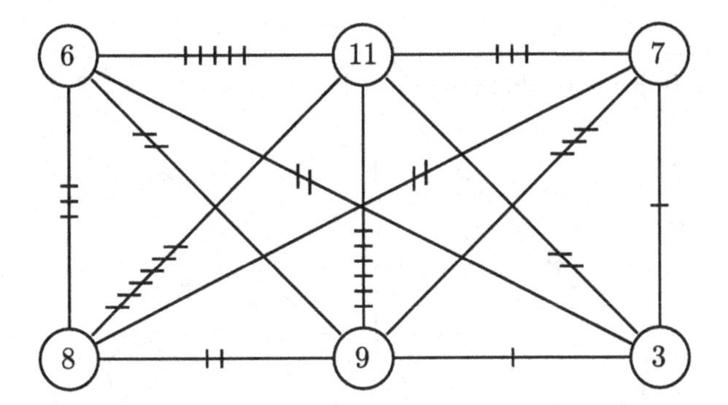

Fig. 3.7. An example of a family $\mathcal{F}$ of six sets (with cardinalities 3, 6, 7, 8, 9 and 11) and the positive-weight edges of $\Omega^w(\mathcal{F})$ (with their weights indicated by cross-hatchings).

The traditional *inclusion-exclusion formula*

$$\left|\bigcup_{i=1}^{n} S_i\right| = \sum_{Q\in\mathcal{Q}(\Omega)} (-1)^{|Q|+1} \left|\bigcap_{S_i\in Q} S_i\right| \tag{3.2}$$

(called a "sieve" in some contexts) gives the exact size of $|\bigcup_{i=1}^{n} S_i|$ if the cardinalities of all the intersections of subsets of $\mathcal{F}$ are known. But in practice, often only the cardinalities of the sets $S_i \in \mathcal{F}$ and the cardinalities of some of their intersections are known—perhaps only the pairwise intersections (as in Fig. 3.7). This section will discusses modifications of (3.2) that only use limited intersection information.

3.2.1. *Several classical set bounds*

Theorem 3.7 will give the *Hunter–Worsley bound* from [10; 22], which uses only $n-1$ of the pairwise intersections of members of $\mathcal{F}$. This can be viewed as a special case of Theorem 3.1, where G is the edgeless graph with vertex set $S_1 \cup \cdots \cup S_n$. (It will also be a special case of Theorem 3.10.)

Theorem 3.7 ([10; 22]). *If T is a spanning tree of $\Omega^w(\mathcal{F})$, then*

$$\left|\bigcup_{i=1}^{n} S_i\right| \leq \sum_{S_i\in V(T)} |S_i| - \sum_{S_iS_j\in E(T)} |S_i \cap S_j|, \tag{3.3}$$

with the inequality strongest when T is a maximum spanning tree.

Figure 3.8 shows one of the maximum spanning trees T for $\Omega^w(\mathcal{F})$ in Fig. 3.7. Since $\sum_{S_i\in V(T)} = 3+6+7+8+9+11 = 44$ and $\sum_{S_iS_j\in E(T)} = 5+6+6+2+3 = 22$, inequality (3.3) says that $|\bigcup_{i=1}^{n} S_i| \leq 44 - 22 = 22$.

Inequality (3.3) is often stated in the probabilistic form [10; 22]

$$\Pr\left(\bigcup_{i=1}^{n} S_i\right) \leq \sum_{S_i\in V(T)} \Pr(S_i) - \sum_{S_iS_j\in E(T)} \Pr(S_i \cap S_j),$$

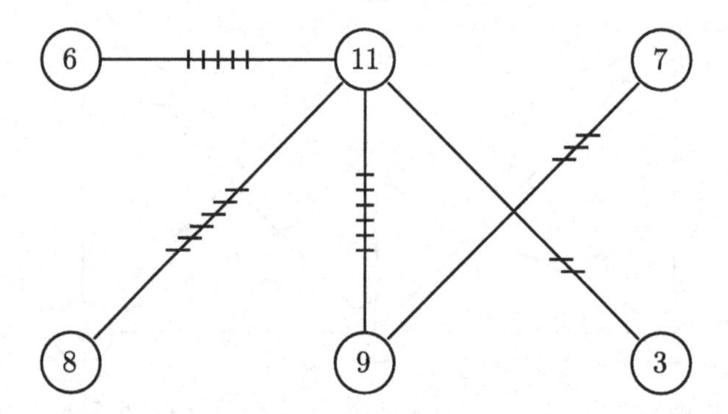

Fig. 3.8. A maximum-weight spanning tree of $\Omega^w(\mathcal{F})$ from Fig. 3.7.

transforming "statements on proportions to probabilities with the uniform distribution on finite sets as the underlying probability space" [7]. In fact, such probabilistic formulations are the real-life applications of all of the inequalities in this section. Galambos and Simonelli's 1996 book [7] is an excellent source of details on all these "Bonferroni-type" probabilistic inequalities and on their many applications.

Replacing the spanning tree T in (3.3) with a spanning star gives a weaker inequality, the *Kounias bound* [11]. Replacing the spanning tree of T in (3.3) with the entire graph $\Omega = \Omega^w(\mathcal{F})$ gives the lower bound

$$\left|\bigcup_{i=1}^{n} S_i\right| \geq \sum_{S_i \in V(\Omega)} |S_i| - \sum_{S_i S_j \in E(\Omega)} |S_i \cap S_j|, \tag{3.4}$$

corresponding to the terms in the inclusion-exclusion formula (3.2) that have $|Q| \leq 2$. Indeed, truncating the summation in (3.2) to the terms with $|Q| \leq k$ gives an upper bound for $|\bigcup_{i=1}^{n} S_i|$ when k is odd and a lower bound when k is even. The very special $k = 1$ case gives the naive upper bound $\sum_{i=1}^{n} |S_i|$ that is often attributed to Boole.

3.2.2. *Two 2-tree versions of the Hunter–Worsley bound*

The class of 2-*trees* [1; 19] is a well-studied generalization of the class of trees. Define 2-trees recursively—starting from K_2 being a 2-tree—as follows:

> If G is any 2-tree with $e \in E(G)$ and if $\Delta \cong K_3$ is vertex-disjoint from G with $e' \in E(\Delta)$, then the graph formed from G and Δ by identifying edges e and e' (along with their endpoints) is another 2-tree.

A *topological* K_4 is a graph H that is isomorphic to a subdivision of K_4 (in other words, H is homeomorphic to K_4). Among several other characterizations [1], a graph is *series-parallel* if and only if it contains no topological K_4. A graph is series-parallel if and only if it is a subgraph of a 2-tree [1].

A simple modification of Kruskal's algorithm— *repeatedly choose an edge of greatest weight so long as no topological K_4 is formed with previously-chosen edges*— will produce all the *maximum spanning 2-trees*—the spanning 2-trees whose sum of edge weights is maximum—of $\Omega^w(\mathcal{F})$. This has the same proof of correctness as the usual Kruskal algorithm, except that now topological K_4 subgraphs are avoided, instead of cycles (which are topological K_3 subgraphs). The graph produced is series-parallel; indeed, it is a spanning 2-tree. (Spanning 2-trees always exist in the complete graph $\Omega^w(\mathcal{F})$.)

One consequence of the recursive definition of 2-tree is that a series-parallel graph G will satisfy $3 \leq 2|V(G)| - |E(G)|$, with equality if and only if G is a 2-tree. This inequality underlies Theorem 3.8, from [17], which will give a bound on $|\bigcup_{i=1}^n S_i|$ in terms of the weights of the $2n-3$ edges of a spanning 2-tree of $\Omega^w(\mathcal{F})$ together with the number of elements that are in unique members of $\mathcal{F}$—denote that number by $\#\text{smpl}(\mathcal{F})$.

Theorem 3.8 ([17]). *If ^2T is a spanning 2-tree of $\Omega^w(\mathcal{F})$, then*

$$\left|\bigcup_{i=1}^n S_i\right| \leq \frac{2}{3}\sum_{S_i\in V(^2T)} |S_i| - \frac{1}{3}\sum_{S_iS_j\in E(^2T)} |S_i \cap S_j| + \frac{\#\text{smpl}(\mathcal{F})}{3}, \tag{3.5}$$

with the inequality strongest when ^2T is a maximum spanning 2-tree.

Proof. Suppose ^2T is a spanning 2-tree of $\Omega^w(\mathcal{F})$. For each element $x \in S_1 \cdots S_n$, let ^2T_x denote the subgraph of ^2T that is induced by those nodes of $\Omega^w(\mathcal{F})$ that, as members of $\mathcal{F}$, contain x. For each x that is in two or more of the subsets S_i, the subgraph ^2T_x will be series-parallel and so will satisfy

$$1 \leq \frac{2}{3}|V(^2T_x)| - \frac{1}{3}|E(^2T_x)|. \tag{3.6}$$

For each element x that is in a unique subset S_i, the subgraph ^2T_x will be a single node and so will satisfy

$$1 = \frac{2}{3}|V(^2T_x)| - \frac{1}{3}|E(^2T_x)| + \frac{1}{3}. \tag{3.7}$$

Summing (3.6) and (3.7) over all $x \in S_1 \cup \cdots \cup S_n$ proves (3.5). □

Figure 3.9 shows one of the maximum spanning 2-trees ^2T for the intersection data given in Fig. 3.7. Suppose you also know that $\#\text{smpl}(\mathcal{F}) = 5$. Since $\sum_{S_i\in V(^2T)} = 44$ and $\sum_{S_iS_j\in E(^2T)} = 3+5+6+2+6+3+3+2+1 = 31$, inequality (3.5) now says that $|\bigcup_{i=1}^n S_i| \leq \frac{2}{3}\cdot 44 - \frac{1}{3}\cdot 31 + \frac{5}{3} = \frac{62}{3}$, and so $|\bigcup_{i=1}^n S_i| \leq 20$.

Theorem 3.9 will give the *Hoover bound* from [9], another upper bound on $|\bigcup_{i=1}^n S_i|$ that is based on spanning 2-trees, except now the cardinalities of $n-2$ triple intersections $S_i \cap S_j \cap S_k$ of members of $\mathcal{F}$ are also required. For any 2-tree ^2T, let $\Delta(^2T)$ denote the set of all triangles of ^2T.

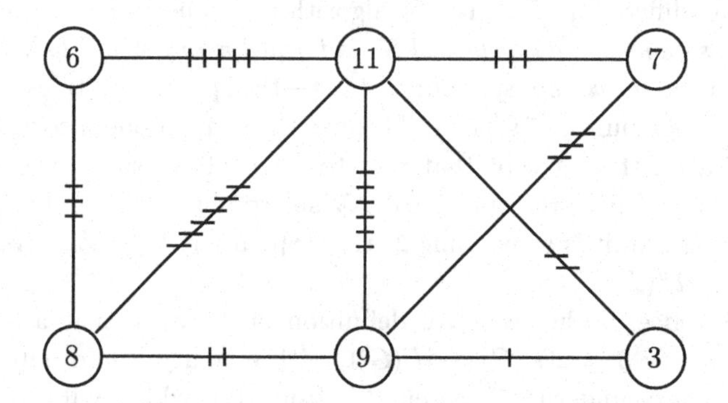

Fig. 3.9. A maximum spanning 2-tree of $\Omega^w(\mathcal{F})$ from Fig. 3.7.

Theorem 3.9 ([9]). *If* ^2T *is a spanning* 2*-tree of* $\Omega^w(\mathcal{F})$*, then*

$$\left|\bigcup_{i=1}^{n} S_i\right| \leq \sum_{S_i \in V(^2T)} |S_i| - \sum_{S_iS_j \in E(^2T)} |S_i \cap S_j| + \sum_{S_iS_jS_k \in \Delta(^2T)} |S_i \cap S_j \cap S_k|. \qquad (3.8)$$

Theorem 3.9 has a direct, graph-based proof in [17] and will also be a special case of Theorem 3.10 (proved below). In the case of the spanning 2-tree ^2T shown in Fig. 3.9 for the intersection data given in Fig. 3.7 with the additional information that the four triangles in ^2T have weights 1, 1, 2, and 3, inequality (3.8) now says that $|\bigcup_{i=1}^{n} S_i| \leq 44-31+(1+1+2+3) = 20$. Reference [17] explores the relationship between inequalities (3.5) and (3.8)—each is sometimes better than the other.

Fig. 3.10 shows one complete data set that is consistent with Fig. 3.7 and the other intersection data mentioned above. (These data are also consistent with $\#\text{smpl}(\mathcal{F}) = 5$, as used with Theorem 3.8.) Here A_i denotes the unique set in $\mathcal{F}$ of cardinality $i \in \{3, 6, 7, 8, 9, 11\}$, and all the nonempty intersections of three or more sets A_i are shown. Using all the intersection data in Figs. 3.7 and 3.10, the inclusion-exclusion formula (3.2) shows that, in fact, $|\bigcup_{i=1}^{n} S_i| = 19$.

$|A_3 \cap A_6 \cap A_9| = 1$ $\quad$ $|A_7 \cap A_9 \cap A_{11}| = 1$ $\quad$ $|A_7 \cap A_8 \cap A_{11}| = 2$

$|A_3 \cap A_9 \cap A_{11}| = 1$ $\quad$ $|A_3 \cap A_6 \cap A_{11}| = 2$ $\quad$ $|A_8 \cap A_9 \cap A_{11}| = 2$

$|A_6 \cap A_8 \cap A_9| = 1$ $\quad$ $|A_6 \cap A_9 \cap A_{11}| = 2$ $\quad$ $|A_6 \cap A_8 \cap A_{11}| = 3$

$|A_3 \cap A_6 \cap A_9 \cap A_{11}| = 1$ $\quad$ $|A_6 \cap A_8 \cap A_9 \cap A_{11}| = 1$

Fig. 3.10. Cardinalities of the nonempty intersections of three or more sets for Fig. 3.7.

3.2.3. *The chordal graph sieve*

Theorem 3.10 will give an elegant modification of the traditional inclusion-exclusion sieve (3.2); it is called the *chordal graph sieve* in Dohmen's 2002 paper [3] (also see the much more abstract setting in his book [4]). The chordal graph sieve uses a spanning chordal subgraph of $\Omega = \Omega^w(\mathcal{F})$, instead of using all of Ω. As an alternative to the proof given in [3]—and instead of a traditional inductive proof using the "perfect elimination ordering" characterization [1,19] of chordal graphs—the proof given below is an intersection graph argument using clique trees, somewhat-similar in spirit to the proof of Theorem 3.1. Let $\mathcal{Q}(G)$ denote the set of all the complete subgraphs of a graph G.

Theorem 3.10 ([3]). *If G is a spanning chordal subgraph of $\Omega^w(\mathcal{F})$, then*

$$\left|\bigcup_{i=1}^{n} S_i\right| \leq \sum_{Q\in\mathcal{Q}(G)} (-1)^{|Q|+1} \left|\bigcap_{S_i\in Q} S_i\right|. \tag{3.9}$$

Proof. Suppose $\mathcal{F} = \{S_1, \ldots, S_n\}$ and G is a spanning chordal graph of $\Omega^w(\mathcal{F})$. Let T be a clique tree for G.

For each node $\mathcal{F}' \subseteq \mathcal{F}$ of T, let $G[\mathcal{F}']$ denote the maxclique of G that is induced by $\mathcal{F}'$. Applying the inclusion-exclusion formula to the complete graph $G[\mathcal{F}']$ gives

$$\left|\bigcup_{S_i\in\mathcal{F}'} S_i\right| = \sum_{Q\in\mathcal{Q}(G[\mathcal{F}'])} (-1)^{|Q|+1} \left|\bigcap_{S_i\in Q} S_i\right|. \tag{3.10}$$

For each edge $\mathcal{F}'\mathcal{F}''$ of T, let $G[\mathcal{F}'\mathcal{F}'']$ denote the complete subgraph of G that is induced by $\mathcal{F}' \cap \mathcal{F}''$. Applying the inclusion-exclusion formula to the complete graph $G[\mathcal{F}'\mathcal{F}'']$ gives

$$\left|\bigcup_{S_i\in\mathcal{F}'\cap\mathcal{F}''} S_i\right| = \sum_{Q\in\mathcal{Q}(G[\mathcal{F}'\mathcal{F}''])} (-1)^{|Q|+1} \left|\bigcap_{S_i\in Q} S_i\right|. \tag{3.11}$$

For each $x \in S_1 \cup \cdots S_n$, let $T(x)$ denote the subgraph of T that consists of those nodes $\mathcal{F}'$ of T such that x is an element of at least one $S_i \in \mathcal{F}'$ and those edges $\mathcal{F}'\mathcal{F}''$ of T such that x is an element of at least one $S_i \in \mathcal{F}' \cap \mathcal{F}''$. (Note that $T(x)$ does not have to be an induced subgraph of T.) Each such $T(x)$ is a forest, and so $1 \leq |V(T(x))| - |E(T(x))|$ holds. Summing those inequalities over all $x \in S_1 \cup \cdots S_n$ shows that

$$\left|\bigcup_{i=1}^{n} S_i\right| \leq \sum_{\substack{\mathcal{F}'\in\\ V(T)}} \left|\bigcup_{S_i\in\mathcal{F}'} S_i\right| - \sum_{\substack{\mathcal{F}'\mathcal{F}''\in\\ E(T)}} \left|\bigcup_{S_i\in\mathcal{F}'\cap\mathcal{F}''} S_i\right|. \tag{3.12}$$

Next, combining inequality (3.12) with the equalities (3.10) and (3.11) gives

$$\left|\bigcup_{i=1}^{n} S_i\right| \leq \sum_{\substack{\mathcal{F}'\in\\ V(T)}} \sum_{\substack{Q\in\\ \mathcal{Q}(G[\mathcal{F}'])}} (-1)^{|Q|+1} \left|\bigcap_{S_i\in Q} S_i\right| - \sum_{\substack{\mathcal{F}'\mathcal{F}''\in\\ E(T)}} \sum_{\substack{Q\in\\ \mathcal{Q}(G[\mathcal{F}'\mathcal{F}''])}} (-1)^{|Q|+1} \left|\bigcap_{S_i\in Q} S_i\right|. \tag{3.13}$$

For each $Q \in \mathcal{Q}(G)$, let T_Q denote the subgraph of T that is induced by the nodes of T that contain Q. Since T is a clique tree of G, each such T_Q is connected and so $|V(T_Q)| - |E(T_Q)| = 1$ holds. Therefore, the right side of (3.13) equals the right side of (3.9), which proves Theorem 3.10. □

For the intersection data given in Figs. 3.7 and 3.10, let G be the spanning chordal subgraph of $\Omega^w(\mathcal{F})$ that is shown in Fig. 3.11. In this example, inequality (3.9) says that $|\bigcup_{i=1}^n S_i| \leq 44 - 30 + 6 = 20$ (grouping together the 6 nodes, then the 9 edges, and then the 4 triangles of G).

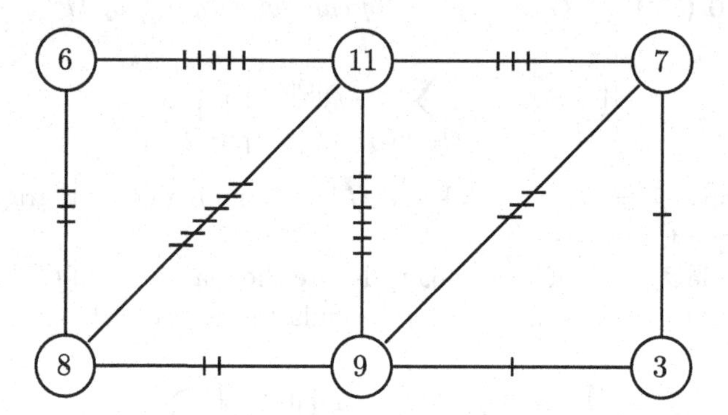

Fig. 3.11. A spanning chordal subgraph of Fig. 3.7 to illustrate Theorem 3.10.

To illustrate the proof of Theorem 3.10, Fig. 3.12 shows the unique clique tree T for the graph G in Figure 3.11. Inequality (3.12) in the proof of Theorem 3.10 says that $|\bigcup_{i=1}^n S_i| < (14+16+16+14) - (13+14+13) = 60 - 40 = 20$. (Since the full inclusion-exclusion formula shows that $|\bigcup_{i=1}^n S_i| = 19 < 20$, this strict inequality corresponds to the existence of an element x that is in the underlying sets $A_7 \cap A_9 \cap A_{11}$ and $A_3 \cap A_7 \cap A_9$ of adjacent nodes of T without being in the underlying set $A_7 \cap A_9$ of the edge between them in T—this means that this $T(x)$ is not connected, and so has $1 < |V(T(x))| - |E(T(x))|$.)

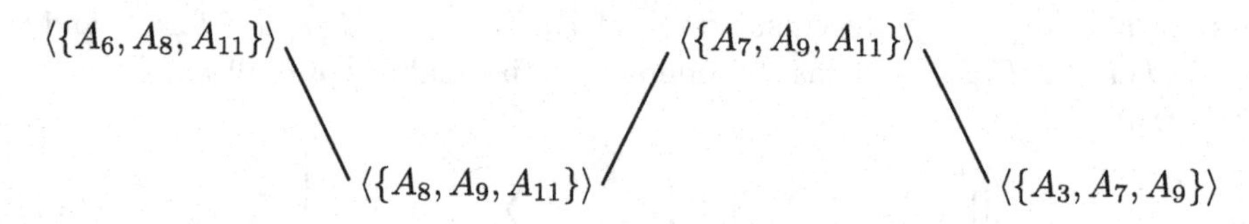

Fig. 3.12. The clique tree T for the graph in Fig. 3.11.

As mentioned above, Theorems 3.7 and 3.9 are special cases of Theorem 3.10. Reference [3] discusses several other special cases of Theorem 3.10 that had previously appeared in the literature.

The following example will show the necessity of requiring the spanning subgraph G in Theorem 3.10 to be chordal. Take $\mathcal{F}$ to consist of $S_1 = \{a, b, z\}$, $S_2 = \{b, c, z\}$, $S_3 = \{c, d, z\}$, and $S_4 = \{a, d, z\}$, and then take G to be the 4-cycle S_1, S_2, S_3, S_4, S_1 of weight-2 edges of $\Omega^w(\mathcal{F})$. This nonchordal G has exactly eight complete subgraphs (four nodes and four edges), making (3.9) the false inequality $5 \leq 12 - 8 = 4$.

3.2.4. *Economical inclusion-exclusion*

Recall that applying the traditional inclusion-exclusion formula (3.2) to the intersection data given in Figs. 3.7 and 3.10 gives the actual cardinality $|A_3 \cup A_6 \cup A_7 \cup A_8 \cup A_9 \cup A_{11}| = 44 - 38 + 15 - 2 = 19$ (grouping together the 6 vertices, then the 13 edges, then the 12 triangles, and then the 4 *tetrahedra*—meaning the four subgraphs isomorphic to K_4—of the graph in Fig. 3.7). Expanding that grouping, this inclusion-exclusion calculation involves 27 nonzero terms (out of $2^6 - 1 = 31$ possible terms). Many of those nonzero terms end up canceling with each other, and so were not needed in the first place.

References [12; 13] contain additional upper and lower bounds on $|\bigcup_{i=1}^{n} S_i|$, as well as a truly economical version of inclusion-exclusion—Theorem 3.11 below, proved in [12]—that will avoid unnecessary canceling of terms. This will involve a "graph structure" based on $\Omega^w(\mathcal{F})$ that consists of a graph $G = \langle V(G), E(G)\rangle$ together with a distinguished set $C(G)$ of cycles of G—each a set of edges such that each node of G is in an even number of those edges—a distinguished set $P(G)$ of *polyhedra* (3-*polyhedra*)—each a set of cycles in $C(G)$ such that each edge of G is in an even number of those cycles—and so on, as described more carefully below.

The *graph structure* $\langle V(G), E(G), C(G), P(G), \ldots\rangle$ is defined from the complete graph $\Omega^w(\mathcal{F})$. Note that each edge E, cycle C, polyhedron P, and so on of $\Omega^w(\mathcal{F})$—whether or not it is chosen for $\langle V(G), E(G), C(G), P(G), \ldots\rangle$—will have its node set correspond to a subset of $\mathcal{F} = \{S_1, \ldots, S_n\}$; define weights $|E|$, $|C|$, $|P|$, and so on to be the cardinalities of the intersections of all of the S_i's in each of those subsets. The specific steps in the construction are listed below. Note that the sets $E(G)$, $C(G)$, $P(G)$, and so on might not be uniquely determined; any of the possible choices can be taken for the graph structure.

- The nodes in $V(G)$ are precisely $S_1, \ldots, S_n$.
- Choose edges for $E(G)$ from the edges of $\Omega^w(\mathcal{F})$ by:
  first, for each $i \geq 0$ in decreasing order, sequentially choosing edges E with $|E| = i$ such that E does not form a cycle with previously-chosen edges $E_1, E_2, \ldots$ (this is Kruskal's Algorithm, giving a maximum spanning tree of $\Omega^w(\mathcal{F})$);
  then, for each $i \geq 0$ in decreasing order, sequentially choosing edges E with $|E| = i$ such that E does not form a cycle C with previously-chosen edges that have $|E| = |C|$ (thus every cycle C of $G = \langle V(G), E(G)\rangle$ has $|C| < |E|$ for each of its edges E).
- Choose cycles for $C(G)$ from the cycles of $G = \langle V(G), E(G)\rangle$ by:
  first, for each $i \geq 0$ in decreasing order, sequentially choosing cycles C with $|C| = i$ such that C does not form a polyhedron with previously-chosen cycles $C_1, C_2, \ldots$;

then, for each $i \geq 0$ in decreasing order, sequentially choosing cycles C with $|C| = i$ such that C does not form a polyhedron P with previously-chosen cycles that have $|C| = |P|$ (thus every polyhedron P of $\langle V(G), E(G), C(G)\rangle$ has $|P| < |C|$ for each of its cycles C).
- And so on—next choosing (3-)polyhedra, being careful with 4-*polyhedra* (sets of 3-polyhedra with each cycle in an even number of those 3-polyhedra).

Theorem 3.11 ([12]). *If* $\langle V(G), E(G), C(G), P(G), \ldots\rangle$ *is a graph structure for* $\mathcal{F} = \{S_1, \ldots, S_n\}$, *then*

$$\left|\bigcup_{i=1}^{n} S_i\right| = \sum_{X \in V(G)} |X| - \sum_{X \in E(G)} |X| + \sum_{X \in C(G)} |X| - \sum_{X \in P(G)} |X| + \cdots. \quad (3.14)$$

For instance, applying this to the intersection data given in Figs. 3.7 and 3.10 might produce the graph structure shown in Fig. 3.13.

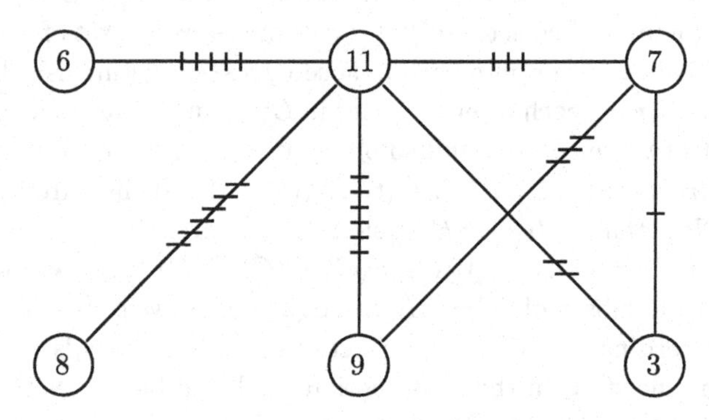

Fig. 3.13. A graph structure from $\Omega^w(\mathcal{F})$ for the example from Figs. 3.7 and 3.10 (including the weight-0 triangle $A_3A_7A_{11}$ and the weight-1 triangle $A_7A_9A_{11}$).

The economical inclusion-exclusion calculation from formula (3.14) for Fig. 3.13 says that $|\bigcup_{i=1}^{n} S_i| = 44 - 26 + 1 - 0 + \cdots = 19$. This calculation requires only 15 nonzero terms, as compared to the 27 nonzero terms required in the traditional inclusion-exclusion formula (3.2). (Reference [12] gives another example; also see [16].)

References

1. A. Brandstädt, V.B. Le, J.P. Spinrad, *Graph Classes: A Survey.* (Society for Industrial and Applied Mathematics, Philadelphia, 1999).
2. H.-J. Bandelt, H.M. Mulder, Distance-hereditary graphs, *J. Combin. Theory,* Ser. B **41** 182–208, (1986).
3. K. Dohmen, Bonferroni-type inequalities via chordal graphs, *Combin. Probab. Comput.* **11** 349–351, (2002).

4. K. Dohmen, *Improved Bonferroni inequalities via abstract tubes. Inequalities and identities of inclusion-exclusion type*, *Lecture Notes in Mathematics*, **1826**. (Springer-Verlag, Berlin, 2003).
5. F.F. Dragan, V.I. Voloshin, Incidence graphs of biacyclic hypergraphs, *Discrete Appl. Math.* **68** 259–266, (1996).
6. E. Gioan, C. Paul, Dynamic distance hereditary graphs using split decomposition, *Lecture Notes in Comput. Sci.* **4835** 41–51, (2007).
7. J. Galambos, I. Simonelli, *Bonferroni-Type Inequalities with Applications*. (Springer, New York, 1996).
8. M.C. Golumbic, A.N. Trenk, *Tolerance Graphs*. (Cambridge, 2004).
9. D.R. Hoover, Subset complement addition upper bounds—an improved inclusion-exclusion method, *J. Statist. Plann. Inference* **24** 195–202, (1990).
10. D. Hunter, An upper bound for the probability of a union, *J. Appl. Prob.* **13** 597–603, (1976).
11. E.G. Kounias, Bounds for the probability of a union, with applications, *Ann. Math. Statist.* **39**, 2154–2158, (1968).
12. T.A. McKee, Graph structure for inclusion-exclusion inequalities, *Congr. Numer.* **125** 5–10, (1997).
13. T.A. McKee, Graph structure for inclusion-exclusion equalities, *Congr. Numer.* **133** 121–126, (1998).
14. T.A. McKee, Subgraph trees in graph theory, *Discrete Math.* **270** 3–12, (2003).
15. T.A. McKee, Restricted circular-arc graphs and clique cycles, *Discrete Math.* **263** 221–231, (2003).
16. T.A. McKee, Clique representations of graphs, *Congr. Numer.* **170** 185–192, (2004).
17. T.A. McKee, Spanning (2-)trees of intersection graphs and Hunter–Worsley-type set bounds, *Util. Math.* **68** 97–102, (2005).
18. T.A. McKee, Graphs with complete minimal k-vertex separators, *Ars Combin.*, to appear.
19. T.A. McKee, F.R. McMorris, *Topics in Intersection Graph Theory*. (Society for Industrial and Applied Mathematics, Philadelphia, 1999).
20. T.M. Przytycka, E. Zotenko, Graph theoretical approaches to delineate dynamics of biological processes. In eds. I. Mandoiu and A. Zelikovsky, *Bioinformatics Algorithms: Techniques and Applications*, pp. 29–54. (Wiley, Hoboken N. J., 2008).
21. E.R. Scheinerman, Characterizing intersection classes of graphs, *Discrete Math.* **55** 185–193, (1985).
22. K.J. Worsley, An improved Bonferroni inequality and applications, *Biometrika* **69** 297–302, (1982).
23. E. Zotenko, K.S. Guimarães, R. Jothi, T.M. Przytycka, Decomposition of overlapping protein complexes: A graph theoretical method for analyzing static and dynamic protein associations, *Algor. Molecular Bio.* **1** #7, (2006).

Chapter 4

Axiomatic Characterization of Location Functions

F.R. McMorris, Henry Martyn Mulder, Rakesh V. Vohra

Department of Applied Mathematics, Illinois Institute of Technology
Chicagi, IL 60616 USA
mcmorris@iit.edu

Econometrisch Instituut, Erasmus Universiteit
P.O. Box 1738, 3000DR Rotterdam, Netherlands
hmmulder@ese.eur.nl

Kellogg Business School, Northwestern University
Evanston, IL, USA
r-vohra@kellogg.northwestern.edu

The general problem in location theory deals with functions that find sites on a graph (discrete case) or network (continuous case) in such a way as to minimize some cost (or maximize some benefit) to a given set of clients represented by vertices on the graph or points on the network. The axiomatic approach seeks to uniquely distinguish, by using a list of intuitively pleasing axioms, certain specific location functions among all the arbitrary functions that address this problem. In this chapter we survey results for three popular location functions (center, mean, median) in both the discrete and continuous cases. Some new results are presented for the median function on median networks.

Introduction

A problem often encountered in the provision of a service is how best to locate a service facility so as to optimize efficiency and accessibility. This problem has received a great deal of attention, as indicated by the reference lists of [6; 17; 22]. Typically, the problem is formulated as an optimization problem in which a facility has to be optimally located at some point on a network of roads. Optimality is defined in terms of the utility of clients. Client utility (or more precisely disutility) is modelled as some monotone function of the distance the client has to travel to reach the facility.

Another way to formulate location problems is to use the perspective of consensus functions. A consensus function is a model to describe a rational way to obtain consensus among a group of agents or clients. The input of the function consists of certain information about the agents, and the output concerns the issue about which consensus should be reached. The rationality of the process is guaranteed by the fact that the consensus function satisfies certain "rational" rules or "consensus axioms" (see e.g. Powers, this volume, for another discussion of consensus). For a location function the input is the location of the clients and the output are the locations that satisfy the optimality criterion. Following Arrow [1], cf. [4], we characterize various location functions in terms of axioms they satisfy. Holzman [7] was the first to study location problems from this persective. In this chapter we survey the axiomatic approach as is stands in the literature as well as describe some new steps.

The possibility or impossibility of axiomatic characterization depends on three factors. First of course the optimality criterion, that is, the specific location function under study. Secondly, the structure of the network on which the clients are situated and the location is to be found. Trees and median structures play a major role here. Finally, it depends on whether the structure is continuous or discrete, or unordered versus ordered. For instance, in the continuous case the edges of the tree are line segments and clients as well as the service facility may be located on interior points. In the discrete case, the tree is just a graph, and clients and facilities are required to be located at vertices only. Finally, in the ordered case, the tree may be considered to be a meet-semilattice with a universal lower bound form which the tree "grows upward" only.

The location functions, for which an axiomatic characterization has been obtained, are the center function, the median function, and the mean function. The first chooses a location so as to minimize the maximum distance any client is from the location. The paradigm for the center function is the Fire Station Problem *FSP*: how to find a location such that (in)flammable objects can be reached in short enough time. The median function chooses a location that minimizes the sum of client distances to itself. The mean function minimizes the sum of squared distances. The paradigm for the median and mean function is the Distribution Center Problem *DCP*: how to find a location for a distribution center that minimizes the cost for stocking up the warehouses located at given points in the network.

This chapter is organized as follows. First we discuss the center function, then the median function, and finally the mean function. Note that this is not in chronological order. For each function we present first the discrete case, then the continuous case, and finally, when available, the ordered case. Trees appear in all cases. For the median function results on a broader class are available: median graphs in the discrete case, median semilattices in the discrete ordered case. In the case of the median function we extend existing results on continuous trees to the new case of continuous cube-free median networks.

4.1. The model

Our terminology is chosen so at to best the purposes of this chapter. For the terminology and notation on graphs we refer the reader to Mulder (this volume). Here we give only the essentials. In this chapter $G = (V, E)$ denotes a finite, connected, simple, loopless graph. The *length* of a path is the number of edges on the path. For any two vertices u and v of G, the *distance* $d_G(u, v)$ between u and v is the length of a shortest u, v-path, or a u, v-*geodesic*. If no confusion arises, we will just write $d(u, v)$ instead of $d_G(u, v)$. The *interval* between u and v is the set

$$I_G(u, v) = \{ \ w \mid d(u, w) + d(w, v) = d(u, v) \ \}.$$

So $I_G(u, v)$ consists of all vertices "between" u and v. Again, we write $I(u, v)$ if no confusion will arise. Recall that a *tree* $T = (V, E)$ is a connected, cycle-free graph. A *median graph* is a graph G such that $|I(u, v) \cap I(v, w) \cap I(w, v)| = 1$, for any three vertices u, v, w of G. Trees, hypercubes, and grid graphs are typical examples of median graphs. These are the models for the discrete case. For all necessary notations and results we refer to Mulder (this volume).

For the continuous case we choose our terminology such that it stresses the similarities and dissimilarities with the discrete case. So we start with a graph and edges are treated as continuous arcs with a length, and interior points are also possible locations. A *network* $N = (G, \lambda) = (V, E, \lambda)$ consists of a graph $G = (V, E)$ and a mapping λ that turns edge uv into an *arc* of *length* $\lambda(uv)$. It follows from this definition that there is at most one arc between any two vertices. Formally, λ maps the vertices onto distinct points of some euclidian m-space, and maps edge uv onto a curve of length $\lambda(uv)$ with extremities u and v. We require that the curves do not intersect in interior points. We call the elements of an arc *points.* So there are points of two types: points in V are vertices, and points on the interior of an arc are *interior points.* For any two points p and q on the same arc, the length $\lambda(pq)$ of the subarc between p and q is just the length of the part of the curve between p and q. A *path* R joining two points p on arc uv_1 and q on arc $v_k w$ is either a subarc or a sequence $p \to v_1 \to v_2 \ \ldots \ \to v_k \to q$ with vertices $v_1, v_2, \ \ldots \ , v_k$, where $v_i v_{i+1}$ is a arc, for $i = 1, \ldots, k$, such that each vertex occurs at most once in the sequence. Since there is at most one arc between any two vertices, this definition of a path uniquely determines the arcs that are used to get from p to q. The length of the path is

$$\lambda(pv_1) + \lambda(v_1 v_2) + \ \ldots \ \lambda(v_{k-1} v_k) + \lambda(v_k q).$$

A shortest path is a path of minimum length. The *distance* $\delta(p, q)$ between two points p and q is the length of a shortest p, q-path. We assume that there are no redundant arcs, that is, each arc uv is the unique shortest path between u and v. This implies the absence of multiple arcs (which were excluded anyway by our definition). Moreover, for each arc uv, any other path between u and v has length greater than $\lambda(uv)$. This assumption is a necessary condition for some results below, but we shall see that it is not a serious restriction of our model. The graph $G = (V, E)$ is the *underlying graph* of the network. In general there is no relation between $\delta(u, v)$ and $d(u, v)$, except that, because of the irredundancy of arcs, we have $\delta(u, v) = \lambda(uv)$ if and only if $d(u, v) = 1$. The *segment* between two points p and q in N is the set

$$S(p, q) = \{ r \mid \delta(p, r) + \delta(r, q) = \delta(p, q)\}.$$

Since there are no redundant arcs, $S(p, q)$ is the subnetwork of N consisting of all shortest paths between p and q.

In the discrete case vertices of degree 2, incident with exactly two edges may occur and cannot be ignored. In the continuous case vertices of degree 2 do not play a special role, so we may just turn it into an interior point, thus merging the two arcs involved into one arc. Hence we may assume that in a tree network the vertices have either degree 1 or degree at least 3.

A *profile* on a set X is a finite sequence $\pi = x_1, x_2, \ldots, x_k$ of elements in X, with $|\pi| = k$ the *length* of the profile. Note that, a profile being a sequence, multiple occurrences of the same element are allowed. For the location functions considered here, the order of the sequence is irrelevant. When k is odd, we call π an *odd profile*, and when k is even, we call π an *even profile.* Denote by X^* the set of all profiles on X. A *consensus function* on a set X is a function $L : X^* \to 2^X - \emptyset$ that returns a nonempty subset of X for each profile on X. Because multiple occurrences in π are allowed, more than one client may be at the same point, or, if the clients are weighted, we can replace a client of weight j by j copies of this client in π. In the discrete case the set X is the set of vertices V of graph G. In the continuous case X is the set of all points of network N. A *location function* is a consensus function, for which the defining criterion is phrased in terms of the distances to the elements of the profile.

4.2. Basic Axioms

Let us first consider obvious and natural axioms that one would want in any rational consensus procedure on graphs or networks. Let $L : X^* \to 2^X - \emptyset$ be a location function on a graph $G = (V, E)$, in which case $X = V$, or a network $N = (V, E, \lambda)$, in which case X is the set of all points of N. The first axiom is Anonymity: the order of the profile does not play a role. There are simple consensus functions that are not anonymous. For example, if there is a dictator amongst the clients in the profile, then he can not hide his identity. All the location functions considered here are anonymous.

(A) Anonymity: for any profile $\pi = x_1, x_2, \ldots, x_k$ on X and any permutation σ of $\{1, 2, \ldots, k\}$, we have $L(\pi) = L(\pi^\sigma)$, where $\pi^\sigma = x_{\sigma(1)}, x_{\sigma(2)}, \ldots, x_{\sigma(k)}$.

Another natural axiom, again violated by dictatorship, is that, if all clients are located at the same point, then that point should be selected.

(U) Unanimity: $L(x, x, \ldots, x) = \{x\}$. for all $x \in X$.

Sometimes a weaker but equally natural axiom is sufficient.

(F) Faithfulness: $L(x) = \{x\}$, for all $x \in X$.

The next axioms are specifications of the following idea. If profile π agrees on output x and profile ρ does as well, then the concatenation $\pi\rho$ of the two profiles should agree on x as well. The idea for this type of consistency appears in a paper of Young [25], where it is used to axiomatize Borda's rule for voting procedures. We need two types of consistency here.

(C) Consistency: If $L(\pi) \cap L(\rho) \neq \emptyset$ for profiles π and ρ, then $L(\pi\rho) = L(\pi) \cap L(\rho)$.

(QC) Quasi-consistency: If $L(\pi) = L(\rho)$ for profiles π and ρ, then $L(\pi\rho) = L(\pi)$.

Because $L(\pi) \neq \emptyset$, for any π, it follows that (C) implies (QC) trivially. It is an easy exercise to prove that (QC) and (F) imply (U).

No combination of (F), (U) or (C) (or (QC)) is sufficient to pin down a particular consensus function. Hence we consider axioms that specify how a consensus function must behave on profiles of length 2. We present two: Middleness and Betweenness, for FSP and DCP, respectively. Since both axioms involve distances, we need to distinguish between the discrete and the continuous case. So L is a consensus function of a graph G, discrete case, or a network N, continuous case.

(Mid) Middleness: [Discrete] Let u, v be two not necessarily distinct vertices in V. If $d(u, v)$ is even, then $L(u, v)$ is the set of all vertices z with $d(u, z) = d(z, v) = \frac{1}{2}d(u, v)$. If $d(u, v)$ is odd, then $L(u, v)$ is the set of all vertices z with $d(u, z) = d(v, z) + 1$ or $d(u, z) = d(v, z) - 1$.

(Mid) Middleness: [Continuous] Let x, y be two not necessarily distinct points in X. Then $L(x, y)$ is the set of all points v with $\delta(x, v) = \delta(v, y) = \frac{1}{2}\delta(x, y)$.

Note that if there is a unique shortest path between x and y, as is the case in trees, then $L(x, y)$ is a single point in the continuous case, and $L(u, v)$ is a single vertex or an edge in the discrete case. The next axiom fits DCP: any point between x and y minimizes the sum of the distances to x and y.

(B) Betweenness: [Discrete] $L(u, v) = I(u, v)$, for all $u, v \in V$.

(B) Betweenness: [Continuous] $L(x,y) = S(x,y)$, for all $x, y \in X$.

Note that betweenness implies that $L(x,x) = \{x\}$. From this it follows that (B) and (C) imply (F), and hence (U). Similarly, (Mid) and (C) imply (F), and hence (U). In both cases we may replace (C) by (QC). We will see below that for specific cases additional axioms are needed.

4.3. The Center Function

Let $\pi = x_1, x_2, \ldots, x_k$ be a profile on a graph $G = (V, E)$. A *central vertex* of π is a vertex x minimizing

$$\max\{\ d(x, x_i) \mid 1 \leq i \leq k\ \}.$$

The *center* of π is the set consisting of all central vertices of π. If π consists of all vertices exactly once, then we call the center of π the *center of G*.

A classical result is that the center of a tree consists of one vertex or two adjacent vertices. This result dates back to 1869 and was proved by C. Jordan [8], although at that time the notion of graph did not exist. In the 1860's Jordan studied automorphisms of mathematical structures. As an example he discussed "assemblages of lines" in the plane with a tree-like structure, and proved that each automorphism fixes substructures, which we now would call the center and the centroid of the tree. The concept of graph was introduced only in 1878 by James Joseph Sylvester in a letter to Nature [23], see [19; 20] for a discussion of the origins of graph theory. Sylvester envisaged a great future for graph theory as a universal science, being a common basis for such diverse sciences as logic, chemistry, kinematics, and algebra. His vision did not come true, but we still owe him for the concept. He did not prove anything worthwhile on graphs. That he left to others.

Let $\pi = x_1, x_2, \ldots, x_k$ be a profile on a network $N = (V, E, \lambda)$. A *central vertex* of π is a point x minimizing

$$\max\{\ \delta(x, x_i) \mid 1 \leq i \leq k\ \}.$$

The *center* of π is the set consisting of all central points of π.

The *Center Function* $Cen : X^* \to 2^X - \emptyset$ on X is defined by $Cen(\pi)$ being the center of π. McMorris, Roberts, and Wang [16], see also [21] give an axiomatic characterization of the center function in the discrete case. If we examine the proof in [21] closely, we see that the same proof holds for the continuous case.

It is easy to see that the center function satisfies (A), (Mid), and (QC), but these are not sufficient to characterize the center function. So we need some specific axioms for this case. Observe that in the definition of a central vertex, multiple occurrences of a vertex in a profile do not influence the outcome. Define the *support* $\{\pi\}$ of profile π to be the set of all points that occur at least once in π. The center function obviously satisfies

(PI) Population Invariance: If $\{\pi\} = \{\rho\}$ then $L(\pi) = L(\rho)$.

The next axiom applies on trees (tree networks) only. For a profile π on a tree (network) T, we define $T(\pi)$ to be the smallest subtree (network) of T containing $\{\pi\}$. Note that all pendant vertices (endpoints) of $T(\pi)$ occur in π. For an element x we denote by $\pi \setminus x$ the profile obtained by removing all occurrences of x from π. Note that, if x is not in π, then $\pi \setminus x = \pi$. The next axiom obviously is also satisfied by the center function.

(R) Redundancy: Let L be a consensus function on a tree (network) T.
If $x \in T(\pi \setminus x)$ then $L(\pi \setminus x) = L(\pi)$.

The discrete case, proved in [16], is then:

Theorem 4.1. *Let L be a consensus function on a tree $T = (V, E)$. Then L is the center function Cen if and only if L satisfies (Mid), (PI), (R), and (QC).*

Adapting the proof of Theorem 4.1 in [21] to the continuous case we get:

Theorem 4.2. *Let L be a consensus function on a tree network $T = (V, E, \lambda)$. Then L is the center function Cen if and only if L satisfies (Mid), (PI), (R), and (QC).*

Extending these characterizations beyond trees is difficult. To see why, let H be any graph. Add to H four new vertices u_l, v_l, v_r, u_r. Make v_l and v_r adjacent to all vertices of H and u_l adjacent to v_l and u_r adjacent to v_r. The resulting graph G has as its center the subgraph H. Since the center can be arbitrary we cannot expect a clean characterization on arbitrary graphs or networks. Developing a characterization for special classes of graphs or networks might be more promising.

4.4. The median Function

For the Distribution Center Problem, the simplest solution is to find a location that minimizes the sum of the distances to the clients. This is known as the absolute median, or median for short. A median vertex of a tree minimizes the sum of the distances to the trivial profile, so to all vertices in the tree. It is easily seen that a median vertex is just a centroidal vertex. So the characterization of the median of a tree as one vertex or two adjacent vertices is, with hindsight, also due to C. Jordan, see [8].

Let $\pi = x_1, x_2, \ldots, x_k$ be a profile on a graph $G = (V, E)$. A *median vertex* of π is a vertex x minimizing $D(x, \pi) = \sum_{1 \le i \le k} d(x, x_i)$. The *median set* of π is the set of all median vertices of π.

Let $\pi = x_1, x_2, \ldots, x_k$ be a profile on a network $N = (V, E, \lambda)$. A *median point* of π is a point x minimizing $\Delta(x, \pi) = \sum_{1 \le i \le k} \delta(x, x_i)$. The *median set* of π is the set of all median points of π.

The *Median Function* $Med : X^* \to X$ on a graph G or a network N is defined by $Med(\pi)$ being the median set of profile π. Clearly, the median function satisfies Anonymity and Betweenness. The next Lemma probably belongs to folklore.

Lemma 4.1. *The Median Function Med on a graph satisfies* (C).

Proof. Let π and ρ be profiles such that $Med(\pi) \cap Med(\rho) \neq \emptyset$. Choose any element x in $Med(\pi) \cap Med(\rho)$ and any element y in $Med(\pi\rho)$. Then we have

$$D(x, \pi\rho) = D(x, \pi) + D(x, \rho) \leq D(y, \pi) + D(y, \rho) = D(y, \pi\rho) \leq D(x, \pi\rho),$$

where the first $\leq$ follows from $x \in Med(\pi) \cap Med(\rho)$, and the second $\leq$ follows from $y \in Med(\pi\rho)$. This implies that we have equality throughout, which means that $x \in Med(\pi\rho)$ and $y \in Med(\pi) \cap Med(\rho)$. This proves Consistency. □

The analogue for the continuous case is proved in the same way. So the median function satisfies three of the basic axioms: (A), (B), and (C). What makes the median function exceptional in our story is that these axioms characterize the median function in some important instances. The first result of this nature is the characterization on tree networks due to Vohra [24]. In that paper the following axiom was used.

(Ca) Cancellation: Let π be a profile with support $\{x, y\}$ such that x and y occur an equal number of times in π. Then $L(\pi) = S(x, y)$.

Note that Betweenness is a special instance of Cancellation. It follows immediately that (A), (B) and (C) imply (Ca). So we may rephrase the characterization in [24] as:

Theorem 4.3. *Let L be a consensus function on a tree network N. Then $L = Med$ if and only if L satisfies (A), (B), and (C).*

In Subsection 4.4.3 we will extend this result to cube-free median networks. For an easier understanding we present the discrete case first.

4.4.1. *The Median Function on Median Graphs*

In Mulder (this volume) median graphs were introduced as an important generalization of trees and hypercubes. This suggests that one might try to generalize results on trees to median graphs. A *median graph* is a graph G such that $|I(u, v) \cap I(v, w) \cap I(w, u)| = 1$, for any triple of vertices u, v, w. Clearly trees (as well as hypercubes) are median graphs. The following characterization of median graphs suggests strongly that any result on the median function for trees might extend to median graphs.

Theorem 4.4. *A graph G is a median graph if and only if $|Med(u, v, w)| = 1$, for any three vertices u, v, w of G.*

It follows from results in Mulder (this volume) that all odd profiles have a single median vertex in a median graph. We now review some important results on median graphs from [14]. For a profile $\pi = x_1, x_2, \ldots, x_k$, we denote by $\pi - x_i$ the vertex-deleted profile with just the i-th element deleted from π.

Theorem 4.5. *Let G be a median graph, and let $\pi = x_1, x_2, \ldots, x_k$ be an odd profile. Then $Med(\pi) = \cap_{1 \leq i \leq k} Med(\pi - x_i)$.*

A subset W of V in a graph $G = (V, E)$ is *convex* if $I(u, v) \subseteq W$, for any two vertices u, v in W. Note that the intersection of convex sets is again convex. The *convex closure* $Con[U]$ of a subset U of V is the smallest convex set containing U.

Theorem 4.6. *Let G be a median graph, and let $\pi = x_1, x_2, \ldots, x_k$ be an even profile. Then $Med(\pi) = Con[\cup_{1 \leq i \leq k} Med(\pi - x_i)]$.*

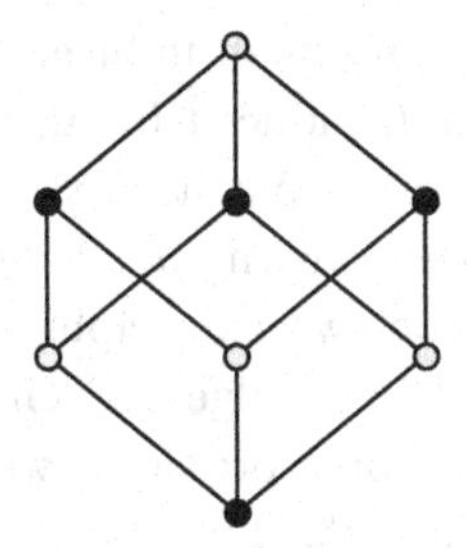

Fig. 4.1. The 3-cube Q_3 with a profile

Theorem 4.5 and induction on the length of the profiles suffice to characterize $Med(\pi)$ on median graphs when π is odd in terms of (A), (B) and (C). This argument does not extend to even profiles. The difficulty is that consistency may have no bite. To see why, consider Fig. 4.1. The black vertices represent a profile π of length 4 on the *cube* Q_3. Clearly, $M(\pi) = V$. However, we cannot find any combination of subprofiles of π that give intersecting median sets. Surprisingly, this example constitutes the bottleneck to easy extensions the characterization.

Call a median graph *cube-free* if Q_3 does not occur in the median graph. In [14] the following strong theorem is obtained, by which we can circumvent induction in the case of even profiles.

Theorem 4.7. *Let G be a cube-free median graph, and let π be an even profile of length $2k$. Then there exists a permutation σ of $\{1, 2, \ldots, 2k\}$ with $\pi^\sigma = x_1, x_2, \ldots, x_{2k}$ such that*

$$Med(\pi) = \bigcap_{1 \leq i \leq k} I(x_{2i-1}, x_{2i}).$$

Given these theorems, one can now easily prove that on cube-free median graphs the obvious axioms (A), (B), (C) suffice to characterize Med.

Theorem 4.8. *Let G be a cube-free median graph and let L be a consensus function on G. Then $L = Med$ if and only if L satisfies (A), (B), and (C).*

For arbitrary median graphs the example in Fig. 4.1 forms an obstacle for similar easy results. Hence some 'heavy-duty' axioms were introduced to do the job, see [14] for the first one.

(K) Convexity: Let $\pi = x_1, x_2, \ldots, x_k$ be a profile of length $k > 1$ on G. If $\bigcap_{1 \leq i \leq k} L(\pi - x_i) = \emptyset$ then $L(\pi) = Con[\bigcup_{1 \leq i \leq k} L(\pi - x_i)]$.

Theorem 4.9. *Let L be a consensus function on a median graph G. Then $L = Med$ if and only if L satisfies (A), (B), (C), and (K).*

An important feature of median graphs is the following. Let uv be an edge in a median graph G. Let G_u^{uv} be the subgraph induced by all vertices closer to u than to v, and let G_v^{uv} be the subgraph induced by all vertices closer to v than to u. It turns out that these two subgraphs partition the vertex set of G in a very nice way. We call such a partition a *split* with *split halves* G_v^{uv}, G_v^{uv}. Both subgraphs are convex, the edges between the two subgraphs form a matching F_{uv}, and for any edge xy in F_{uv} with x in G_u^{uv} the edge xy defines the same split as uv, that is, $G_x^{xy} = G_u^{uv}$, and $G_y^{xy} = G_v^{uv}$. For a profile π, we denote by π_u^{uv} the subprofile of π consisting of all elements in G_u^{uv}. For even profiles the median set $Med(\pi)$ satisfies the following property, see [14]: if $|\pi_u^{uv}| = |\pi_v^{uv}|$, then $u \in Med(\pi)$ if and only if $v \in Med(\pi)$. If we turn this into an axiom, see [11], then we get another characterization of the median function on median graphs.

($\frac{1}{2}$-Co) $\frac{1}{2}$-Condorcet Let uv be an edge of a graph G, and let π be a profile with $|\pi_u^{uv}| = |\pi_v^{uv}|$. Then $u \in L(\pi)$ if and only if $v \in L(\pi)$.

Theorem 4.10. *Let L be a consensus function on a median graph G. Then $L = Med$ if and only if L satisfies (F), (C), and ($\frac{1}{2}$-Co).*

Note the omission of anonymity and betweenness. These axioms do not follow from (F), (C), and ($\frac{1}{2}$-*Co*) in general, yet in median graphs they are implied by (F), (C), and ($\frac{1}{2}$-*Co*).

4.4.2. *The t-Median Function on Median Semilattices*

The ordered case displays some significant and unexpected differences with the ordered case. A full discussion exceeds the scope of this chapter. Instead, we give only the essential details and use graphs to explain what is going on.

The $\frac{1}{2}$-Condorcet property of median graphs is a consequence of the following characterization of median sets in median graphs. Recall that a split in a median graph does not depend on the edge chosen between the two sides of the split. So we may refer to a split in a median graph G as a pair of subgraphs G_1, G_2. For a profile π on G, we denote by π_i the subprofile of all elements in G_i. Then the median set of π is always on the side of the split where the majority of π is located, see [14]:

Theorem 4.11. *Let G be a median graph, and let M be the median function on G. Then $Med(\pi) = \bigcap\{ G_1 \mid G_1$ is a splithalve with $|\pi_1| > |\pi_2| \}$, for any profile π on V.*

Barthélémy and Janowitz [3] introduced a stronger criterion for being a median. We reformulate this for median graphs: instead of requiring a simple majority on the side of the split, one requires a larger portion of π to be on the side where the median is to be found. Let t be a rational number with $\frac{1}{2} \leq t < 1$. Then the *t-Median Function* $M_t : V^* \to 2^V - \emptyset$ on a median graph $G = (V, E)$ is defined by

$$M_t(\pi) = \bigcap\{ G_1 \mid G_1 \text{ is a splithalve with } |\pi_1| > t|\pi|\}.$$

The analogue of the ($\frac{1}{2}$-*Co*) axiom of the previous subsection for t-medians is then

(t-Co) t-Condorcet: [Discrete] u is in $L(\pi)$ if and only if v is in $L(\pi)$, for each profile π and each split G_u^{uv}, G_v^{uv} with $|\pi_u^{uv}| = t|\pi|$.

An isometric subgraph of a hypercube is called a *partial cube.* Median graphs are partial cubes. So are all even cycles, whereas only the 4-cycle is a median graph. It was proved in [13] that the t-median function M_t is t-Condorcet on a partial cube G. Thus, one might conjecture that a consensus function on a median graph that is faithful, consistent, and t-Condorcet should be the t-median function M_t. Surprisingly, this is *not* the case, as the following impossibility result from [13] shows.

Theorem 4.12. *Let $G = (V, E)$ be a median graph with $|V| \geq 3$, and let t be a rational number with $\frac{1}{2} < t < 1$. Then there does not exist a consensus function $L : V^* \to 2^V - \{\emptyset\}$ on G satisfying (F), (C), and (t-Co).*

The reason for this surprise is that M_t is *not* consistent. However it does satisfy two weaker types of consistency: subconsistent and subquasi-consistent.

(SC) Subconsistency: If $L(\pi) \cap L(\rho) \neq \emptyset$ for profiles π and ρ, then $L(\pi\rho) \subseteq L(\pi) \cap L(\rho)$.

(SQC) Subquasi-consistency: If $L(\pi) = L(\rho)$ for profiles π and ρ, then $L(\pi) \subseteq L(\pi\rho)$.

Now we turn to the ordered case. Let $G = (V, E)$ be a median graph, and let z be any vertex of G. We define the *partial order* $\leq_z$ on V by $u \leq_z v$ if $u \in I(z, v)$.

Loosely speaking, we take z as the bottom and let G 'grow upwards'. Then the ordered set $(V, \leq_z)$ is a *median semilattice.* This is a distributive meet semilattice with the *coronation property*, that is, if the pairwise joins $u \vee v$, $v \vee w$, $w \vee u$ exist for u, v, w then the join $u \vee v \vee w$ exists. Median semilattices have been studied widely in the literature. Given a finite semilattice $(V, \leq)$, the *covering graph* $G = (V, E)$ is defined by uv being an edge if v covers u, that is, $u < v$ but there is no element w with $u < w < v$. From our viewpoint it is important to know that the covering graph of a median semilattice $(V, \leq)$ is a median graph, see e.g. [18].

The following notions are rather technical. One might speed-read these technicalities. We will clarify what the axiom and the theorems mean in terms of graphs.

Let $(V, \leq)$ be a finite semilattice with universal lower bound z. An element x is an *atom* if it covers the universal lower bound z. Atoms are precisely the neighbors of z in the covering graph.

An element s if *join-irreducible* if $s = x \vee y$ implies that $s = x$ or $s = y$. We can find the join-irreducible elements in a median semilattice $(V, \leq)$ as follows. Let z be the universal lower bound, and let G_1, G_2 be a split of the underlying median graph G with z in G_1. Then the vertex in G_2 closest to x is a join-irreducible. This induces a one-to-one correspondence between the join-irreducibles in $(V, \leq)$ and the splits in G.

Let $(V, \leq)$ be a finite distributive semilattice, S be the set the join-irreducible elements of $(V, \leq)$, and $\pi = x_1, x_2, \ \ldots \ , x_k$ be a profile on V. Then the *index* of a join-irreducible element $s \in V$ with respect to π is

$$\gamma(s, \pi) = \frac{|\{ \, i \mid s \leq x_i \, \}|}{k}.$$

In terms of the underlying median graph let G_1, G_2 be the split of s, that is, z is in G_1 and s is the vertex in G_2 closest to z. Then $\gamma(s, \pi)$ is π_2, that is, the elements of π on the side of s. Note that G_2 consists of all elements w in $(V, \leq)$ with $s \leq w$. Now let

$$\alpha_t(\pi) = \bigvee \{ \, s \mid s \in S \text{ with } \gamma(s, \pi) > t \, \}.$$

In graph terms the element $\alpha_t(\pi)$ is the vertex in $M_t(\pi)$ closest to z. The other elements of $M_t(\pi)$ we get by taking the meets of $\alpha_t(\pi)$ with the join-irreducible elements with index exactly t. Thus the *t-Median Function*, M_t, on $(V, \leq)$ is defined by

$$\begin{aligned} M_t(\pi) = &\{\alpha_t(\pi)\} \cup \\ &\cup \{\alpha_t(\pi) \vee s_1 \vee \ldots \vee s_k \mid \gamma(s_i, \pi) = t, \ i = 1, \ldots, k, \text{ provided the join exists}\}. \end{aligned}$$

The t-Condorcet axiom for a consensus function L on the semilattice $(V, \leq)$ is phrased as follows.

(t-Co) t-Condorcet: [Ordered] If s is join-irreducible element in $(V, \leq)$ covering w_s and $\gamma(s, \pi) = t$, then $x \vee s$ is in $L(\pi)$ if and only if $x \vee w_s$ is in $L(\pi)$, provided

$x \vee s$ exists.

In [15] the following characterization of the t-Median Function is given.

Theorem 4.13. *Let $(V, \leq)$ be a distributive meet semilattice in which all join-irreducibles are atoms, and let t be a rational number with $\frac{1}{2} \leq t < 1$. Let L be a consensus function on $(V, \leq)$. Then $L = M_t$ if and only if L satisfies (F), (C), and $(t\text{-}Co)$.*

In [13] a similar result was proved for the case of median semilattices.

Theorem 4.14. *Let $(V, \leq)$ be a finite median semilattice, and let t be a rational number with $\frac{1}{2} \leq t < 1$. Let L be a consensus function on $(V, \leq)$. Then $L = M_t$ if and only if L satisfies (F), (C), and $(t\text{-}Co)$.*

4.4.3. *The Median Function on Cube-free Median Networks*

It is natural to ask if the results in Subsection 4.4.1 extend to the network case. For the cube-free case this can be done under certain conditions. Our discussion will underscore this fact. We use the relation between median graphs and median networks from [2]. While cube-free median networks are still rather special, the class is rich enough to encompass one category of real world network structure: the grid equipped with the Manhattan metric (or city-block norm). However, as we will see, we have to sacrifice something: client locations must be confined to vertices of the network.

We need some additional concepts. Let W be a subset of V. Then $\langle W \rangle_G$ denotes the subgraph of G induced by W, that is, the subgraph with W as its vertex set and all edges with both ends in W as its edge set. Furthermore $\langle W \rangle_N$ denotes the subnetwork of N induced by W, that is, the subnetwork with W as its vertex set and all arcs with both ends in W as its set of arcs. Note that $I(u, v)$ is a *subset* of V, so we may consider the subgraph $\langle I(u, v) \rangle_G$ of G or the subnetwork $\langle I(u, v) \rangle_N$ of N induced by $I(u, v)$. The subgraph $\langle I(u, v) \rangle_G$ may consist of more than the geodesics between u and v, viz. if there exists some edge between vertices, say x and y, in $I(u, v)$ with $d(u, x) = d(u, y)$, so that we also have $d(x, v) = d(y, v)$. Such an edge will be called a *horizontal edge* in $I(u, v)$. Note that, if G is bipartite, then such horizontal edges do not exist.

A *m-cycle* in N, with $m \geq 3$, is a closed path with m arcs, or more precisely, a sequence $v_1 \to v_2 \to \ \ldots \ \to v_m \to v_{m+1}$ with $v_1, v_2, \ \ldots \ , v_m$ distinct vertices and $v_1 = v_{m+1}$, such that $v_i v_{i+1}$ is a arc, for $i = 1, 2, \ldots, m$. A *rectangle* in N is a 4-cycle such that non-adjacent (i.e. opposite) arcs have equal length.

A network N is a *median network* if

$$|S(u, v) \cap S(v, w) \cap S(w, u)| = 1,$$

for any three vertices u, v, w in N. A characterization of median networks was given in [2]:

Theorem 4.15. *A network N is a median network if and only if its underlying graph G is a median graph and all 4-cycles in N are rectangles.*

The usefulness of this theorem lies in the fact that we can make use of the rich structure theory for median graphs, see e.g. [18; 9; 14] and Mulder (this volume).

The *cube network* is the network whose underlying graph is the cube Q_3, see Fig. 4.1, in which all 4-cycles are rectangles. This means that in the figure parallel arcs have the same length. Our main result holds for cube-free median networks. For this case we consider a slightly different type of consensus function: $L : V^* \rightarrow 2^X - \emptyset$, where V is the set of vertices of network N and X is the set of all points of N. So client positions are restricted to vertices, whereas the facility may still be located at any point. We will see below that we need this restriction is necessary to get results.

Tree networks have the characterizing property that, for any two points p and q, there exists a unique path connecting them. Trivially, this unique path is the shortest p, q-path. For any three points p, q, r in a tree network, the three paths between the pairs of p, q, r have a unique common point, which is necessarily a vertex (unless one of the three points is an interior point on the path between the other two). A striking difference between networks and graphs arises with respect to this property.

Theorem 4.16. *Let N be a network. Then N is a tree network if and only if*

$$S(p,q) \cap S(q,r) \cap S(r,p) \neq \emptyset,$$

for any three points p, q, r in N.

Proof. Let N be a tree. Then $S(p,q) \cap S(q,r) \cap S(r,p)$ consists of the unique point (vertex) lying simultaneously on the shortest paths between the three pairs of p, q, and r.

Assume that N is not a tree. If N is disconnected, we choose p and q in different components, whence $S(p,q) = \emptyset$, so that, for any r, we have three points for which the corresponding segments have empty intersection. Now suppose that N contains cycles. Let C be a cycle of minimal length t. Minimality of t implies that C is an *isometric* cycle in N, i.e., the distance along the cycle between any two points p and q on C equals their distance $\delta(p,q)$ in N. Now choose two distinct points q and r on C such that $\delta(q,r) < \frac{1}{2}t$. Then $\delta(q,r)$ is the length of the shortest of the two arcs on C between q and r. Let p be the point on the other arc with equal distance to q and r. Then $\delta(p,q)$ is the length of the shorter arc on C between p and q, and $\delta(p,r)$ is the length of the shorter arc on C between p and r. Now, if $S(p,q) \cap S(q,r)$ contained a point x different from q, then a shortest p, x-path together with a shortest x, r-path and the shorter arc of C between r and p would contain a cycle shorter than C, which contradicts the minimality of C. So $S(p,q) \cap S(q,r) = \{q\}$. Similarly, any two of the segments intersect only in their common endpoint. Since

p, q, r are distinct points on C, the intersection of all three segments is empty. This settles the proof of the Theorem. □

In the discrete case we get even more than the median graphs when we require $I(p,q) \cap I(q,r) \cap I(r,p) \neq \emptyset$. Because of the uniqueness of the point on the three paths between the pairs of p, q, r, we may refine the property in Theorem 4.16 to obtain the following characterization.

Corollary 4.1. *Let N be a network. Then N is a tree if and only if only if $S(p,q) \cap S(q,r) \cap S(r,p)$ is a unique point, for any three points p, q, r in N.*

If $S(p,q) \cap S(q,r) \cap S(r,p) \neq \emptyset$, for some points p, q, r, then, by (C) and (B), we have $Med(p,q,r) = S(p,q) \cap S(q,r) \cap S(r,p)$. If N is a network with cycles, then let C be any cycle of minimal length, and let p, q, r be points as in the proof of Theorem 4.16. Now $S(p,q) \cap S(q,r) \cap S(r,p) = \emptyset$, whereas $Med(p,q,r) = \{q,r\}$. So we cannot determine $Med(p,q,r)$ from the three segments using consistency. This is the main reason forcing us to restrict ourselves to profiles consisting of vertices only. As we know, in the graph case the situation is quite different: we consider vertices only anyway, and median graphs come in view.

In a median network, we have the following property: For any arc vw and any vertex u, either $v \in S(u,w)$ or $w \in S(u,v)$. This follows from the median property and the irredundancy of arcs (whence $S(v,w) \cap V = \{v,w\}$). This property was called the *bottleneck property* in [2], where the following Lemma was proved.

Lemma 4.2. *Let $N = (V, E, \lambda)$ be a network with the bottleneck property. Then*

$$I(u,v) = S(u,v) \cap V,$$

for any two vertices u, v in V.

The proof in [2] contains a minor but repairable gap. An examination of the proof reveals that one can prove more. First we need some notation: $I_1(u,v)$ is the set of all neighbors of u in $I(u,v)$. Clearly,

$$I(u,v) = \{u\} \cup [\bigcup_{x \in I_1(u,v)} I(x,v)].$$

Lemma 4.2 can be strengthened. Instead of giving a full proof here by extending the one in [2], we just use Lemma 4.2 as a starting-point, and restrict ourselves to completing the proof.

Lemma 4.3. *Let $N = (V, E, \lambda)$ be a network with the bottleneck property. Then $S(u,v)$ is the subnetwork of N induced by $I(u,v)$, for any two vertices u and v in V.*

Proof. We use induction on the length $n = d(u,v)$ of the intervals $I(u,v)$ in G. If $d(u,v) = 0$, then $u = v$, so that $S(u,u) = \{u\} = I(u,u)$. If $d(u,v) = 1$, then uv is an edge in G and a arc in N, and we have $I(u,v) = \{u,v\}$ and $S(u,v)$ is the arc uv. So assume that $n \geq 2$. First we observe that there are no horizontal edges in $I(u,v)$. For otherwise, suppose xy is a horizontal edge in the interval, so that $d(u,x) = d(u,y) < n$. Then, by induction, $S(u,x) = \langle I(u,x)\rangle_N$ and $S(u,y) = \langle I(u,y)\rangle_N$. So we have $x \notin S(u,y)$ and $y \notin S(u,x)$, which contradicts the bottleneck property. So all edges in $\langle I(u,v)\rangle_G$ are on u,v-geodesics. By Lemma 4.2, any shortest u,v-path in N starts with a arc ux with x in $I(u,v)$. So it starts with an edge ux, where x is a neighbor of u in $I(u,v)$. Hence we have

$$S(u,v) = \bigcup_{x \in I_1(u,v)} [S(u,x) \cup S(x,v)],$$

which, by induction is equal to

$$\bigcup_{x \in I_1(u,v)} [S(u,x) \cup \langle I(x,v)\rangle_N].$$

Since there are no horizontal edges in $I(u,v)$, the assertion now follows. □

In fact we have proved more. In networks satisfying the bottleneck property, for any two vertices u,v, each u,v-geodesic in G can be obtained from a shortest u,v-path in N by ignoring the lengths of the arcs. Conversely, each shortest u,v-path in N can be obtained from the corresponding u,v-geodesic in G by assigning the appropriate lengths to the arcs. This more informal version of Lemma 4.3 is the one we will use.

A consequence of Lemma 4.3 involves the notion of convexity. Let W be a subset of vertices. Then $\langle W\rangle_N$ is *convex* in N if $S(u,v) \subseteq \langle W\rangle_N$, for any two vertices u,v in W. Similarly, $\langle W\rangle_G$ is *convex* in G if $I(u,v) \subseteq W$, for any two vertices u,v in W. In a network N with the bottleneck property $\langle W\rangle_N$ is convex in N if and only if $\langle W\rangle_G$ is convex in the underlying graph G. Moreover, $\langle W\rangle_N$ can be obtained from $\langle W\rangle_G$ by assigning the appropriate lengths, and, vice versa, $\langle W\rangle_G$ can be obtained from $\langle W\rangle_N$ by ignoring lengths of arcs. These observations imply that all results for median graphs that can be proved using the concepts of distance, geodesic, and convexity have their counterparts for median networks provided we restrict ourselves to profiles consisting of vertices only. So, in the sequel *the median function on a network $N = (V,E,\lambda)$ is a consensus function with profiles on V and with the set of all points X as set of possible outcomes*:

$$Med_N : V^* \to 2^X - \{\emptyset\}.$$

Let N be a median network, and let uv be an arbitrary arc in N. Let G be the underlying median graph of N. Denote by W_u^{uv} the set of vertices strictly closer to u than to v. By Lemma 4.3, we can use the same notation for the underlying median graph. Hence W_u^{uv} is the vertex set of G_u^{uv}. It follows from the structure theory for median graphs in [18] that the sets W_u^{uv} and W_v^{uv} are convex. Moreover, for any other arc xy between the two sets with $x \in W_u^{uv}$ and $y \in W_v^{uv}$, it turns out that $W_x^{xy} = W_u^{uv}$ and $W_y^{xy} = W_v^{uv}$. So arc xy defines the same sides as arc uv. Finally, for any shortest u,x-path $u \to u_1 \to \ldots \to u_k \to x$ in the u-side there exists a shortest v, y-path $v \to v_1 \to \ \dots \ \to v_k \to y$ in the v-side such that $u_i v_i$ is a arc, for $i = 1, \ \dots \ , k$. From the rectangle property we deduce that all these arcs have the same lengths, in particular uv and xy have the same length. This property is typical of the Manhattan metric.

Recall that $M_G(\pi)$ consists of all vertices u such that a majority (not necessarily strict) of the profile is closer to u than any of its neighbors. Note also that the sets $M_G(\pi)$ are convex in median graphs.

Theorem 4.17. *Let $N = (V, E, \lambda)$ be a median network, let $G = (V, E)$ be its underlying median graph, and let π be a profile on V. Then*

$$M_N(\pi) = \langle M_G(\pi) \rangle_N.$$

Proof. By Lemma 4.3, we know that $M_N(\pi) \cap V = M_G(\pi)$. So we only have to check interior points. Let uv be any arc of N, and let p be an interior point of uv. By the definition of W_u^{uv}, and the bottleneck property, the distance from v to any vertex in W_u^{uv} can be measured via u. Hence the distance from p to any vertex in W_u^{uv} can also be measured via u. The same holds if we interchange the roles of u and v. This implies

$$\Delta(p, \pi) = \Delta(p, \pi_u) + \Delta(p, \pi_v) =$$

$$= |\pi_u|\lambda(pu) + \Delta(u, \pi_u) + |\pi_v|\lambda(pv) + \Delta(v, \pi_v).$$

From this equality we deduce that if uv is a arc with $|\pi_u| = |\pi_v| = \frac{1}{2}|\pi|$, then $\Delta(p, \pi) = \Delta(u, \pi) = \Delta(v, \pi)$. Hence, either the entire arc uv is in $M_N(\pi)$ or none of it is in $M_N(\pi)$. Finally, if $|\pi_u| > |\pi_v|$, then $\Delta(p, \pi) > \Delta(u, \pi)$. Hence p is not in $M_N(\pi)$. These observations and the facts on median sets in median graphs preceding the theorem complete the proof. □

By Lemma 4.3, we get an analogue of Theorem 4.7 for networks.

Theorem 4.18. *Let N be a cube-free median network, and let π be a profile on V of even length $2m$. Then there exists a permutation $y_1, y_2, \ \dots \ , y_{2m-1}, y_{2m}$, such that*

$$M_N(\pi) = \bigcap_{i=1}^{m} S(y_{2i-1}, y_{2i}).$$

Now we are ready to prove the main result of this section.

Theorem 4.19. *Let N be a cube-free median network, and let $L : V^* \to 2^X - \emptyset$ be a consensus function on N where X is the set of all points in N. Then $L = Med_N$ if and only if L satisfies (A), (B) and (C).*

Proof. Let L be a consensus function on N satisfying (A), (B), and (C). We use induction on the length of the profiles to prove that $L(\pi) = Med_N(\pi)$, for all profiles π on V.

By (C) and (B), we have $L(x) = L(x) \cap L(x) = L(x,x) = S(x,x) = \{x\} = Med_N(x)$. Now let $\pi = x_1, x_2, \ldots, x_k$ be a profile of length $k > 1$. If k is even, then, by Theorem 4.18, we can write $\pi = y_1, y_2, \ldots, y_{2m}$ such that $Med_N(\pi) = \bigcap_{i=1}^{m} S(y_{2i-1}, y_{2i})$. By (B) and (C), we conclude that $L(\pi) = Med_N(\pi)$.

If k is odd, then, by Theorem 4.5, we have $Med_N(\pi) = \bigcap_{i=1}^{k} Med_N(\pi - x_i)$. Hence, by the induction hypothesis, we have $Med_N(\pi) = \bigcap_{i=1}^{k} f(\pi - x_i)$. Since $Med_N(\pi) \neq \emptyset$, axiom (A) and repeated use of (C) gives $Med_N(\pi) = f(x_1, \ldots, x_1, x_2, \ldots, x_2, \ldots, x_k, \ldots, x_k)$ with the i-th element x_i appearing exactly $k-1$ times in f. Using (A), we have $Med_N(\pi) = L(\pi, \ldots, \pi)$ with π appearing exactly $k-1$ times in L. Hence, by (C), we deduce that $L(\pi) = Med_N(\pi)$. □

At first glance it might be thought that one could just "lift" Theorem 4.3 up to median networks. This is true in the case that the range X of the consensus function L is V as well. Then Theorem 4.18 would suffice to prove this result. But in Theorem 4.19 we include interior points in the range of L as well.

As noted above, we excluded redundant arcs to ensure that the unique shortest path between a pair of adjacent vertices was the arc connecting them. Here we outline why this entails no great loss.

First we give a precise definition of redundant arcs. Let N be a connected network. Consider two vertices u and v. If there exists a shortest u,v-path with more than one arc, then any arc with ends u and v is a *redundant* arc. If there is no such shortest path, then each shortest u,v-path is an arc with ends u and v. Take one such arc, and call this arc the irredundant arc between u and v. All other arcs between u and v are *redundant arcs.* The *reduced network* $\bar{N}$ is the network obtained from N by deleting all redundant arcs. We will argue that $Med_{\bar{N}}(\pi) \subseteq Med_N(\pi)$.

Observe first that the distance between any pair of points in $\bar{N}$ is the same as their distance in N. Therefore, if $Med_N(\pi)$ contains no point interior to a redundant arc, then $Med_N(\pi) = Med_{\bar{N}}(\pi)$.

Now let p be an interior point of a redundant arc e incident to the vertices u and v, say, with $\delta(u,p) \leq \delta(p,v)$. Take a shortest u,v-path P in $\bar{N}$, and let p' be the point on P with $\delta(u,p) = \delta(u,p')$. Then it is straightforward to check that $\Delta(p,\pi) \geq \Delta(p',\pi)$. This implies that, if p lies in $Med_N(\pi)$, then p' lies in $Med_{\bar{N}}(\pi)$ as well as $Med_N(\pi)$. It is again straightforward to deduce from this fact that $Med_{\bar{N}}(\pi) \subseteq Med_N(\pi)$.

4.5. The Mean Function on Trees

The third function we consider is the mean function. Instead of the sum of distances as optimality criterion a euclidian measure is used: the square-root of the sum of the squares of the distances. Because we minimize, we may omit taking the square-root. So let N be a network and let $\pi = x_1, x_2, \ldots, x_k$ be a profile on N. A *mean point* of π is a point x minimizing

$$\sum_{1 \leq i \leq k} [\delta(x, x_i)]^2.$$

The *mean* of π is the set of mean points of π. The *Mean Function Mean* on N is the function $Mean : X^* :\rightarrow 2^X - \emptyset$ with $Mean(\pi)$ being the mean of π. Historically this was the first location function to be characterized axiomatically, see Holzman [7]

If N is a tree network, then any profile has a unique mean point, so $Mean$ is single-valued. Holzman introduced two axioms for the tree case. Both involve specifying how the selected location changes as one of the clients moves to another position. So we introduce the following notation. Let $\pi = x_1, x_2, \ldots, x_k$ be a profile on N, and let y_i be some point. Denote by $\pi[x_i \rightarrow y_i]$ the profile obtained from π by replacing x_i by y_i, that is, client i moves the new position y_i.

(Li) Lipschitz: Let $\pi = x_1, x_2, \ldots, x_k$ be a profile of length k. Then $\delta(L(\pi), L(\pi[x_i \rightarrow y_i]) \leq \frac{1}{k}\delta(x_i, y_i)$.

(Inv) Invariance: Let $\pi = x_1, x_2, \ldots, x_k$ be a profile. Let y_i be a point in the branch of $L(\pi)$ that contains x_i with $\delta(y_i, L(\pi)) = \delta(x_i, L(\pi))$. Then $L(\pi[x_i \rightarrow y_i]) = L(\pi)$.

Theorem 4.20. *Let L be a single-valued consensus function on a tree network N. Then $L = Mean$ if and only if L satisfies (U),(Inv), and (Li).*

In [24] the invariance axiom was replaced by consistency.

Theorem 4.21. *Let L be a single-valued consensus function on a tree network N. Then $L = Mean$ if and only if L satisfies (U),(C), and (Li).*

Again, the discrete case is quite different from the continuous case. As in the case of the center function, the discrete case is not single-valued.

4.6. Concluding Remarks

In [5] another interesting instance of axiomatic characterization is discussed. It involves a nondecreasing, nonnegative, differentiable, strictly convex function f, which is used to 'weigh' the distances. Define $Med_f(\pi)$ to be the set of vertices x minimizing $\sum_{1 \leq i \leq k} f(\delta(x, x_i))$, for profile $\pi = x_1, x_2, \ldots, x_k$. $Med_f(\pi)$ is characterized by the axioms (A), (C), plus three axioms called Continuity, Tree Independence, and Population Monotonicity. These three additional axioms do not imply our axioms above.

So far axiomatic characterizations have been successful on trees and tree networks, and in the case of the Median Function also on median graphs and cube-free median networks. The discrete case for the Mean Function is in preparation, see

[10]. But there are still many open questions. For instance, on which graphs is the Median Function characterized by the three basic axioms (A), (B), and (C) only? Or, given a class of graphs, what are the location functions satisfying (A), (B), (C), or what extra axioms do we need to characterize Med?

References

1. K.J. Arrow, *Social Choice and Individual Values*, no. 12 in Cowles Commission for Research in Economics:Monographs, Wiley, New York, first ed. 1951.
2. H.J. Bandelt, Networks with Condorcet solutions, *European J. Operational Research* **20** (1985) 314–326.
3. J.P. Barthélemy, M.F. Janowitz, A formal theory of consensus, *SIAM J. Discrete Math.* **4** (1991) 305–322.
4. W.H.E. Day, F.R. McMorris, *Axiomatic Consensus Theory in Group Choice and Biomathematics*, Frontiers in Applied Math., SIAM, Philadelphia, 2003.
5. D.P. Foster, R.V. Vohra, An axiomatic characterization of a class of location functions in tree networks, *Operations Research* **46** (1998) 347–354.
6. P. Hansen, M. Labbé, D. Peeters, J.-F. Thisse, *Single facility location on networks*, Annals of Discrete Mathematics **31** (1987) 113 – 145.
7. R. Holzman, An axiomatic approach to location on networks, *Math. Oper. Res.* **15** (1990) 553 – 563.
8. C. Jordan, Sur les assemblages de lignes, *J. Reine Angew. Math.* **70** (1869) 193–200.
9. S. Klavžar, H.M. Mulder, *Median graphs: characterizations, location theory, and related structures*, J. Combin. Math. Combin. Comput. **30** (1999) 103–127.
10. F.R. McMorris, H.M. Mulder, O. Ortega, *Axiomatic characterization of the mean function on trees*, manuscript.
11. F.R. McMorris, H.M. Mulder, R.C. Powers, The median function on median graphs and semilattices, *Discrete Appl. Math.* **101** (2000) 221– 230.
12. F.R. McMorris, H.M. Mulder, R.C. Powers, The median function on distributive semilattices, *Discrete Appl. Math.* **127** (2003) 319–324.
13. F.R. McMorris, H.M. Mulder, R.C. Powers, The t-median function on graphs, *Discrete Appl. Math.* **127** (2003) 319–324.
14. F.R. McMorris, H.M. Mulder, F.S. Roberts, *The median procedure on median graphs*, Discrete Appl. Math. **84** (1998) 165 –181.
15. F.R. McMorris and R.C. Powers, The median procedure in a formal theory of consensus, *SIAM J. Discrete Math.* **14** (1995) 507–516.
16. F.R. McMorris, F.S. Roberts, C. Wang, The center function on trees, *Networks* **38** (2001) 84–87.
17. P.B. Mirchandani, R.L. Francis, *Discrete location theory*, John Wiley & Sons, New York, 1990.
18. H.M. Mulder, *The Interval Function of a Graph*, Mathematical Centre Tracts 132, Mathematisch Centrum, Amsterdam, 1980.
19. H.M. Mulder, To see the history for the trees, *Congressus Numerantium* **64** (1988) 25–43.
20. H.M. Mulder, Die Entstehung der Graphentheorie, in: K. Wagner, R. Bodendiek, *Graphentheorie III*, Wissenschaftsverlag, Mannheim, 1992, pp. 296-313, 381-383.
21. H.M. Mulder, K.B. Reid, M.J. Pelsmajer, Axiomization of the center function on trees, *Australasian J. Combin.* **41** (2008) 223–226.
22. S. Nickel, J. Puerto, *Location Theory: A Unified Approach*, Springer, Berlin, 2005.

23. J.J. Sylvester, Chemistry and Algebra, *Nature* **17** (1878) 284.
24. R. Vohra, *An axiomatic characterization of some locations in trees*, European J. Operational Research **90** (1996) 78 – 84.
25. H.P. Young, *An axiomatization of Borda's rule*, Journal of Economic Theory **9** (1974) 43 – 52.

Chapter 5

Median Graphs. A Structure Theory

Henry Martyn Mulder*

Econometrisch Instituut, Erasmus Universiteit
P.O. Box 1738, 3000DR Rotterdam, Netherlands
hmmulder@ese.eur.nl

Median graphs are a common generalization of trees and hypercubes in the following sense: for any three vertices u, v, w there exists a unique x lying on a shortest path between any pair of u, v, w. The origins of median graphs have to be found within pure mathematics: they arose from semilattices, ternary algebras, and Helly hypergraphs. Form the view point of mathematics they form a very interesting class: a rich structure theory has been developed, and there is still much more to come. Another interesting feature is that median graphs appear in different guises in many other mathematical areas. But equally important is that median graphs and median-type structures have many applications in such diverse fields as: mathematical biology, psychology, chemistry, economics, literary history, location theory, voting theory, and the like. This chapter provides an introduction into the structure theory of median graphs. It surveys median structures within mathematics and presents a concise overview of the many applications.

Introduction

Two well-known classes of graphs are that of the trees and that of the hypercubes. Recall that the n-dimensional hypercube Q_n, n-cube for short, has the 0,1-vectors of length n as its vertices, and two vertices are joined by an edge if, as 0,1-vectors, they differ in exactly one coordinate. When we add an ordering $\leq$ on the vertices of the n-cube, with $u \leq v$ whenever the 0,1-vector of v has a one in every coordinate where the 0,1-vector of u has a one, then this ordered graph is precisely the Hasse diagram of the n-dimensional Boolean lattice. Papers on both classes are abundant in the literature.

At first sight these classes seem to be quite different. But appearances are deceptive. There is a striking feature that both trees and hypercubes have in common: if we take any three vertices u, v, w, then there exists a unique vertex x, called the *median* of u, v, w, that lies simultaneously on a shortest path between each pair of

*This chapter is dedicated to my good friend and colleague Buck McMorris on the occasion of his 65-th birthday.

the three. In a tree this is obvious, because there is only one path between any two vertices. In the hypercube vertex x is determined as follows: each coordinate of x takes the value of the majority amongst the values of the corresponding coordinate of u, v, w.

Then the question arises: what are the graphs that share this property with trees and hypercubes. The graphs defined by this property are the *median graphs.* These graphs where introduced independently by various authors. First by Avann under the name of *unique ternary distance graphs*, see [1; 2]. The focus here was on distributive semi-lattices, and the graphs basically went almost unnoticed. Second by Nebeský, see [66], where the focus was on the relation with median algebras. Finally by Schrijver and the author, see [64], where the focus was on the relation with Helly hypergraphs. The first paper on median graphs with a focus only on the graphs is [56], where the Expansion Theorem was proved, see Section 5.2. The results on median graphs below are presented from a perspective that was developed much later. It is based on the idea mentioned above where median graphs arise as a common generalization of trees and hypercubes. In [60] the following "Metaconjecture" was proposed, which may be utilized as a guiding principle for finding nice problems on median graphs.

Metaconjecture. *Let $\mathcal{P}$ be a property that makes sense, which is shared by the trees and the hypercubes. Then $\mathcal{P}$ is shared by all median graphs.*

The point here, of course, is that the property should make sense. First, it does not make sense to observe such trivialities as: all trees and hypercubes are connected, hence all median graphs are connected; they are connected by definition. An example that does not work is: the center of a tree is a vertex or an edge, hence a hypercube, the center of a hypercube is the hypercube itself, therefore the center of a median graph should always be a hypercube. This is not true: take the 2 by 1 grid and attach a vertex at two opposite ends of the grid. Now the center consists of the path on the four vertices of the original 2 by 1 grid that are not involved in these attachments. A stronger version of this "conjecture", the *Strong Metaconjecture*, asks for properties $\mathcal{P}$ shared by trees and hypercubes that actually characterize median graphs. Both the Metaconjecture and the Strong Metaconjecture have produced new interesting results on median graphs. Below we present many examples where the (Strong) Metaconjecture does work, although some of these results precede the actual formulation of the Metaconjecture.

Median graphs allow a rich structure theory as is shown in the literature. But equally important, there are many interesting generalizations of median graphs. Also median structures are abundant in many other mathematical disciplines. And last but not least, median graphs and median structures have many applications in such diverse areas as evolutionary theory, chemistry, literary history, location theory, consensus theory, and computer science. This chapter serves as an introduction into the theory of median structures and their applications with a focus on median graphs.

5.1. Definitions and Preliminaries

In this Chapter $G = (V, E)$ denotes a finite, *connected*, simple, loopless graph. For any two vertices u and v of G, the *distance* $d_G(u, v)$ between u and v is the length of a shortest u, v-path, or a *u, v-geodesic*. If no confusion arises, we will just write $d(u, v)$ instead of $d_G(u, v)$. The *interval* between u and v is the set

$$I_G(u, v) = \{\ w \mid d(u, w) + d(w, v) = d(u, v)\ \}.$$

So it consists of all vertices "between" u and v. Again, we write $I(u, v)$ if no confusion will arise. A subset W of V is *convex* if $I(x, y)$ is contained in W, for any x, y in W. A subgraph H of G is *convex* if it is induced by a convex set in G. For a set $W \subseteq V$, the *convex closure* $Con[W]$ is the smallest convex set containing W. Because of finiteness, it can also be obtained by applying the following extension on W until no new set is found: the *extension* of a subset S of V is $ext(S) = \bigcup_{x,y \in S} I(x, y)$. A basic property of convex sets is that the family of convex sets is closed under taking arbitrary intersections. Loosely speaking, this is the defining property of a *convexity* in abstract convexity theory. In the infinite case a second defining property is that the union of a nested family of convex sets is again convex. In the sequel we will not distinguish between a subset W of V and the subgraph of G induced by W. The notions of interval and convexity in a graph probably are already part of folklore for a couple of decades. Notation and terminology were fixed in [58], where they were studied systematically for the first time.

For u, v, w in G, we write

$$I(u, v, w) = I(u, v) \cap I(v, w) \cap I(w, v).$$

The sets $I(u, v, w)$ can be empty: Take any isometric odd cycle C in G of length $2k + 1$, let v, w be adjacent vertices on C and let u be the vertex on C with $d(u, v) = d(u, w) = k$. Then $I(u, v, w) = \emptyset$. See 5.1 for examples. If $I(u, v, w) \neq \emptyset$ for all triples u, v, w in G, then G is called a *modular graph.* Clearly, modular graphs are bipartite. A *median graph* G is a special instance of a modular graph. It has the defining property that $|I(u, v, w)| = 1$, for all triples of vertices u, v, w in G. Obviously, a convex subgraph of a median graph is again a median graph.

Let W be a subset of V in the graph $G = (V, E)$, and let u be a vertex of G. A *gate* for u in W is a vertex x in W such that $x \in I(u, w)$, for every vertex w in W. A set W is *gated* if every vertex has a unique gate in W. Note that the gate for u in the gated set W is the vertex x in W closest to u. A nice property of median graphs is that the gated sets are precisely the convex sets. For a set W we denote by $\langle W \rangle$ the *gated closure* of W being the smallest gated set containing W.

A *profile* on G is a sequence $\pi = x_1, x_2, \ \dots\ , x_k$ of vertices of G. The *length* of the profile is $|\pi| = k$. Note that, a profile being a sequence, multiple occurrences of vertices are allowed. We call π an *odd* profile when k is odd, and *even* when k is

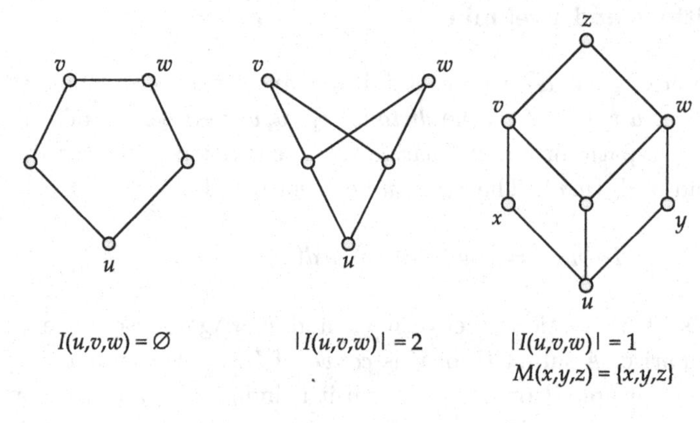

Fig. 5.1.

even. The distance of a vertex v to the profile π is

$$D(v,\pi) = \sum_{i=1}^{k} d(v, x_i).$$

The ordering in the profile can be ignored when only the distance to the profile matters. A *median* or *median vertex* of π is a vertex x minimizing the distance $D(x,\pi)$ to the profile, and the *median set* $M(\pi)$ of π is the set of all medians of π. Since G is connected, a median set is always non-empty. The profile consisting of all vertices of V just once is called the *trivial profile*, and is denoted by V. We write $D(x)$ for $D(x,V)$. The median set $M(G)$ of G is just the set $M(V)$ of vertices x minimizing $D(x)$. Clearly we have $M(u) = \{u\}$ and $M(u,v) = I(u,v)$ in any connected graph G. Note that, if $I(u,v,w) \neq \emptyset$, then $M(u,v,w) = I(u,v,w)$. Hence in a median graph profiles of length 3 have a singleton as median set. Actually, this characterizes median graphs, see [26]. It is also an immediate consequence of results in [58].

Theorem 5.1. *A graph G is a median graph if and only if $|M(u,v,w)| = 1$, for any triple of vertices u,v,w of G.*

For two graphs $G_1 = (V_1, E_1)$ and $G_2 = (V_2, E_2)$, the *Cartesian product* $G_1 \Box G_2$ is the graph with vertex set $V_1 \times V_2$, where two vertices $(u_1,u_2), (v_1,v_2)$ are adjacent whenever they have equality in one coordinate and adjacency in the other. The Cartesian product of more graphs is defined likewise. We write G^2 for $G \Box G$, and G^k for the product of k copies of G. With this notation we have $Q_n = K_2^n$. A *Hamming graph* is the Cartesian product of complete graphs, see [58; 59], whence the hypercube is a special instance of a Hamming graph. For a beautiful, in-depth treatise on the theory of various important types of graph products including the Cartesian product see [40].

For two graphs $G_1 = (V_1, E_1)$ and $G_2 = (V_2, E_2)$, the *union* $G_1 \cup G_2$ is the graph with vertex set $V_1 \cup V_2$ and edge set $E_1 \cup E_2$, and the *intersection* $G_1 \cap G_2$ is the graph with vertex set $V_1 \cap V_2$ and edge set $E_1 \cap E_2$. We write $G_1 \cap G_2 \neq \emptyset$ when $V_1 \cap V_2 \neq \emptyset$. The graph $G_1 - G_2$ is the subgraph of G_1 induced by the vertices in G_1 but not in G_2. A *proper cover* of a connected graph G consists of two subgraphs G_1 and G_2 such that $G_1 \cap G_2 \neq \emptyset$ and $G = G_1 \cup G_2$. Note that this implies that there are no edges between $G_1 - G_2$ and $G_2 - G_1$. If both G_1 and G_2 are convex, we say that G_1, G_2 is a *convex cover.* Every graph admits the *trivial cover* G_1, G_2 with $G_1 = G_2 = G$. This cover trivially is convex. On the other hand a cycle of length at least four does not have a convex cover with two proper subgraphs.

5.2. The Expansion Theorem

Any tree can be constructed from smaller trees in various ways. First we consider the following construction. Let T' be a tree, and let T'_1 and T'_2 be two proper subtrees sharing one vertex u' that cover T'. Loosely speaking, we now pull the two subtrees apart, by which vertex u' is doubled, and then we insert a new edge between the two copies of u'. Thus we obtain a larger tree T, which we say is the expansion of T' with respect to the proper cover T'_1, T'_2. Clearly, any tree can be obtained from the *one-vertex graph* K_1 by a succession of such expansions. Next we construct the hypercube Q_n by an expansion from a smaller hypercube. Take the hypercube $H' = Q_{n-1}$ and the trivial cover H'_1, H'_2 with $H'_1 = H'_2 = H'$. Now we pull the two hypercubes H'_1 and H'_2 apart obtaining two disjoint copies of Q_{n-1} and insert an edge between each pair of corresponding vertices, thus getting $H = Q_n$. Again it is clear that we can obtain any hypercube from K_1 by a sequence of such expansions. In view of the Metaconjecture, what is the property $\mathcal{P}$ that we are tracking here? It turns out that we need convex covers for our expansion. We give the precise definitions. For an illustration of the definitions and notations see Figure 5.2.

Let G' be a connected graph and let G'_1, G'_2 be a convex cover of G' with $G'_0 = G'_1 \cap G'_2$. For $i = 1, 2$, let G_i be an isomorphic copy of G'_i, and let λ_i be an isomorphism from G'_i to G_i. We write $G_{0i} = \lambda_i[G'_0]$ and $u_i = \lambda_i(u')$, for u' in G'_0. The *convex expansion* of G' with respect to the convex cover G'_1, G'_2 is the graph G obtained from the disjoint union of G_1 and G_2 by inserting an edge between u_1 in G_{01} and u_2 in G_{02}, for each u' in G'_0. We denote the set of edges between G_{01} and G_{02} by F_{12}. Note that F_{12} induces an isomorphism between G_{01} and G_{02}. We say that λ_i *lifts* G'_i up to G_i. For any subgraph H' of G' we abuse notation and write $\lambda_i[H']$ for $\lambda_i[H' \cap G'_i]$. So λ_i lifts the part of H' lying in G'_i up to G_i. The proof of the next Proposition is straightforward. The key in the proof is the following Lemma.

Lemma 5.1. *Let G be the expansion of G' with respect to the convex cover G'_1, G'_2 with lift maps λ_1, λ_2. Then the intervals in G_i are obtained by lifting*

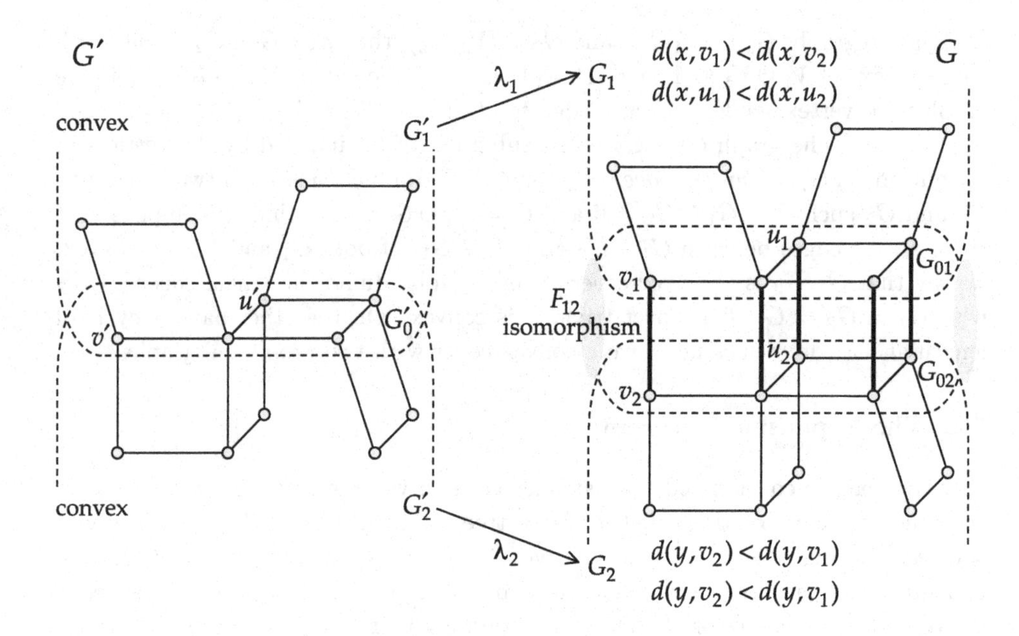

Fig. 5.2. Expansion

up the corresponding interval in G'_i, and, for v in G_1 and w in G_2, the interval $I_G(v,w)$ is obtained as follows: let $v = \lambda_1(v')$ and $w = \lambda_2(w')$, then $I_G(v,w) = \lambda_1[I_{G'}(v',w')] \cup \lambda_2[I_{G'}(v',w')]$.

Proposition 5.1. *A graph G obtained by successive convex expansions from K_1 is a median graph.*

The Strong Metaconjecture now asks whether we get all median graphs in this way, so that we have a characterization. We need some more notation, see Figure 5.2 for an illustration. Let G be a median graph, and let v_1v_2 be an edge in G. Let G_1 be the subgraph induced by all vertices closer to v_1 than to v_2 and let G_2 be the subgraph induced by all vertices closer to v_2 than to v_1. Since G is bipartite, it follows that G_1, G_2 vertex-partition G. Let F_{12} be the set of edges between G_1 and G_2, and let G_{0i} be the subgraph induced by the ends of F_{12} in G_i, for $i = 1, 2$. Then it is proved in [56] (although not exactly in that order) that the following facts hold:

(i) F_{12} is a matching as well as a cutset (minimal disconnecting edge set),
(ii) the subgraphs G_1, G_2, G_{01}, G_{02} are convex subgraphs of G,
(iii) the obvious mapping of G_{01} onto G_{02} defined by F_{12} (i.e. $u_1 \to u_2$, for any edge u_1u_2 in F_{12} with u_{0i} in G_{0i}) is an isomorphism,
(iv) for every edge u_1u_2 in F_{12} with u_i in G_{0i}, the subgraph G_1 consists of all vertices of G closer to u_1 than u_2, and the subgraph G_2 consists of all vertices

of G closer to u_2 than u_1.

We call such a partition G_1, G_2 of G a *split*. Note that any edge in F_{12} defines the same split. The subgraphs G_1 and G_2 are the *sides* of the split. If we are in u_1 of an edge u_1u_2 of F_{12}, then G_1 is the *side* of u_1 and G_2 is the *opposide* of u_1. We use this neologism as an homage to the guide at a boat trip that Buck McMorris and my family took on the river Zaan north of Amsterdam along the many windmills there. The guide did not speak any foreign language, but for the English speaking tourists he had a leaflet, in which 'opposide' was used to point at interesting features on the other side of the river.

Using the above notation it is clear that a convex expansion of a median graph G' with respect to the convex cover G'_1, G'_2 results in a median graph G with split G_1, G_2. But now we also have the converse operation: by contracting the edges in F_{12}, that is, identifying the ends of each edge in F_{12} and then delete the edges of F_{12}, we obtain the *contraction* G' of G with respect to the split G_1, G_2. To illustrate this, just go from right to left in Figure 5.2. The *contraction map* κ is defined as follows: restricted to G_i it is precisely λ_i^{-1}, for $i = 1, 2$. When this far, it is straightforward to prove that, if G is a median graph, then G' is again a median graph. Thus we get the following characterization of median graphs, see [56; 58].

Theorem 5.2 (Expansion Theorem). *A graph G is a median graph if and only if it can be obtained from K_1 by successive convex expansions.*

Because it allows induction on the number of vertices, but also on the number of splits, this theorem is a very powerful tool in understanding the structure of median graphs. We present a number of such structural features in the next section.

First we have another look at how to obtain larger trees from smaller ones, and hence from K_1. A simpler way to achieve this is: adding a pendant vertex. How can we reformulate this property such that we can use the (Strong) Metaconjecture? It must be such that we can apply the same procedure on hypercubes. Well, adding a pendant vertex to a tree T' by making the new vertex adjacent to v in T' can be achieved by expansion with respect to the convex cover $T', \{v\}$, where the one covering subgraph is the whole tree and the other is a 'subtree', in this case consisting of a single vertex. The expansion for the hypercubes is of a similar type, the one covering subgraph is the whole hypercube, the other is a convex subgraph (being the whole hypercube again). We call a convex expansion of G a *peripheral expansion* if it is performed with respect to the convex cover G_1, G_2, where $G = G_1$ and G_2 is a convex subgraph of G. For the peripheral contraction we define a *peripheral subgraph* of a median graph G to be a subgraph G_2 such that it is part of a split G_1, G_2 with $G_2 = G_{02}$. Using the Expansion Theorem we can prove that a median graph always contains a peripheral subgraph, see [60]. Note that this is not trivial. Then we have another instance of the possibilities of the Strong Metaconjecture.

Theorem 5.2 (Peripheral Expansion Theorem). *A graph G is a median graph if and only if it can be obtained from K_1 by successive peripheral expansions.*

The Expansion Theorem is a trivial corollary of the Peripheral Expansion Theorem. But to find a peripheral subgraph in a median graph one has to start with an arbitrary split G_1, G_2 and then prove that there exists a split H_1, H_2 such that G_1 is a proper subgraph of H_1 and H_2 a proper subgraph of G_2, see [60]. This is the reason that we did not start with the simple case of adding pendant vertices in the tree case, but with the case of arbitrary expansion with respect to two proper subtrees sharing one vertex.

A basic proof technique when using the Expansion Theorem is as follows. One or more contractions on the median graph G are performed to obtain a smaller median graph G', on which we apply the appropriate induction hypothesis. Then we perform the corresponding expansions in reverse order on G' so that we regain G. During this process a vertex v is contracted to a unique vertex v' in G'. When we recover G from G' by expansions, then v' is lifted up in each expansion to the appropriate side until we regain v. The sequence of vertices and expansions that we obtain in this way from v' to v is called the *history* of v (with respect to the expansions involved). Similarly, if $\pi = x_1, x_2, \ldots, x_k$ is a profile on G, then π is contracted to a profile $\pi' = x'_1, x'_2, \ldots, x'_k$ on G', where x'_i is the contraction of x_i, for $i = 1, 2, \ldots, k$. Thus we define the *history* of π in the obvious way, and, similarly, we define the history of a subset of V or a subgraph of G. If v' is a vertex of G' and we lift v' up to a vertex z in an expansion of G', then we call z a *descendant* of v'. Hence, if we know which lifts are applied on v' in the expansions to regain G from G', then we know the history of all descendants of v'.

5.3. The Armchair

In Economics (and Philosophy) the concept of Armchair Theorizing exists: by only sitting in their armchair and looking at the world economists can come up with new theories and insights, see [80]. Buck McMorris introduced me to this concept in mathematics: now sitting in our armchair after proving some heavy duty theorems, we let these do the work and come up with nice and new results. This approach is much more solid than that within Economics. The results below are an example of the use of the Armchair. We use the Expansion Theorem and the ideas and notations developed in its proof.

A feature that follows immediately from the structure described above is the following. A cutset coloring of a connected graph is a proper coloring of the edges (adjacent edges have different colors) such that each color class is a cutset (a minimal disconnecting edge-set). Of course, most graphs will not have a cutset coloring, whereas even cycles of length at least six have more than one. If we want to cutset color the edges of a graph, then, in an induced 4-cycle $u \to v \to w \to x \to u$, opposite edges must have the same color. For, if the edge uv gets a color, then vw

and ux must get another color. So u, x are on one side and v, w are on the other side of the cutset color of uv, and thus wx must get the same color as uv. We call this the 4-*cycle property* of cutset colorings. It follows from (i) to (iv) above that, in any cutset coloring of the median graph G, the set F_{12} must be a color class. By induction on the number of colors we get the next corollary [56; 58].

Corollary 5.1. *A median graph is uniquely cutset colorable, up to the labelling of the colors.*

Note that, using the 4-cycle property, we find a split and its sides in a median graph without computing any distances. The Strong Metaconjecture does not apply here. Take the 3-cube Q_3 and delete one vertex x. Now the three neighbors of the deleted vertex do not have a unique median. On the other hand, the 4-cycle property gives a unique cutset coloring. The 4-cycle property and the facts **(i)**, **(ii)**, **(iii)**, **(iv)** given in the previous section have been the basis for various recognition algorithms for median graphs, see for instance [43; 41].

Let G_1, G_2 be a proper cover of a graph G. Then we say that G is the *amalgamation* of G_1 and G_2 along the subgraph $G_1 \cap G_2$. If the cover is convex, then we say that it is a *convex amalgamation.* Note that in this case we amalgamate along a convex subgraph of the two covering subgraphs. A tree is the convex amalgamation of two smaller trees along a vertex. A hypercube is the 'amalgamation' with respect to the trivial cover of itself. The following application of the Strong Metaconjecture follows easily from the structural characterizations above.

Theorem 5.3. *Let G be a graph that is not a hypercube. Then G is a median graph if and only if it can be obtained from two smaller median graphs by convex amalgamation.*

A tree can be considered as built from 1-dimensional hypercubes, the edges, glued together along 0-dimensional hypercubes, i.e. vertices. A hypercube has only one hypercube as building stone. Thus we get the following characterization, see [14], which is easily deduced from the previous theorem, but also from the Expansion Theorem.

Theorem 5.4. *A graph G is a median graph if and only if it can be obtained from a set of hypercubes by convex amalgamations.*

One should read this theorem carefully. It does not state that in each step we amalgamate a new hypercube from our initial set of hypercubes to the graph constructed so far. It may be necessary to construct two graphs by amalgamations, and then amalgamate these two to obtain a larger median graph. For example, take the 2 by 2 grid (the Cartesian product of two copies of the path of length 2). The initial set of hypercubes consists of four 4-cycles. First we construct two 2 by 1 grids by amalgamating two 4-cycles along an edge. Then we amalgamate these 2 by 1 grids along a path of length 2, see 5.3. Thus we amalgamate always along a convex subgraph. If we would have added a 4-cycle in each step, then we can only amalgamate along a convex subgraph up to three of the four 4-cycles. To add the fourth one we have to amalgamate along a path of length 2, which is not convex in the 4-cycle.

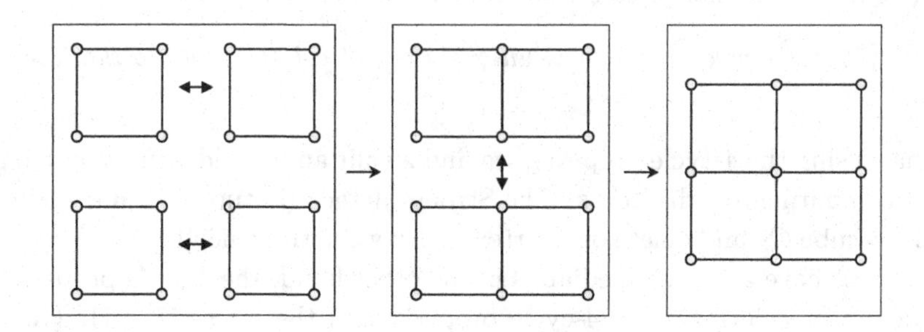

Fig. 5.3. Amalgamation

Another early result on median graphs is the following characterization, see [57; 58]. An *isometric subgraph* H of G is a such that distances in H equal those in G:

$$d_H(u,v) = d_G(u,v) \text{ for any } u,v \text{ in } H.$$

Theorem 5.5. *A graph G is a median graph if and only if it can be isometrically embedded in a hypercube Q such that the median in Q of any three vertices in G lies also in G.*

A simple corollary is that the hypercubes are the only regular median graphs.

A less easy consequence of the Armchair is the following result, see [14], which again can be explained using the Metaconjecture. Every automorphism of a tree fixes its center and its centroid, that is, the center is mapped onto the center and the centroid is mapped onto the centroid. This is a classical result of C. Jordan [44]. He studied trees, or actually, tree-like line structures in the plane, precisely for this purpose: the behavior of the automorphisms on these structures. A consequence of these results is that each automorphism fixes an edge, that is a subcube. Each automorphism of a hypercube fixes the whole hypercube, hence again a subcube.

Theorem 5.6. *Each automorphism of a median graph fixes some subcube, i.e. some regular median subgraph.*

5.4. Median Sets in Median Graphs

In [36] Goldman proved the classical result on how to find the median set of a profile on a tree using majority rule, see [37; 82] for origins of this idea. Rephrased in our terminology: if π is a profile on a tree T and T_1, T_2 is a split of T, then $M(\pi)$ is on the side where the majority of the elements of the profile are located. In the case of an even profile something else might happen: if half of the profile is in T_1 and the other half is in T_2, then both ends of the edge between T_1 and T_2 are in the median set of the profile. Note that in case the profile is even, the median set might induce a path. A similar result holds for hypercubes. Now in the even case the median set might be a subcube. These facts may be considered as inspiration for this section. We need some extra notation. Let G be a median graph with split G_1, G_2, and let π be a profile on G. Then π_i is the subprofile of π consisting of all elements of π contained in G_i, for $i = 1, 2$. The majority rule for median graphs reads as follows and was proved first in [53].

Theorem 5.7. *Let G be a median graph with split G_1, G_2, and let G' be the contraction with respect to this split. Let π be a profile on G. If $|\pi_1| > |\pi_2|$, then $M(\pi') \subseteq G'_1$ and $M(\pi) = \lambda_1[M(\pi')] \subseteq G_1$. If $|\pi_1| = |\pi_2|$, then $M(\pi) = \lambda_1[M(\pi')] \cup \lambda_2[M(\pi')]$, and u_1 is in $M(\pi)$ if and only if u_2 is in $M(\pi)$, for any edge u_1u_2 in F_{12}.*

An easy consequence of this theorem is the following result, which implies immediately that median sets are convex.

Theorem 5.8. *Let G be a median graph and let π be a profile on G. Then $M(\pi) = \bigcap_{G_1, G_2 \text{ split with } |\pi_1| > |\pi_2|} G_1$.*

The majority rule can also be formulated in a different form, see [61], which then is called the *Majority Strategy.* If we are at a vertex u in a tree T and uv is an edge, then we move to v if at least half of the profile is at the side of v. Note that, if half of the profile is at the side of u and the other half at the side of v, then we can move back and forth between u and v. We park and erect a sign at a vertex where we get stuck, in case there is always a majority on one side, or we find a path, of which all vertices are visited at least twice and all other vertices at most once. In the latter case we park and erect a sign at each vertex visited at least twice. It turns out that the median set is the set of vertices with a sign. Again such a move to majority also works in hypercubes. The Strong Metaconjecture suggests the Majority Strategy on graphs given below. This idea arose in Louisville, Kentucky, while the author was visiting F.R. McMorris. We were driving to the University of Louisville along Eastern Parkway. At some stretch there is a beautiful median on Eastern Parkway, with green grass and large trees. Along this median there were traffic signs that read: "Tow away zone. No parking on the median at any time". We use the notation π_{wv}: it is the subprofile consisting of all elements of π closer to w than to v.

Majority Strategy

Input: A connected graph G, a profile π on G, and an *initial vertex* in V.

Output: The set of vertices where signs have been erected.

• Start at the initial vertex.

- If we are in v and w is a neighbor of v with $|\pi_{wv}| \geq \frac{1}{2}|\pi|$, then we *move* to w.
- We move only to a vertex already visited if there is no alternative.
- We stop when
  (i) we are stuck at a vertex v *or*
  (ii) we have visited vertices at least twice, and, for each vertex v visited at least twice and each neighbor w of v, either $|\pi_{wv}| < \frac{1}{2}|\pi|$ or w is also visited at least twice.
- We park and erect a sign at the vertex where we get stuck or at each vertex visited at least twice.

Do we always find the median set using the Majority Strategy? The answer is no, a simple example suffices. Take the complete graph K_3 with vertices u, v, w and let $\pi = u, v, w$. Now, for each edge xy there is only one vertex closer to y than to x, viz. y itself. So we do not move from x to y. This means that, being at x we are stuck at x, and only find x, whereas $M(\pi)$ is the whole vertex set. We find *one* median vertex but not all. Having applied the Strong Metaconjecture already a couple of times the first equivalence in following theorem does not come as a surprise, the main result in [61].

Theorem 5.9. *Let G be a graph. Then the following statements are equivalent:*

(i) G is a median graph.
(ii) The Majority Strategy produces $M(\pi)$ in G, for each profile π.
(iii) The Majority Strategy produces the same set from any initial position v in G, for each profile.

Statement *(iii)* in the theorem came as a bonus and was not foreseen in any way.

We conclude with another interesting feature of median sets in median graphs, in which the median set of π is related to the vertex-deleted profiles: if $\pi = x_1, x_2, \ldots, x_k$, then $\pi - x_i$ is the profile obtained from π by deleting x_i.

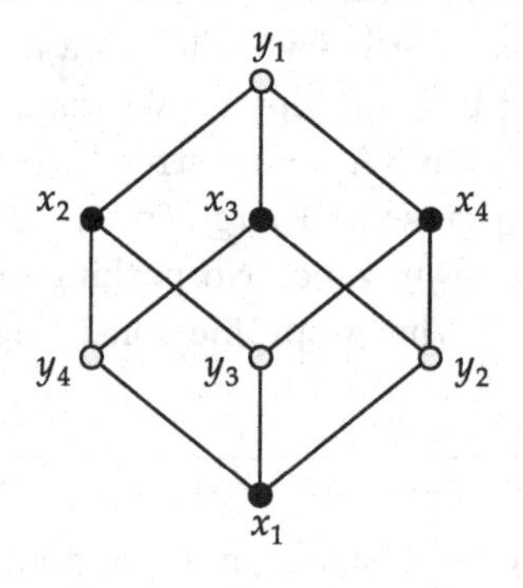

Fig. 5.4.

Theorem 5.10. *Let G be a median graph, and let $\pi = x_1, x_2, \ldots, x_k$ be a profile on G. If π is odd, then $M(\pi) = \bigcap_{i=1}^{k} M(\pi - x_i)$. If π is even, then $M(\pi) = Con[\bigcup_{i=1}^{k} M(\pi - x_i)]$.*

In Fig. 5.4 the profile $\pi = x_1, x_2, x_3, x_4$ is given as the black vertices. We have $M(\pi - x_i) = \{y_i\}$, and $M(\pi) = Con[\cup_{i=1}^{4} M(\pi - x_i)] = Con[\{y_1, y_2, y_3, y_4\} = V$.

5.5. Median Graphs in the Mathematical Universe

Although median graphs are a very nice common generalization of trees and hypercubes, they have a very special structure. So it seems that the class of median graphs is quite esoteric and resides just somewhere in a remote corner of the universe of all graphs. But again appearances may be deceptive, as we shall see in this section.

Let $G = (V, E)$ be a (connected) graph. The *subdivision* $\widehat{G}$ of G is obtained by inserting a new vertex on each edge of G, thus subdividing each edge into two new edges. Otherwise stated, the vertex set of $\widehat{G}$ is $V \cup E$, and for any edge $e = uv$ in G, the vertex e in $\widehat{G}$ is joined to the vertices u and v of $\widehat{G}$. The graph G^* is obtained from $\widehat{G}$ by adding a new vertex z and making it adjacent to each vertex in V of $\widehat{G}$. Note that, if G is a triangle, then G^* is precisely Q_3 minus a vertex. This graph is *not* a median graph. So if G contains triangle, then G^* cannot be a median graph. On the other hand, if G is triangle-free, then it is a simple exercise to prove that G^* is a median graph. Just identify in G^* a vertex $e = uv$ with the set $\{u, v\}$, identify a vertex u in V with the set $\{u\}$, and identify z with $\emptyset$. Then the unique median of the sets A, B, C is the set $(A \cap B) \cup (B \cap C) \cup (C \cap A)$. Thus we have the following theorem, see [41].

Theorem 5.11. *A connected graph G is triangle-free if and only if G^* is a median graph.*

Let $\mathcal{O}$ be the class of connected, triangle-free graphs. Let $\mathcal{M}_2^*$ be the class of median graph just constructed from the connected, triangle-free graphs. We call a median graph *cube-free* if it does not contain a Q_3. Let $\mathcal{M}_2$ be the class of cube-free median graphs. This class plays a prominent role in McMorris, Mulder & Vohra (this volume). Finally let $\mathcal{M}$ be the class of all median graphs. From the proof of Theorem 5.11 in [41], it also followed that the mapping ϕ defined by $G \to G^*$ is a one-to-one correspondence between the class $\mathcal{O}$ and the class $\mathcal{M}_2^*$. Then we have

$$\mathcal{O} \longmapsto \mathcal{M}_2^* \subsetneq \mathcal{M}_2 \subsetneq \mathcal{M} \subsetneq \mathcal{O}.$$

So, essentially, there are as many connected triangle-free graphs as there are median graphs in the Universe of All Graphs. Surely, we cannot consider triangle-free graphs esoteric. Theorem 5.11 was used in [41] to show that the complexity of recognizing median graphs is related to that of triangle-free graphs. This implies that the algorithm in [35] seems to be best possible.

5.6. Median Structures

Median structures can be found in many different guises. Here we present the main ones: in various algebraic terms, in terms of set functions, in terms of hypergraphs, in terms of convexities, in terms of geometries, and in terms of conflict models. In the case of trees and hypercubes appropriate axioms can be found in the literature, we do not review these here. A rich literature can be found on each of these median structures and their respective contexts. We restrict ourselves to the references pertaining to the origins and some recent exemplary ones.

5.6.1. *Ternary algebras*

A *ternary algebra* (V, m) consists of a set V and a ternary operator

$$m : V \times V \times V \to V.$$

The *underlying graph* $G_m = (V, E_m)$ of (V, m) is defined by

$$uv \in E_m \iff u \neq v \text{ and } m(u, x, v) \in \{u, v\} \text{ for all } x \in V.$$

A ternary algebra (V, M) is a *median algebra* if it satisfies a certain set of algebraic axioms. They were first introduced as *ternary distributive semilattices* by Avann [1]. Median algebras were discovered independently a couple of times. The second discovery of median algebras was by Sholander [75], named by him median semilattice. In 76; 77] Sholander proved the equivalence of these median algebras with the structures given below in Sections 5.6.2 - 5.6.3. The third independent discovery of median algebras was by Nebeský in [65] under the name of normal graphic algebra (or simple graphic algebra in [66]). Each of these authors had a different set of axioms to characterize median algebras . We present here only those of Nebeský. Let u, v, w be arbitrary elements of V.

$(n1)$ $m(u, u, v) = u$;
$(n2)$ $m(u, v, w) = m(w, v, u) = m(v, u, w)$;
$(n3)$ $m(m(u, v, w), w, x) = m(u, m(v, w, x), w)$.

It turns out that (V, m) is a median algebra if and only if its graph G_m is a median graph. Conversely, let $G = (V, E)$ be a median graph, and define $m(u, v, w)$ to be the median of the triple u, v, w. Then (V, m) is a median algebra. Using different terminology, this was first proved by Avann [2], and later independently by Nebeský [66].

The term median algebra seems to be independently introduced by Evans [34], Isbell [42] and Mulder [58]. Many other axiom systems for median algebras can be found in the literature. Some more recent references are [32; 45].

5.6.2. *Betweenness*

A *betweenness structure* (V, B) consists of a set V and a *betweenness relation*

$$B \subseteq V \times V \times V$$

satisfying at least the following conditions:

$(b1)$ $(u, u, v) \in B$, for any u and v,
$(b2)$ if $(u, v, w) \in B$ then $(w, v, u) \in B$,

In such a broad sense betweenness structures were introduced in [48]. Various types of betweenness structures with additional axioms exist in the literature, see e.g. [74; 55]. If $(u, v, w) \in B$, then we say that "v is between u and w". The graph $G_B = (V, E_B)$ of (V, B) is given by

$$uv \in E_B \iff u \neq v \text{ and } (u, x, v) \in B \text{ only if } x \in \{u, v\},$$

so there is nothing strictly between u and v.

A betweenness structure (V, B) is a *median betweenness structure* if it satisfies the extra axioms:

$(mb1)$ for all $u, v, w \in V$, there exists x such that (u, x, v), (v, x, w), $(w, x, u) \in B$,
$(mb2)$ if $(u, v, u) \in B$, then $v = u$,
$(mb3)$ if $(u, v, w), (u, v, x), (w, y, x) \in B$, then $(y, v, u) \in B$.

Then (V, B) is a median betweenness structure if and only if its graph G_B is a median graph. Conversely, let $G = (V, E)$ be a median graph with interval function I, and let B be the betweenness relation on V defined by $(u, v, w) \in B$ if $v \in I(u, w)$. Then (V, B) is a median betweenness structure. This gives a one-to-one correspondence between the median betweenness structures (V, B) and the median graphs with vertex set V. The median betweenness structures were introduced by Sholander [76] and proven to be equivalent with his median algebras from Section 5.6.1.

5.6.3. *Semilattices*

A *semilattice* is a partially ordered set $(V, \leq)$ in which any two elements u, v have a *meet* (greatest lower bound) $u \wedge v$. If u and v have a *join* (least upper bound), then we denote it by $u \vee v$. For $u \leq v$, we define the *order interval* to be $[u, v] = \{w \mid u \leq w \leq v\}$. As usual, the *covering graph* $G_{\leq} = (V, E_{\leq})$ of $(V, \leq)$ is given by

$$uv \in E_{\leq} \iff u \neq v \text{ and } u \leq w \leq v \text{ if and only if } w \in \{u, v\}.$$

A semilattice $(V, \leq)$ is a *median semilattice* if satisfies the following axioms:

$(\ell 1)$ every order interval is a distributive lattice,
$(\ell 2)$ if $u \vee v$, $v \vee w$ and $w \vee u$ exist, then $u \vee v \vee w$ exists for any $u, v, w \in V$.

Then $(V, \leq)$ is a median semilattice if and only if its covering graph $G_{\leq}$ is a median graph. Note that two different median semilattices can have the same median graph as covering graph. Conversely, let $G = (V, E)$ be a median graph with interval function I and let z be a fixed vertex in G. We define the ordering $\leq_z$ on V by $u \leq_z v$ if $u \in I(z, v)$. Then $(V, \leq_z)$ is a median semilattice with universal lower bound z. This yields a one-to-one correspondence between the median semilattices and the pairs (G, z), where G is a median graph and z is a vertex of G.

Median semilattices were introduced by Sholander [77] and proven to be equivalent with his median algebras, median betweenness structures and median segments (see Section 5.6.7). The equivalence with median graphs follows from the result of Avann [2] on median algebras and median graphs. A direct proof of the above result is given in Mulder [58]. For a recent reference see [52].

5.6.4. *Hypergraphs and convexities*

A *copair hypergraph* $(V, \mathcal{E})$ consists of a set V and a family $\mathcal{E}$ of nonempty subsets of V such that $A \in \mathcal{E}$ implies $V - A \in \mathcal{E}$. As usual, its graph $G_{\mathcal{E}} = (V, E_{\mathcal{E}})$ is given by

$$uv \in E_{\mathcal{E}} \iff u \neq v \text{ and } \cap\{A \in \mathcal{E} \mid u, v \in A\} = \{u, v\}.$$

A copair hypergraph $(V, \mathcal{E})$ is a *maximal Helly copair hypergraph* if it satisfies the conditions:

$(h1)$ $\mathcal{E}$ has the Helly property,
$(h2)$ if $A \notin \mathcal{E}$, then $\mathcal{E} \cup \{A, V - A\}$ does not have the Helly property.

Then a copair hypergraph $(V, \mathcal{E})$ is maximal Helly if and only if its graph $G_{\mathcal{E}}$ is a median graph. Conversely, let $G = (V, E)$ be a median graph, and let $\mathcal{E}$ consists of the vertex sets of the split sides in G. Then $(V, \mathcal{E})$ is a maximal Helly copair hypergraph. This result was proven by Mulder and Schrijver in [64] using the Expansion Theorem, see also [58]. Note that, if we take the closure $\mathcal{E}^*$ of $\mathcal{E}$ by taking all intersections and add V, then $\mathcal{E}^*$ consists precisely of the convex sets of $G_{\mathcal{E}}$. Thus we get an alternative formulation of the above result as follows.

A *convexity* $(V, \mathcal{C})$ consists of a set V and a family $\mathcal{C}$ of subsets of V that is closed under taking intersections. Its graph $G_{\mathcal{C}} = (V, E_{\mathcal{C}})$ is given by

$$uv \in E_{\mathcal{C}} \iff u \neq v \text{ and } \bigcap\{A \in \mathcal{C} \mid u, v \in A\} = \{u, v\}.$$

The above result reads then: in a convexity $(V, \mathcal{C})$ the family of convex sets $\mathcal{C}$ has the Helly property and the separation property S_2 if and only if its graph $G_{\mathcal{C}}$ is a median graph (cf. [79]).

5.6.5. *Join geometries*

A *join geometry* $(V, \circ)$ consists of a set V and a *join operator* $\circ : V \times V \to \mathcal{P}(V)$ satisfying

($j1$) $u \circ u = \{u\}$,
($j2$) $u \in u \circ v$,
($j3$) $u \circ v = v \circ u$,
($j4$) $u \circ (v \circ w) = (u \circ v) \circ w$, for all $u, v, w \in V$.

Its graph $G_\circ = (V, E_\circ)$ is given by

$$uv \in E_\circ \iff u \neq v \text{ and } u \circ v = \{u, v\}.$$

For subsets U, W of V, we write $U \circ W$ for the union of all $u \circ w$ with $u \in U$ and $w \in W$. If $U = \{u\}$, we write $u \circ W$ instead of $U \circ W$. A set C in a join geometry $(V, \circ)$ is *convex* if $C \circ C = C$. Join geometries were introduced and extensively studied by Prenowitz and Jantosciak [72]. A join geometry $(V, \circ)$ is a *join space* if it satisfies the following conditions:

($S4$) (Kakutani separation property) if C, D are disjoint convex sets in $(V, \circ)$, then there is a convex set $H \subset V$ such that $C \subseteq H$ and $D \subseteq -H$ and $V - H$ is also convex,

(JHC) (Join-hull commutativity) if C is a convex set then $Con(\{u\} \cup C) = u \circ C$, for u in V.

Then the convex sets in a join space $(V, \circ)$ have the Helly property if and only if its graph $G_\circ$ is a median graph. Conversely, let $G = (V, E)$ be a median graph with interval function I, and define the join operator $\circ$ on V by $u \circ v = I(u, v)$. Then $(V, \circ)$ is a join space with convex sets having the Helly property. This result is essentially due to Van de Vel [78], see also [68; 12; 79]. For a recent reference see [28].

5.6.6. *Conflict models*

A *conflict model* $(X, \leq, A)$ consists of a set X, a partial ordering $\leq$ of X and a set of edges A such that (X, A) is a graph. As usual, a subset Y of X is an *ideal* whenever $x \in Y$ and $y \leq x$ implies $y \in Y$, and Y is *independent* whenever there are no edges in (X, A) between vertices of Y. One can construct a graph $G = (V, E)$ from a conflict model $(X, \leq, A)$ as follows: the vertex-set V of G consists of all independent ideals of $(X, \leq, A)$, and we connect two vertices Y and Z by an edge whenever they differ in one element (i.e. have symmetric difference of size 1). This graph is a median graph, see [18].

Now the problem arises how to construct a conflict model $(X, \leq, A)$ from a median graph $G = (V, E)$ such that G can be reconstructed as above from $(X, \leq, A)$.

Barthélémy and Constantin [18] gave a nice construction, which amounts to the following. Let $G = (V, E)$ be a median graph, and let z be a fixed vertex of G. For a split G_1, G_2 defined by the edge uv with u in G_1, we write W_{uv} for the vertex set of G_1 and W_{vu} for the vertex set of G_2, and call W_{uv}, W_{vu} a *split* as well. The elements of X are the splits W_{uv}, W_{vu}. We say that two different splits $\{W_{uv}, W_{vu}\}$ and $\{W_{ab}, W_{ba}\}$ *cross* if the four intersections $W_p \cap W_q$, where $p \in \{uv, vu\}$ and $q \in \{ab, ba\}$ are nonempty. Now let $x = \{W_{uv}, W_{vu}\}$ and $y = \{W_{ab}, W_{ba}\}$ be two distinct non-crossing elements of X. We put

$x \leq y$ if, say, $z \in W_{uv} \cap W_{ab}$, and $W_{uv} \subseteq W_{ab}$,
$xy \in A$ if x and y are incomparable with respect to $\leq$.

Note that crossing elements of X are incomparable as well as non-adjacent. Barthélémy and Constantin [18] proved that the so obtained conflict model $(X, \leq, A)$ reproduces G by the above construction. These conflict models $(X, \leq, A)$ constructed from median graphs have the following additional property

($p1$) if $xy \in A$ and $x \leq u$, $y \leq v$, then $uv \in A$.

Barthélémy and Constantin [18] named conflict models satisfying ($p1$) *sites*. In computer science, they are known as *conflict event structures*, cf. [30]. The main result of [18] reads then as follows: there is a one-to-one correspondence between the sites and the pairs (G, z), where G is a median graph and z a vertex of G.

5.6.7. *Transit Functions*

For the purpose of this Section a *transit function* on a set V is a set-valued function $R : V \times V \to 2^V - \emptyset$ satisfying the two axioms:

($t1$) $u \in R(u, v)$, for any u, v,
($t2$) $R(u, v) = R(v, u)$, for any u, v,

See Section 5.7.5 for more information on this concept. The graph $G_R = (V, E_R)$ of the transit function is defined by

$$uv \in E_R \iff u \neq v \text{ and } R(u, v) = \{u, v\}.$$

A transit function R on V is a *median transit function* if it satisfies the extra axioms:

($s1$) there exists z such that $R(u, v) \cap R(u, w) = R(u, z)$, for all u, v, $w \in V$;
($s2$) if $R(u, v) \cap R(u, w) = R(u, v)$, then $R(x, u) \cap R(x, w) \subseteq R(v, x)$, for any $x \in V$;
($s3$) if $R(u, v) \cap R(u, w) = R(u, u)$, then $R(u, u) \cap R(v, w) = \{u\}$.

Then R is a median transit function if and only if its graph G_R is a median graph. Conversely, let $G = (V, E)$ be a median graph with interval function I. Then I

is a median transit function. This gives a one-to-one correspondence between the median transit functions I on V and the median graphs with vertex set V.

Median transit functions R with the axioms ($s1$), (s2) and (s3) were introduced by Sholander [76], which he called *median segments*, and were proven to be equivalent to his median semilattices from Section 5.6.3. The equivalence with median graphs follows from Avann [2]. In Mulder and Schrijver [64] a different set of axioms and direct proofs of the relation between median graphs and median transit functions were given. There they were called *median interval structures*, cf. Mulder [58].

In [79] a function satisfying ($t1$) and ($t2$) was called an *interval operator*. The perspective in this book is that of convexities and separation properties similar to those in topology. The reason why they are called transit function here is explained in Section 5.7.5.

5.7. Generalizations

From early on, see [58], generalizations of median graphs have been considered. We have encountered a couple in the previous section, in the form of various discrete structures. But here the focus is on the generalizations of the graphs.

5.7.1. *Quasi-median Graphs*

The first generalization was already developed along with the first in-depth study of median graphs in [58]. It is easiest to explain using the idea of expansion. In Section 5.2 the expansion is performed with respect to a proper cover with *two* convex subgraphs. We extend the definition of proper cover to k subgraphs. Let $G = (V, E)$ be a connected graph, and let $G_0 = (V_0, E_0)$ and $G_1 = (V_1, E_1), G_2 = (V_2, E_2), \dots, G_k = (V_k, E_k)$ be subgraphs of G. We say that $G_1 = (V_1, E_1), G_2 = (V_2, E_2), \dots, G_k = (V_k, E_k)$ is a *proper k-cover* of G if

$$G = \cup_{i=1}^{k} G_i \text{ and } G_i \cap G_j = G_0, \text{ for } 1 \leq i <\leq j.$$

The cover is called convex if all subgraphs involved are convex. The *quasi-median expansion* with respect to this cover is obtained in the obvious way. Loosely speaking: take disjoint copies of $G_1, G_2, \dots, G_k$ and join the k copies of u by edges, for each u in G_0. So in the expansion each vertex u in G_0 is replaced by a k-clique, and G_0 is replaced by a graph isomorphic to $G_0 \Box K_k$. Clearly, for $k \geq 3$, the resulting graphs is not bipartite. For $k = 2$, we call the expansion *binary*. With this new terminology, a graph is a median graph if and only if it can be obtained from K_1 by a succession of binary convex expansions. If we do not restrict the value of k, then we call the resulting graph a *quasi-median graph*. This was not the definition in [58]. There the search was for a generalization in terms of the defining property of median graphs.

The thing that worked was to replace the unique median for a triple of vertices u, v, w by a triangular structure. A triple x, y, z is called *triangular* if

$$I(x, y) \cap I(y, z) = \{y\}, I(y, z) \cap I(z, x) = \{z\}, I(z, x) \cap I(x, y) = \{x\}.$$

It is an *equilateral triangle* if $d(x, y) = d(y, z) = d(z, x)$. The size of the equilateral triangle is the value $d(u, v)$. An equilateral triangle of size 0 is just a triple x, x, x, which may be taken as a unique vertex x. A *quasi-median* for u, v, w is an equilateral triangle x, y, z of minimal size such that x, y lie on a shortest u, v-path, y, z lie on a shortest v, w-path, and z, u lie on a shortest w, u-path. Note that the ordering of u, v, w and x, y, w is relevant. If we want to make this explicit, we could write (x, y, z) being a quasi-median of (u, v, w). Then a *quasi-median graph* is a graph in which any three vertices have a unique quasi-median with two extra conditions. These are rather technical but necessary. We state these here for the sake of completeness. The first one is as follows. Let $K_4 - e$ denote the graph obtained from K_4 by deleting an edge. Then $K_4 - e$ is forbidden as induced subgraph in a quasi-median graph. Clearly, every triple in $K_4 - e$ has a unique quasi-median, but this graph cannot be obtained from K_1 by expansions. The second condition is that an induced C_6 in a quasi-median graph has as convex closure either $Q_3 = K_2^3$ of K_3^2. If we use this as definition, then one of the mains results in [58] reads as follows.

Theorem 5.12. *A graph G is a quasi-median graph if and only if it can be obtained from K_1 by a succession of quasi-median expansions.*

Again, these graphs may look quite exotic, but we shall see in Section 5.8.4 that there are non-trivial applications of these graphs. Note that a quasi-median (x, x, x) of size 0 basically is just a single vertex. In this sense a quasi-median graph is rightly a non-bipartite generalization of median graphs. It was the first such class to be discovered, but not the simplest. For a simpler one we refer to Section 5.7.4.

Having a closer look at quasi-median graphs we see that the role of convex subgraphs in median graphs is now played by gated subgraphs. Then the Armchair provides us with two expected results: a quasi-median graph can be obtained from smaller ones by amalgamation along a common gated subgraph, and a quasi-median graph can be built from Hamming graphs as building stones by amalgamation along gated subgraphs.

Given a quasi-median graph $G = (V, E)$, we can define a ternary algebra in the following way: if (x, y, z) is the quasi-median of (u, v, w), then for the ternary operator q we define

$$(u, v, w) = x, (v, w, u) = y, (w, u, v) = z.$$

As in Section 5.6.1 we can characterize a ternary algebra coming from a quasi-median graph by a set of axioms. There are many different sets of axioms available, see e.g. [58; 13]. Brešar had a different nice approach using the imprint function.

Let u, v, w be vertices of G, then the *imprint* $i(u, v, w)$ of u on v, w is the gate of u in $\langle\{u, v\}\rangle$. It turns out that in a quasi-median graph $i(u, v, w)$ is precisely x in the quasi-median (x, y, z) of (u, v, w). The idea of imprint function provides is with possibilities for different generalizations, see [19]. For recent results and uses of quasi-median graphs see [20; 38].

5.7.2. *Expansions*

In [60] the idea of expansion was generalized as follows. We start with a proper cover $G_1 = (V_1, E_1), G_2 = (V_2, E_2), \ldots, G_k = (V_k, E_k)$ of a connected graph G with G_0 as *common subgraph* of any of the two covering subgraphs. Let $\mathcal{P}$ be some property shared by $G_1 = (V_1, E_1), G_2 = (V_2, E_2), \ldots, G_k = (V_k, E_k)$. Note that we might also require a condition $\mathcal{Q}$ on G_0. If $\mathcal{P}$ is the property convex, then $\mathcal{Q}$ is automatically the property convex as well. Now we take disjoint copies of $G_1 = (V_1, E_1), G_2 = (V_2, E_2), \ldots, G_k = (V_k, E_k)$ and insert edges between the respective copies of G_0 according to some rule ρ. The resulting graph is then the $\mathcal{P}, \rho$*-expansion* of G with respect to the given cover. If $k = 2$, then we call the expansion *binary*. Now a graph is a median graph if and only if it can be obtained from K_1 by successive binary, convex, convex, Cartesian expansion, where Cartesian means that the rule ρ is that the new edges make a Cartesian product $G_0 \Box K_2$.

Another important example is when we take property $\mathcal{P}$ to be *isometric*, that is, distances in the subgraphs are equal to those in the whole graph. A *partial cube* is an isometric subgraph of a hypercube. Note that any even cycle is a partial cube. Median graphs are another instance. Partial cubes were first studied by Djokovic [31], but not yet under the name partial cube. His characterization involved the relation Θ on the edges of G: let $e = uv$ and $f = xy$ be two edges, then $e\Theta f$ if

$$d(u, x) + d(v, y) \neq d(u, y) + d(x, v).$$

Note that in an induced 4-cycle opposite edges are in relation Θ. This relation is reflexive and symmetric. By Θ^* we denote the transitive closure of Θ. The characterization by Djokovic of partial cubes was:

Theorem 5.13. *A connected graph G is a partial cube if and only if G is bipartite and $\Theta^* = \Theta$.*

A characterization involving expansions is due to Chepoi [25]

Theorem 5.14. *A graph G is a partial cube if and only if it can be obtained from K_1 by successive binary, isometric, Cartesian expansions.*

Note that the common subgraph G_0 does not need to be isometric, because the intersection of two isometric subgraphs can still be anything.

Partial cubes have been extensively studied in the last decade, notably by Imrich, Klavžar, Brešar and their co-authors, and by Polat, see e.g. [46; 70; 71]. Several papers on partial cubes were inspired by properties of median graphs, see e.g. [49].

In [39] two other interesting generalizations of median graphs are characterized with special instances of expansion: *almost median graphs* and *semi-median graphs*. We omit details.

5.7.3. *Retracts*

A subgraph H of a subgraph$G$ is a *retract* of G if there exists a mapping $r : V(G) \to V(H)$ such that each edge of G is mapped either onto an edge or a vertex of H. So it fixes every vertex of H, and it may shrink distances. Using peripheral expansions one can prove the following characterization of median graphs due to Bandelt [5].

Theorem 5.15. *A graph G is a median graph if and only if it is a retract of a hypercube.*

Not surprisingly, quasi-median graphs also have a retract characterization, due to Wilkeit [81]:

Theorem 5.16. *A graph G is a quasi-median graphs if and only if it is the retract of a Hamming graph.*

Retracts form an interesting area in graph theory, but retracts can also be defined for other structures than graphs, for instance (partially) ordered sets.

5.7.4. *Median-type Graphs*

By now there is an abundance of generalizations of median graphs, each with its own merits. First we concentrate on those where the condition is a generalization of the idea of a median vertex of a triple of vertices. The simplest non-bipartite case is that of pseudo-median graphs. A *pseudo-median* of a triple of vertices (u, v, w) is a triangle, i.e. a K_3, on x, y, z such that edge xy is on a shortest u, v-path, edge uz is on a shortest v, w-path, and edge zx is on a shortest w, u-path. A *pseudo-median graph* is a graph G, in which each triple of vertices has a unique median or a unique pseudo-median. These graphs share nice properties with median graphs, see [11; 12]. For instance, there is an amalgamation characterization. The regular ones are characterized. And each automorphism of a pseudo-median graph has a fixed regular pseudo-median subgraph.

Another generalization is dropping the uniqueness of medians and the like. A *modular graph* is a graph in which any three vertices have at least one median. They occur for instance when considering retracts of bipartite graphs, see [9]. A *pseudo-modular graph* is a graph in which any three vertices have either a median or a pseudo-median (so not necessarily unique). Again these graphs may seem rather exotic, but surprisingly they are a common generalization of two seemingly unrelated classes: that of the median graphs and that of the distance hereditary graphs, see [10]. But also retracts of reflexive graphs are pseudo-modular, see [9].

A quite broad generalization is that of weakly median graphs, see [7; 8; 24]. It involves three conditions that one or another play a role in many characterizations of median graphs, see [48]. Therefore they are given here in full. A connected graph is a *weakly median graph* if it satisfies the following three conditions. These conditions were already present in [58], although only implicitly.

(T) Triangle condition: for any three vertices u, v, w with $1 = d(v, w) < d(u, v) = d(u, w)$ there exists a unique common neighbor x of v and w such that $d(u, x) = d(u, v) - 1$.

(Q) Quadrangle condition: for any four vertices u, v, w, z with $d(v, z) = d(w, z) = 1$ and $2 = d(v, w) \leq d(u, v) = d(u, w) = d(u, z) - 1$, there exists a unique common neighbor x of v and w such that $d(u, x) = d(u, v) - 1$.

(TQ) Meshed condition: for any three vertices u, v, w such that v and w are at distance 2 and have some common neighbor z with $2d(u, z) > d(u, v)+d(u, w)$, there exists a unique common neighbor x of v and w with $2d(u, x) < d(u, v)+d(u, w)$.

If we drop uniqueness in the above conditions then of course we get the *weakly modular graphs.* There is a host of graph classes that are a superclass of the median graphs and a subclass of the weakly median (modular) graphs. Moreover, one may drop one or two of the above conditions to get even broader classes. It seems that a lot is still to be done here. And nice results still might be obtained. We let these things stand here.

5.7.5. *Transit Functions*

The idea of transit function was already proposed by the author in 1998 at the Cochin Conference on "Graph Connections", but it was published only in 2008, see [62], see also Section 5.6.7 for more information on transit functions. Transit functions are a way to have a broad perspective on how to move around in a graph, or another discrete structure, say (partially) ordered sets. As we will see, the interval function I of a graph G is a special instance of a transit function.

A *discrete structure* (V, σ) consists of a finite set V and a 'structure' σ on V. Prime examples of discrete structures on V are a graph $G = (V, E)$, where $\sigma = E$ is the edge set of the graph, and a partially ordered set $(V, \leq)$, where $\sigma = \leq$ is a partial ordering on V. We denote the power set of V by 2^V.

A *transit function* on a discrete structure (V, σ) is a function $R : V \times V \to 2^V$ satisfying the three transit axioms

($t1$) $u \in R(u, v)$, for any u, v,

($t2$) $R(u, v) = R(v, u)$, for any u, v,

($t3$) $R(u, u) = \{u\}$, for any u.

Axioms as these, which are phrased in terms of the function only, will be called *transit axioms.* In Section 5.6.7 axiom ($t3$) was not included because it followed from axiom ($s3$). One might drop this axiom altogether, but then some uninteresting degenerate cases evolve. Note that the structure σ is not involved in the definition of a transit function. But, of course, the focus of our interest will be on the interplay between the function R and the structure σ. Usually, R will just be defined in terms of σ. Here we restrict ourselves to graphs. Then we are interested in the interrelations between the two structures E and R on V. Basically, a transit function on a graph G describes how we can get from vertex u to vertex v: via vertices in $R(u,v)$. The *underlying graph* $G_R = (V, E_R)$ of transit function R is defined by

$$uv \in E_R \iff u \neq v \text{ and } R(u,v) = \{u,v\}.$$

The idea behind the introduction of this concept was the following. Having a transit function of one type, for instance the interval function I, what can we say about another transit function R if we carry over specific properties of I to R. The following example might clarify this approach. To characterize the median graphs one considers $I(u,v,w) = I(u,v) \cap I(v,w) \cap I(w,u)$ and puts a condition on this set. Now take the *induced path function* J defined by

$$J(u,v) = \{\ w \mid w \text{ lies on some induced } u,v\text{-path }\}.$$

What happens if we impose conditions on $J(u,v,w) = J(u,v) \cap J(v,w) \cap J(w,u)$? In the case of I, the set $I(u,v,w)$ is empty more often than not. But for $J(u,v,w)$ it is the opposite: in most cases it will be a large chunk of vertices. A first observation is that $J(u,v,w)$ is empty for any triple of distinct vertices if and only if the graph is complete. A second observation is that $|J(u,v,w)| = 1$ for any triple of vertices u,v,w if and only if each block (i.e. maximal 2-connected subgraph) is a K_2 or C_4. The interesting case arises when we require $|J(u,v,w)| \leq 1$. Then we get a non-trivial class of graphs called the *svelte graphs*, see [55]. Another example is provided in [22]. For a systematic approach and many ideas and possibilities for future research see [62]. See also [23] for interesting cases. Changat and his co-authors and students have studied various transit functions inspired by [62]: for instance, the induced path function, the longest path function, and the triangle-path function. A recent example is [63].

5.8. Applications

In the previous section an application of median graphs in computer science in the guise of conflict models was already given. By now there are many and diverse applications of median graphs. Not surprisingly, these are often generalizations of existing applications of trees or hypercubes. We shortly survey the most important ones.

5.8.1. *Location Theory*

By definition, median graphs are interesting from the viewpoint of finding optimal locations. From Theorem 5.8 it follows that odd profiles always have a single median. The converse is also true: given any odd number $2k + 1$, if all profiles of length $2k + 1$ have a unique median, then the graph is a median graph. Slackening the conditions in the Majority Strategy we get other interesting strategies, for instance the Plurality Strategy: we may move from v to its neighbor w if there are at least as many elements in the profile closer to w than to v. For this strategy see [4]. The center in a tree is an edge, the center of the hypercube is the whole cube. What is the center in a median graph? We have not yet come up with the correct property $\mathcal{P}$ to apply the Metaconjecture. Hence this problem is still open: it seems to be non-trivial. For a generalization of the centroid of a tree that applies to median graphs see [69].

We present only one other use of median graphs in finding optimal locations. In [26] Chung, Graham and Saks formulated a dynamic search problem on graphs. Let $G = (V, E)$ be a connected graph, and assume that there is some specific information located at every vertex. An operator is located at a vertex u in the graph and handles quests for information. For the sake of simplicity, we assume that a quest for information at vertex v just comes in as a quest for vertex v. The operator can do two things, both of which have costs attached to it:

(*a*) retrieve vertex v from position u, which costs $d(u, v)$,
(*b*) move from position u to position w, which costs $d(u, w)$.

The aim of the operator is, given a sequence of quests, to find a sequence of positions in the graph from which he retrieves the requested information such that the total costs of moving and retrieving is minimized. Otherwise formulated, let $Q = q_1, q_2, \ldots, q_n$ be a quest sequence, and let $P = p_0, p_1, p_2, \ldots, p_n$ be a position sequence with initial position p_0 , both consisting of vertices of G, then the i-th step consists of moving from position p_{i-1} to position p_i , with cost $d(p_{i-1}, p_i)$, and retrieving quest q_i from position p_i , with cost $d(p_i, q_i)$. Hence the total costs of the i-th step are $d(p_{i-1}, p_i) + d(p_i, q_i)$. With this terminology the *Dynamic Search Problem* reads: given initial position p_0 and quest sequence $Q = q_1, q_2, \ldots, q_n$ in a connected graph G, find a position sequence $P = p_0, p_1, p_2, \ldots, p_n$ minimizing $\sum_{1 \leq i \leq n} [d(p_{i-1}, p_i) + d(p_i, q_i)]$. A position sequence attaining this minimum is called an *optimal position sequence*. In practice, the operator can not just wait until all quests have come in, so at any time he will only have partial knowledge of the quest sequence. If he only knows one quest at a time, then nothing smarter can be done than just retrieving the information. So theory will not be of any help. If the operator has foreknowledge of the next k quests in position p_i , then the situation is different. Now the best thing to do is to choose p_{i+1} from $M(p_i, q_{i+1}, q_{i+2}, \ldots, q_{i+k})$. For, if no other quests come in, the operator will minimize the costs of retrieving the remaining information by staying in p_{i+1}. We will call this strategy the *median strategy.* Now there might be a problem: with hindsight after having dealt with all the quests, it may turn out that the median strategy, although being optimal during short stretches of time, does not produce an optimal position sequence with respect to the entire sequence Q. This raises the question on which graphs the median strategy for the short term is also optimal for the long term. In [26] the following interesting theorem was proved.

Theorem 5.17. *Let G be a connected graph. Then the median strategy with foreknowledge of the next two quests in each step yields an optimal position sequence for any initial position and any quest sequence if and only if G is a median graph.*

In [27] it was proven that the quasi-median graphs are the graphs where dynamic search with foreknowledge of k quests is always optimal. Then the maximum number of subgraphs in the covers used should be k.

5.8.2. *Consensus Functions*

A consensus function is a model to describe a rational way to obtain consensus among a group of agents or clients. The input of the function consists of certain information about the agents, and the output concerns the issue about which consensus should be reached. The rationality of the process is guaranteed by the fact that the consensus function satisfies certain "rational" rules or "consensus axioms". More formally, let V be a set, and let V^* be the set of all profiles on V. A *consensus function* is a function $f : V^* \to 2^V - \emptyset$. The problem of the rationality is then to characterize f in terms of axioms on f. For more information on consensus theory, see Powers (this volume), and also [29].

A typical instance of a consensus function is the median function $M : V^* \to 2^V - \emptyset$ on a graph or a partially ordered set, where $M(\pi)$ is just the median set of π. The input is the location of the agents in the graph or ordered set, and the output are the vertices (or points) that minimize the distance sum to the location of the agents. For axiomatic characterizations of location functions see McMorris, Mulder & Vohra (this volume). We omit further treatment of this topic here.

5.8.3. *Chemistry*

In recent years a rather unexpected occurrence of median graphs in nature was discovered, e.g. in chemical substances, see [73; 51]. The benzene molecule consists of six carbon atoms and six hydrogen atoms. There bonds are depicted in Figure 5.5. Each H-atom is singly bonded to a unique C-atom. The six C-atoms are bonded in a hexagon. This accounts for three bonds per C-atom. A C-atom is 4-valent. The six remaining free bonds are realized by three pairwise bonds as in Figure 5.5. These extra bonds are then a perfect matching of the six C-atoms. There are two possibilities for the perfect matching, which produce two states. In the case

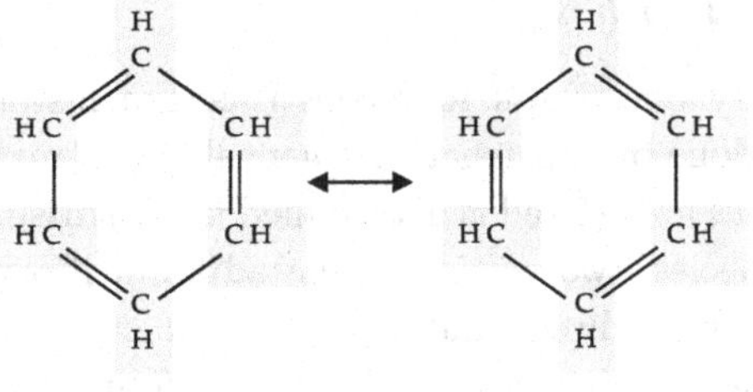

Fig. 5.5. Benzene: its two states

of benzene, these states are identical. Usually the benzene molecule is depicted without the letters signifying the atoms. This structure is called the *benzene ring* and was discovered by Kekulé in 1865.

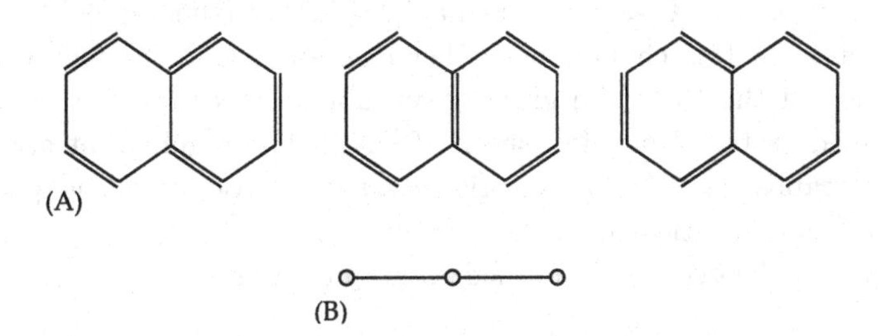

Fig. 5.6. The three naphthalene states and their graph

A *benzenoid* is a chemical substance, of which the molecule consists of benzene rings formed as a hexagonal substructure of the hexagonal grid in the plane, see Figure 5.6 for the example of naphthalene . The H-atoms attached to the perimeter of the structure are not depicted. Again the free bonds are grouped together to form a perfect matching. Note that we restrict ourselves here to the case that such a perfect matching is possible. Such structures are known in the chemical literature as Kekulean. As we can see in the figure, there are several perfect matchings possible. These are the *states* of the molecule. A chemical question now is how can we analyze the states of the molecule? For this purpose the *resonance graph* was introduced: its vertices are the perfect matchings, and two vertices are adjacent if, as perfect matchings, they differ in exactly one hexagon. Then these graphs are studied. It turns out that the resonance graphs of Kekulean structures are median graphs with some specific extra properties. This was first observed in [73] due to a remark of Tomaž Pisanski, and proven in [50; 51].

5.8.4. *Biology and Psychology*

In evolutionary theory one wants to reconstruct the history of the evolution of species. This can be done by constructing a genealogical tree for a family of species. How closely two species are related is determined by the dissimilarities. The vertices in the tree are the species, two vertices being adjacent if one of the corresponding species is assumed to be a direct mutant of the other. It turns out that it is not possible to explain the full history with the species found sofar. So we hope to find new species in the future that fill in the gaps. To make the tree as complete as possible so-called *missing links* are introduced: species to be found, these will then be presented by virtual or *latent vertices*. What we want to avoid is to introduce a new missing link for every unexplained gap. So we would like to construct a family tree with as few latent vertices as possible. In [21] Buneman developed a method how to construct such trees given the available data. Classification based on similarities/dissimilarities plays an important role in biomathematics, see [29], and in psychology as well. This is a well-developed area, see also Estabrook (this volume) for the use of other mathematical disciplines in biomathematics. See for instance [33] for the use of trees and median graphs in taxonomy.

In the context of this chapter one other example should be mentioned. It is a generalization of the Buneman graph given above. Now the object of study is the analysis of data based in information of DNA. The graphs that arise are now quasi-median graphs, see [38]. So, surprisingly, such exotic graphs as quasi-median graphs have a 'manifestation' in nature.

An example of the occurrence of median graphs in psychology is [17].

5.8.5. *Literary History*

One of the applications that appeals to me the most is the following. As we have seen above, in evolutionary theory there exist missing links that need to be included to get a full picture of evolutionary history. In literary history a similar phenomenon exists. To get a copy of a medieval manuscript one had to transcribe the manuscript. This was done mostly in monasteries. So one gets a transcription history for each text. Because sometimes manuscripts were partially destroyed or unreliable it happened quite often that two or more manuscripts of the same text were used to make a new transcription of the text. So instead of a genealogical tree one gets a structure that has cycles in it. Due to various reasons, quite often fires, complete manuscripts have been lost. So a complete transcription history cannot be reconstructed. Because of the cycles Buneman's theory of the previous section does not apply anymore. In [16] Barthélémy proved a very nice result. The median graphs provided precisely the graphs for this problem.

5.8.6. *Economics and Voting Theory*

In Economics voting strategies have been a point of focus already for many decades. One such voting procedure is the following. Take a connected graph $G = (V, E)$, and take two vertices u and v. Let $\pi = x_1, x_2, \ldots, x_k$ be a profile of clients on G, who have to chose between u and v. Now each client x_i casts a vote: the vertex closest to it. If $d(x_i, u) = d(x_i, v)$, then x_i abstains. Question: what vertices win over all other vertices? For instance, on what graphs do we find the median set of π? Take the profile $\pi = x_1, x_1, x_2, x_3, x_4$ in Fig. 5.4, where vertex x_1 occurs twice. Now y_1 wins over x_1, but $Med(\pi) = \{x_1\}$. So on a 3-cube the voting procedure does not produce the median set. But Bandelt [6] proved that a graph is a cube-free median graph if and only if this voting procedure always finds the median set for any profile. He also considered some related voting procedures.

An important feature of a voting procedure is whether it can be manipulated or not by one or more of the voters. A way to manipulate the voting is to *not* cast your true preference in voting. A voting procedure is called *strategy-proof* if no one voter can manipulate the outcome by not casting his/her true preference. We focus here on voting relevant for optimal location. A voting procedure on a graph $G = (V, E)$ is a function $L : V^k \to V$ that assigns a single value to each profile of length k on G. By $\pi[x_j \to w]$ we denote the profile obtained from $\pi = (x_1, \ldots, x_j, \ldots, x_k)$ by replacing x_j by w. So $\pi[x_j \to w] = (x_1, \ldots, x_{j-1}, w, x_{j+1}, \ldots, x_k)$, for $1 < j < k$, and $\pi[x_1 \to w] = (w, x_2, \ldots, x_k)$, and $\pi[x_k \to w] = (x_1, \ldots, x_{k-1}, w)$. A single-valued location function L on G is *strategy-proof* if, for each $\pi = (x_1, \ldots, x_k)$ and for each j with $1 \leq j \leq k$, we have

$$d(x_j, L(\pi)) \leq d(x_j, L(\pi[x_j \to w]))$$

for all w in V. Thus, voter i will never be able to improve (from her/his point-of-view) the result of applying the location function by reporting anything other than his/ger true preference x_i.

Now the question is, what functions L are strategy-proof on G given the structure of G. Clearly the answer depends on L as well as G. So far in economics the focus has been on the continuous analogue. In this case, the interior points on edges are also taken into account: such interior points can be profile elements as well as outcomes of L. A path is then just a segment of the real line. When familiar with median structures, one is not surprised that these arise in the study of strategy-proofness of voting procedures. We give two relevant references, viz. [67; 15]. The discrete case will be studied more extensively in [54].

There exists a rich literature in Economics on this concept of strategy-proofness. Note that this is by no means an instance of Armchair Theorizing: real, hard work has been done.

5.9. Concluding Remarks

In the scope of this chapter we could only give an idea of the richness of the structure of median graphs. We could only touch many topics, generalizations, and applications. Also the bibliography necessarily is just a small selection of the existing literature. There is still much to be done on median graphs, median structures, their generalizations and applications. Recently especially Klavžar, Brešar, Imrich, and their respective co-authors have been very productive in this area, see e.g. [3; 47]. For a more in-depth treatment of a specific problem area in median graph theory we refer the reader to McMorris, Mulder & Vohra (this volume).

References

1. S.P. Avann, Ternary distributive semi-lattices, *Bull. Amer. Math. Soc.* **54** (1948) 79.
2. S.P. Avann, Metric ternary distributive semi-lattices, *Proc. Amer. Math. Soc.* **12** (1961) 407–414.
3. K. Balakrishnan, B. Brešar, M. Changat, W. Imrich, S. Klavžar, M. Kovše, A.R. Subhamathi, On the remoteness function in median graphs, *Discrete Appl. Math.* **157** (2009) 3679–3688.
4. K. Balakrishnan, M. Changat, H.M. Mulder, The Plurality Strategy on Graphs, *Australasian J. Combin.* **46** (2010) 191-202.
5. H.J. Bandelt, Retracts of hypercubes, J. Graph Theory 8 (1984) 501–510.
6. H.J. Bandelt, Networks with Condorcet solutions, *European J. Operational Research* **20** (1985) 314–326.
7. H.J. Bandelt, V. Chepoi, The algebra of metric betweenness. I. Subdirect representation and retraction, *European J. Combin.* **28** (2007) 1640–1661.
8. H.J. Bandelt, V. Chepoi, The algebra of metric betweenness. II. Geometry and equational characterization of weakly median graphs. *European J. Combin.* **29** (2008) 676–700.
9. H.J. Bandelt, A. Dählmann, H. Schtte, Absolute retracts of bipartite graphs, *Discrete Appl. Math.* **16** (1987) 191–215.
10. H.J. Bandelt, H.M. Mulder, Distance-hereditary graphs, *J. Combin. Theory Ser. B* **41** (1986) 182–208.
11. H.J. Bandelt, H.M. Mulder, Regular pseudo-median graphs. *J. Graph Theory* **12** (1988) 533–549.
12. H.J. Bandelt, H.M. Mulder, Pseudo-median graphs: decomposition via amalgamation and Cartesian multiplication, *Discrete Math.* **94** (1992) 161–180.
13. H.J. Bandelt, H.M. Mulder, E. Wilkeit, Quasi-median graphs and algebras, *J. Graph Theory* **18** (1994) 681–703.
14. H.J. Bandelt and M. van de Vel, A fixed cube theorem for median graphs, *Discrete Math.* **67** (1987) 129–137.
15. S. Barberà, An introduction to strategy-proof social choice functions, *Social Choice Welfare* **18** (2001) 619–653.
16. J.-P. Barthélemy, From copair hypergraphs to median graphs with latent vertices, *Discrete Math.* **76** (1989) 9–28.
17. J.-P. Barthélemy, Median graphs and tree analysis of dichotomous data, an approach to qualitative factor analysis, *J. Math. Psychology*, **33** (1989) 452–472.

18. J.-P. Barthélémy and J. Constantin, Median graphs, parallelism and posets, *Discrete Math.* **111** (1993) 49–63.
19. B. Brešar, On the natural imprint function of a graph, *Europ. J. Combinatorics* **23** (2002) 149-161.
20. B. Brešar, S. Klavžar, R. Škrekovski, Quasi-median graphs, their generalizations, and tree-like equalities, *European J. Combinatorics* **24** (2003) 557-572.
21. P. Buneman, The recovery of trees from measures of dissimilarity, in: F.R. Hodson, D.G. Kendall and P. Tautu (Eds), *Mathematics in historical and archeological sciences*, Edinburgh University Press, Edinburgh, 1971, pp. 387–395.
22. M. Changat, S. Klavžar, H.M. Mulder, The all-paths transit function of a graph, *Czechoslovak Math. J.* **51** (2001) 439–448.
23. M. Changat, H.M. Mulder, G. Sierksma, Convexities related to path properties on graphs, *Discrete Math.* **290** (2005) 117–131.
24. M. Chastand, N. Polat, On geodesic structures of weakly median graphs. II. Compactness, the role of isometric rays, *Discrete Math.* **306** (2006) 1846–1861.
25. V. Chepoi, *d*-Convexity and isometric subgraphs of Hamming graphs, *Cybernetics*, **1** (1988) 6–9.
26. F.R.K. Chung, R.L. Graham, M.E. Saks, Dynamic search in graphs, in *Discrete algorithms and complexity* (H. Wilf, ed.), Academic Press, New York, 1987, 351–387.
27. F.R.K. Chung, R.L. Graham, M.E. Saks, A dynamic location problem for graphs, *Combinatorica* **9** (1989) 111–132.
28. P. Corsini, Binary relations, interval structures and join spaces, *J. Appl. Math. Comput.* **10** (2002) 209–216.
29. W.H.E. Day, F.R. McMorris, *Axiomatic consensus theory in group choice and biomathematics*, SIAM, Philadelphia, 2003.
30. P. Degano, R. De Nicola , U. Montanari, Partial ordering descriptions and observations of nondeterministic concurrent processes, in *Linear time, branching time and partial order in logics and models of concurrency* (J.W. de Bakker, W.-P. de Roever and G. Rozenberg, eds.), Lect. Notes in Comp. Sci. **354** (1989) 438–466.
31. D. Djoković, Distance preserving subgraphs of hypercubes, *J. Combin. Theory Ser. B* **14** (1973) 263–267.
32. H. Draškovicová, Modular median algebras generated by some partial modular median algebras. *Math. Slovaca* **46** (1996) 405–412.
33. A. Dress, V. Moulton, M. Steel, Trees, taxonomy, and strongly compatible multi-state characters, *Adv. in Appl. Math.* **19** (1997) 1–30.
34. E. Evans, Median lattices and convex subalgebras, in: *Universal Algebra, Colloq. Math. Soc. János Bolyai*, Vol. 29 North-Holland, Amsterdam, 1982, 225–240.
35. J. Hagauer, W. Imrich, S. Klavžar, Recognizing median graphs in subquadratic time, *Theoret. Comput. Sci.* **215** (1999) 123–136.
36. A.J. Goldman, Optimal center location in simple networks, *Transportation Science* **5** (1971) 212 - 221.
37. A.J. Goldman, C.J. Witzgall, A localization theorem for optimal facility location, *Transportation Science* **4** (1970) 406–409.
38. K.T. Huber, V. Moulton, C. Semplec, Replacing cliques by stars in quasi-median graphs, *Discrete Appl. Math.* **143** (2004) 194–203.
39. W. Imrich, S. Klavžar, A convexity lemma and expansion procedures for graphs, *Europ. J. Combin.* **19** (1998) 677-685.
40. W. Imrich, S. Klavžar, *Product graphs*, Wiley Interscience, New York, 2000.
41. W. Imrich, S. Klavžar, H.M. Mulder, Median graphs and triangle-free graphs, *SIAM J. Discrete Math.* **12** (1999) 111–118.

42. J.R. Isbell, Median algebra, *Trans. Amer. Math. Soc.* **260** (1980) 319–362.
43. P.K. Jha, G. Slutzki, Convex-expansion algorithms for recognizing and isometric embedding of median graphs, *Ars Combin.* **34** (1992) 75–92.
44. C. Jordan, Sur les assemblages de lignes, *J. Reine Angew. Math.* **70** (1869) 193–200.
45. Y.B. Jun,K.H. Kim, Fuzzy convex sets in median algebras, *J. Appl. Math. Comput.* **10** (2002) 157–165.
46. S. Klavžar, Hunting for cubic partial cubes in: *Convexity in discrete structures*, Ramanujan Math. Soc. Lect. Notes Ser., 5, Ramanujan Math. Soc., Mysore, 2008, pp. 87–95.
47. S. Klavžar, M. Kovše, Induced cycles in crossing graphs of median graphs, *Discrete Math.* **309** (2009) 6585–6589.
48. S. Klavžar, H.M. Mulder, Median graphs: characterizations, location theory, and related structures, *J. Combin. Math. Combin. Comput.* **30** (1999) 103–127.
49. S. Klavžar, H.M. Mulder, Crossing graphs and parital cubes, *SIAM J. Discrete Math.* **15** (2002) 235–251.
50. S. Klavžar, P. Žigert, Resonance graphs of catacondensed benzenoid graphs are median, *Preprint Set. Univ. Ljubljana IMFM* **38**.694 (2000) 1–9.
51. S. Klavžar, P. Žigert, G. Brinkmann, Resonance graphs of catacondensed even ring systems are median, *Discrete Math.* **253** (2002) 35-43
52. F.R. McMorris, H.M. Mulder and R.C. Powers, The median function on distributive semilattices, *Discrete Appl. Math.* **127** (2003) 319 – 324.
53. F.R. McMorris, H.M. Mulder and F.S. Roberts, The median procedure on median graphs, *Discrete Appl. Math.* **84** (1998) 165–181.
54. F.R. McMorris, H.M. Mulder and F.S. Roberts, Strategy-proof location functions on finite graphs, in preparation.
55. M.A. Morgana and H.M. Mulder, The induced path convexity, betweenness, and svelte graphs, *Discrete Math.* **254** (2002) 349–370.
56. H.M. Mulder, The structure of median graphs, *Discrete Math.* **24** (1978) 197–204.
57. H.M. Mulder, n-Cubes and median graphs, *J. Graph Theory* **4** (1980) 107–110.
58. H.M. Mulder, *The interval function of a graph*, Math. Centre Tracts 132, *Math. Centre*, Amsterdam, Netherlands 1980.
59. H.M. Mulder, Interval-regular graphs, *Discrete Math.* **41** (1982) 253–269.
60. H.M. Mulder, The expansion procedure for graphs, in: R. Bodendiek ed., *Contemporary Methods in Graph Theory*, B.I.-Wissenschaftsverlag, Mannheim/Wien/Zürich, 1990, 459–477.
61. H.M. Mulder, The majority strategy on graphs, *Discrete Applied Math.* **80** (1997) 97–105.
62. H.M. Mulder, Transit functions on graphs (and posets), in: M. Changat, S. Klavžar, H.M. Mulder and A. Vijayakumar eds., *Convexity in Discrete Structures*, pp. 117–130, Ramanujan Math. Soc. Lect. Notes Ser. **5**, *Ramanujan Math. Soc.*, Mysore, 2008.
63. H.M. Mulder, The prefiber convexity and its transit function, in preparation.
64. H.M. Mulder and A. Schrijver, Median graphs and Helly hypergraphs, *Discrete Math.* **25** (1979) 41–50.
65. L. Nebeský, Graphic algebras, *Comment. Math. Univ. Carolinae* **11** (1970) 533–544.
66. L. Nebeský, Median graphs, *Comment. Math. Univ. Carolinae* **12** (1971) 317–325.
67. K. Nehring, C. Puppe, The structure of strategy-proof social choice Part I: General characterization and possibility results on median spaces, *J. Economic Theory* **135** (2007) 269–305.
68. J. Nieminen, The congruence lattice of simple ternary algebras, *Serdica* 8 (1982) 115–122.

69. J. Nieminen, Centrality, convexity and intersections in graphs, *Bull. Math. Soc. Sci. Math. R. S. Roumanie (N.S.)* **28** (1984) 337–344.
70. N. Polat, Netlike partial cubes. I. General properties. *Discrete Math.* **307** (2007) 2704–2722.
71. N. Polat, G. Sabidussi, On the geodesic pre-hull number of a graph, *European J. Combin.* **30** (2009) 1205–1220.
72. W. Prenowitz and J. Jantosciak, *Join Geometries*, Springer, Berlin, 1979.
73. M. Randić, Resonance in catacondensed benzenoid hydrocarbons, *Int. J. Quantum Chem.* **63** (1997) 585-600.
74. E. Sampathkumar, B-systems, in: S. Arumugam, B.D. Acharya, E. Sampathkumar eds., *Graph Theory and its Applications*, Proceedings of the National Workshop, Manonmaniam Sundaranar University, Tirunelveli, *Tata McGraw-Hill*, New Dehli, 1996.
75. M. Sholander, Trees, lattices, order, and betweenness, *Proc. Amer. Math. Soc.* **3** (1952) 369–381.
76. M. Sholander, Medians and betweenness, *Proc. Amer. Math. Soc.* **5** (1952) 801–807.
77. M. Sholander, Medians, lattices, and trees, *Proc. Amer. Math. Soc.* **5** (1954) 808–812.
78. M. van de Vel, Binary convexities and distributive lattices, *Proc. London Math. Soc.* **48** (1984) 1–33.
79. M. van de Vel, *Theory of Convex Structures*, North Holland, Amsterdam, 1993.
80. Wikipedia, *Armchair Theorizing.*
81. E. Wilkeit, Retracts of Hamming graphs, *Disctrete Math.* **102** (1992) 197–218.
82. C.J. Witzgall, Optimal location of a central facility, *Mathematical models and concepts, Report National Bureau of Standards*, Washington D.C., 1965.

Chapter 6

Generalized Centrality in Trees

Michael J. Pelsmajer, K.B. Reid

Department of Applied Mathematics, Illinois Institute of Technology
Chicago, IL 60616 USA
pelsmajer@iit.edu

Department of Mathematics, California State University San Marcos
San Marcos, California 92096-0001 USA
breid@csusm.edu

In 1982, Slater defined and studied path subgraph analogues to the center, median, and branch weight centroid of a tree. We define three families of central substructures of trees, including three types of central subtrees of degree at most D, which yield the center, median, and centroid vertices for $D = 0$ and Slater's path analogues for $D = 2$. We generalize these results for several types of substructures in trees, each yielding three families of subtrees suggested by the three centrality measures involved with the center (eccentricity), the median (distance), and the branch weight centroid (branch weight). We present a theoretical framework in which several previous results sit. We prove that, in a tree, each type of generalized center and generalized centroid is unique and discuss algorithms for finding them. We also discuss the non-uniqueness of our generalized median and an algorithm to find one.

Dedicated to F.R. Buck McMorris on the occasion of his 65th birthday

Introduction

For many purposes one is interested in determining the "middle" of the graph. For instance, already in 1869 Jordan [1] used one concept of the middle in the case of trees in order to determine the automorphism group of a tree. From the viewpoint of applications, an interesting example is the placing of one or more facilities on a network: given a set of clients that has to be serviced by the facilities, the aim is to find a location for the facilities that optimizes certain criteria. Even in the case of trees and only one facility, there is no uniquely determined "middle" of a tree; it very much depends on the problem at hand and the criteria employed. For example, if the facility is a fire station, then a criterion is likely to be: minimize the maximum distance from the facility to the flammable objects. Whereas, if the facility is a distribution center for a set of warehouses or outlet stores, then a criterion is likely to be: minimize the sum of the distances from the facility to all of the warehouses, as that is a determining factor in the cost of servicing the warehouses. Or, if the managers of a distribution center are interested in economizing deliveries, then a criterion might be: minimize the maximum possible number of deliveries to customers from the facility with a single delivery truck before the truck returns to the facility for re-supplying. To date, an abundance of all kinds of centrality notions for trees occur in the literature (see Reid [2]), and many have natural generalizations for arbitrary connected graphs. The two classical examples, the center and centroid, are due to Jordan [1]. The *center* of a tree T is the set of all vertices of T that minimize the maximum distance to all other vertices, i.e., $\{x \in V(T)\colon\ e(x) \le e(y),\ y \in V(T)\}$, where, for $u \in V(T)$, $e(u) = \max\{d(u,v)\colon\ v \in V(T)\}$. The *branch weight centroid* of a tree T (or merely *centroid*) is the set of vertices x of T that minimize the maximum order of a connected component of the subforest of T resulting by deleting x (and all of its incident edges), i.e., $\{x \in V(T)\colon\ bw(x) \le bw(y),\ y \in V(T)\}$, where, for $u \in V(T)$, $bw(u) = \max\{|V(W)|\colon\ W$ is a connected component of $T - u\}$. A third, fundamental middle set is the median of a tree (apparently first mentioned by Ore [3] in 1962). The *median* of a tree T is the set of vertices of T that minimize the sum of the distances — or, equivalently, the average distance — to all other vertices, i.e., $\{x \in V(T)\colon\ D(x) \le D(y),\ y \in V(T)\}$, where, for $u \in V(T)$, $D(u) = \sum\{d(u,v)\colon\ v \in V(T)\}$. The center and the median have natural generalizations to arbitrary connected graphs; generalizations of centroids to arbitrary connected graphs have been given by Slater [4; 5].

Slater [6] generalized these notions, although in a different sense; he considered paths in a tree that minimize appropriate criteria. In this case, the notions of center, centroid, and median lead to three different optimal paths: *central path*, *spine*, and *core*, respectively, in the terminology of [6]. A path in a tree is a subtree of maximum degree at most 2, so a natural next step is to consider subtrees of maximum degree at most D. In Section 2 we define the $\mathcal{T}_D$*-center*, $\mathcal{T}_D$*-centroid*, and $\mathcal{T}_D$*-median*, which generalize centers, centroids, medians, central paths, spines, and cores—and our results generalize known results for these six as well.

Slater's work on path centers and spines stems from Jordan's classic algorithm for finding the center of a tree: delete all leaves, and repeat. In Section 3 we develop a general framework for central substructures that are amenable to this type of approach. This includes $\mathcal{T}_D$-centers, $\mathcal{T}_D$-centroids, work by McMorris and Reid [7] (following Minieka [8]) on subtrees of order k that minimize eccentricity, and the notion of a central caterpillar, suggested by McMorris [9]. Although there have been many papers on generalizations of the center and centroid of a tree (see Reid [2]), none directly generalize central paths or spines, and it seems that the common aspect of centrality measures of this type that we describe has not been explored — or even noticed. Indeed, we present a theoretical framework for many centrality concepts in trees using eccentricity, branch weight, and distance.

In Section 4 we consider the $\mathcal{T}_D$-median. There have been many interesting generalizations of a core [10; 11; 12; 13; 14]. In particular, the *k-tree core* is like our $\mathcal{T}_D$-median, except that instead of bounding the maximum degree, this concept considers subtrees with at most k leaves. Both the 2-tree core and the $\mathcal{T}_2$-median are equivalent to a core. There are linear time algorithms for finding a k-tree core of a tree when k is fixed [10; 11], and we provide the analogous results for $\mathcal{T}_D$-medians.

To summarize our results: We will give linear time algorithms for finding the unique $\mathcal{T}_D$-center, the unique $\mathcal{T}_D$-centroid, and a $\mathcal{T}_D$-median, for any tree when D is fixed. We will also show how to obtain all of the $\mathcal{T}_D$-medians in time that is linear in the sum of the order of T and the number of $\mathcal{T}_D$-medians. Also, for any tree T, there is a small family of subtrees for the center measure and a small family of subtrees for the branch weight measure, such that each generalization of a center and centroid is realized in T by a unique member of the appropriate small family; these families can be produced in linear time. Section 5 includes open questions and possible extensions. There are also linear time algorithms for many of our generalizations, but ultimately the running time depends on the ease of recognizing which of a small family of subtrees is in a desired class of trees.

There are many generalizations that we do not discuss, most importantly where vertices may have weights, edges may have lengths, and/or where edges are treated as continuums on which central substructures may be located arbitrarily. A good source for references to early work concerning these ideas is [15]. More recently, variations have been studied extensively (see, for example, [16]).

6.1. Preliminaries

Let $G = (V, E)$ be a connected graph with vertex set V and edge set E. For any subgraph H of G, we denote its vertex set by $V(H)$. The *order* of a (sub)graph is the number of vertices in the (sub)graph. The *degree* $d(v)$ of a vertex v in G is the number of vertices of G adjacent to V, i.e., $|\{u \in V\colon\ vu \in E\}|$. As usual, K_n denotes the *complete graph of order* n, i.e., every two distinct vertices are joined by an edge. The subgraph of G *induced by* a subset W of V has vertex set W and edge set consisting of all edges of G with both ends in W; it is denoted by $G[W]$. If G is a tree and $W \subseteq V$, the smallest connected subgraph of G whose vertex set contains W is denoted by $G\langle W\rangle$. Of course, $G[W]$ is a subgraph of $G\langle W\rangle$, and the inclusion can be proper (e.g., when W is a subset of order at least 2 of the leaves of a tree T of order at least 3, $T[W]$ has no edges, while the edge set of $T\langle W\rangle$ is the union of the edge sets of all paths between pairs of distinct vertices in W). If $W \subseteq V$, then $G - W$ denotes the subgraph $G[V - W]$. When $W = \{x\}$, $G - \{x\}$ will be abbreviated as $G - x$. The *length* $l(P)$ of a path P in G is the number of edges in P. The *distance* $d(u, v)$ between vertices u and v in connected graph G is the length of a shortest path in G between u and v, i.e., $d(u, v) = \min\{l(P)\colon\ P \text{ is a path between } u \text{ and } v\}$. The *diameter* $dia(G)$ of G is $\max\{d(u, v)\colon\ u, v \in V\}$. The *eccentricity* $e(v)$ of a vertex v in G is the maximum of the distances from v to all the other vertices in G, i.e., $e(v) = \max\{d(v,u)\colon u \in V\}$. The *center* $C(G)$ of G is the set of vertices of G with smallest eccentricity, i.e., $\{x \in V\colon e(x) \leq e(y),\ y \in V\}$. The *distance* $D(v)$ of a vertex v is the sum of the distances from v to all other vertices in G, i.e., $\sum\{d(u,v)\colon u \in V\}$. The *median* $M(G)$ of G is the set of vertices of smallest distance, i.e, $\{x \in V\colon D(x) \leq D(y),\ y \in \text{V}\}$. A vertex in $M(G)$ is a *median vertex.* Note that $\frac{D(v)}{|V|-1}$ is the average distance from v to all other vertices of G, so $M(G)$ is the set of vertices of smallest average distance to the other vertices.

In the case that G is a tree, the *branch weight* $bw(v)$ of a vertex v in G is the order of a largest connected component of $G - v$, i.e., $\max\{|V(W)|\colon W$ a connected component of $G - v\}$. The *branch weight centroid* $B(G)$ (or merely *centroid*) of G is the set of all vertices of smallest branch weight, i.e., $\{x \in V\colon bw(x) \leq bw(y),\ y \in V\}$. More than 140 years ago, Jordan [1] proved that for a tree G, $C(G)$ and $B(G)$ each consist of either one vertex or two adjacent vertices. A particularly nice proof was given by Graham, Entringer, and Székely [17]. Linear time algorithms can be found in [18; 19; 20; 21]. Zelinka [22] showed that for a tree G, $M(G) = B(G)$. However, for every positive integer k there is a tree G so that $d(x,y) \geq k$ for every $x \in M(G) = B(G)$ and $y \in C(G)$. In fact, as shown in [23], given any two graphs H_1 and H_2 and any positive integer k, there exists a connected graph H so that H_1 and H_2 are induced subgraphs of H with $C(H) = V(H_1)$, $M(H) = V(H_2)$, and $\min\{d(x,y)\colon x \in V(H_1),\ y \in V(H_2)\} = k$.

Slater [6] generalized each of these notions to path subgraphs. Suppose X and Y are subgraphs or subsets of vertices of a connected graph G. The distance $d(X,Y)$ between X and Y is the smallest of the distances between vertices of X and vertices of Y, i.e., $d(X,Y) = \min\{d(X,Y)\colon x \in X,\ y \in Y\}$. When $Y = \{y\}$ we abbreviate $d(X,\{y\})$ by $d(X,y)$. So, $d(X,Y) = 0$ if and only if $X \cap Y \neq \emptyset$ (or $V(X) \cap V(Y) \neq \emptyset$ if X, Y are subgraphs). The *eccentricity* $e(X)$ is the maximum of the distances between X and each of the other vertices of G, i.e., $e(X) = \max\{d(X,y)\colon y \in V(G)\}$. A *path center* (or *central path*) of G is a path of shortest length among all paths in G of minimum eccentricity, i.e., a shortest path P in G so that $e(P) \leq e(P')$ for all paths P' of G. The *distance* $D(X)$ of X is the sum of the distances from X to all other vertices of G, i.e., $\sum\{d(X,y)\colon y \in V(G)\}$. A *path median* (or *core*) of G is a path P of minimum distance, i.e., a path P of G so that $D(P) \leq D(P')$ for all paths P' of G. In case T is a tree, the *branch weight* $bw(X)$ of X is the the order of a largest connected component of $G - X$ (or $G - V(X)$, if X is a subgraph). A *path centroid* (or *spine*) of G is a path of shortest length among all paths of minimum branch weight, i.e., a shortest path P in G so that $bw(P) \leq bw(P')$ for all paths P' of G. It is possible in a tree for a path center, a spine, and a core to be distinct [6]. For trees, the path center and spine are unique [18; 6], contain $C(G)$ and $M(G)$ respectively, and there are linear time algorithms for computing each [6]. However, there may be many cores of a tree, and a core need not contain the median, but there are linear time algorithms for finding a core of a tree and the set of vertices that are contained in cores [24].

In this paper, we generalize these central paths of a tree to several other types of central subtrees of a tree.

Definition 6.1. Suppose that G is a tree and $\mathcal{T}_D$ is the set of subtrees of G with maximum degree at most D.
A $\mathcal{T}_D$-*center* of G is a tree of least order in the set $\{T \in \mathcal{T}_D\colon\ e(T) \leq e(T'),\ T' \in \mathcal{T}_D\}$.
A $\mathcal{T}_D$-*median* of G is a tree in the set $\{T \in \mathcal{T}_D\colon\ D(T) \leq D(T'),\ T' \in \mathcal{T}_D\}$.
A $\mathcal{T}_D$-*centroid* of G is a tree of least order in the set $\{T \in \mathcal{T}_D\colon\ bw(T) \leq bw(T'),\ T' \in \mathcal{T}_D\}$.

Note that for $D = 0$, each $\mathcal{T}_D$-center is a single vertex in $C(G)$, and each $\mathcal{T}_D$-median and each $\mathcal{T}_D$-centroid is a single vertex in $M(G) = B(G)$; hence, these are not unique when $|C(G)| = 2$ or $|M(G)| = 2$. The $\mathcal{T}_1$-center is the subtree induced by the center, $G[C(G)]$, and the $\mathcal{T}_1$-centroid is the subtree $G[B(G)]$. If $|M(G)| = 2$ or if $|V(G)| = 1$, then the $\mathcal{T}_1$-median is $G[M(G)]$; otherwise, a $\mathcal{T}_1$-median is the subgraph induced by the unique median vertex x and a vertex adjacent to x in a largest connected component of $G - x$. Since paths are the trees of maximum degree 2, when $D = 2$ the above definitions become the path center, core, and spine of a tree. When $D \geq \max\{d(x)\colon\ x \in V(G)\}$, each of the subtrees is simply G itself.

6.2. Generalized path centers and path centroids

The following theorem is key to understanding our generalizations of the path center such as the $\mathcal{T}_D$-center. To say that a subtree T of a tree G is *minimal* with respect to some property P means that T has property P and no proper subtree of T has property P.

Theorem 6.1. *If G is a tree and k is an integer, $0 \leq k \leq e(C(G))$, then G has a unique minimal subtree with eccentricity at most k.*

Proof. Suppose that G is the tree of order n and k is an integer, $0 \leq k \leq e(C(G))$. For simplicity, denote $e(C(G))$ by e. The conclusion clearly holds for $n \leq 3$, so assume $n \geq 4$ and $e \geq 1$. We describe a "pruning" process, often used to prove that $|C(G)| = 1$ or 2. For a tree T, let $L(T)$ denote the leaves of T, i.e., $\{x \in V(T)\colon\ d(x) = 1\}$. Inductively define a nested family of subtrees of G as follows: $G_0 = G$, and, for $1 \leq k \leq e$, $G_k = G_{k-1} - L(G_{k-1})$. Note that $e(G_k) = k$ and $G_e = G[C(G)]$.

First we observe that for each pair of distinct indices i and j, $0 \leq i < j \leq e$, and for each $x \in L(G_j)$ there is $y \in L(G_i)$ and a path of length $j - i$ given by $x = v_j, v_{j-1}, \ldots, v_i = y$ such that $v_m \in L(G_m)$ for $i \leq m \leq j$. We also recall that if subtree B of a tree A is such that $L(A) \subseteq B$, then $A = B$.

We show that no proper subtree of G_k has eccentricity at most k. This is certainly true for $k = 0$, so assume $k > 0$. Suppose not, i.e., suppose that W is a proper subtree of G_k with $e(W) \leq k$. So, there is $x \in L(G_k)$ so that $x \notin V(W)$.

By the observation above, there is $y \in L(G_0)$ and a path of length k given by $x = x_k, x_{k-1}, \ldots, v_0 = y$ such that $v_m \in L(G_m)$ for $0 \leq m \leq k$. If $v_m \in V(W)$ for some m, $0 \leq m \leq k$, then $v_m \in V(G_k)$ (as W is a subtree of G_k), so $v_m = v_k = x$ since none of $v_{k-1}, v_{k-2}, \ldots, v_0$ is in $V(G_k)$. That implies $x \in V(W)$, a contradiction. So, $v_m \notin V(W)$ for all m, $0 \leq m \leq k$. As $d(x, W) > 0$ and the path of length k given above contains no vertex of W, $d(y, W) = k + d(x, W) > k$. This implies $e(W) > k$, contrary to $e(W) \leq k$. So, G_k is a minimal subtree with eccentricity at most k.

To complete the proof, we prove uniqueness. Suppose W is a subtree of G so that (i) $e(W) \leq k$ and (ii) no proper subtree of W has eccentricity at most k. We claim that either G_k is a subtree of W or W is a subtree of G_k. Suppose not, i.e., suppose that G_k is not a subtree of W and W is not a subtree of G_k. Then there is $x \in L(G_k)$, $x \notin V(W)$, and there is $w \in L(W)$, $w \notin V(G_k)$. Now, every vertex of G not in G_k, such as w, is a leaf of some G_j, $0 \leq j \leq k-1$. By the observation above, there is a path P of length $k-j$ given by $x = x_k, x_{k-1}, \ldots, x_j = w$ where $v_m \in L(G_m)$, $j \leq m \leq k$. So, none of $x_{k-1}, x_{k-2}, \ldots, x_j$ is in $V(G_k)$. If W and G_k contain a common vertex, say $w' \in V(W) \cap V(G_k)$, there is a path Q in G_k between w' and x (as both x and w' are in $V(G_k)$). Note that Q uses none of $x_{k-1}, x_{k-2}, \ldots, x_j$. The path Q followed by path P is a path in G between two vertices w' and w in $V(W)$. Thus, this is the path between w' and w, and consequently, every vertex on that path is in $V(W)$. In particular, $x \in V(W)$, a contradiction to the choice of x. We deduce that subgraphs W and G_k are vertex disjoint. but, $C(G) \subseteq V(G_k)$; so W is completely contained in a connected component R of $G - C(G)$ and $d(C(G), W) > 0$. And, there exist at least two distinct leaves $a \neq b$ in G in distinct component of $G - C(G)$ so that $d(a, C(G)) = d(b, C(G)) = e(C(G)) = e$. At least one, say a, is not in R. Then $e(W) \geq d(a, C(G)) + d(C(G), W) = e + d(C(G), W) > e$, a contradiction to the choice of E (i.e., (i)). We deduce that either G_k is a subtree of W or W is a subtree of G_k. In the former case, the inclusion cannot be proper by assumption (ii) and the fact that $e(G_k) \leq k$. In the latter case, the inclusion cannot be proper by assumption (i) and the first part of this proof concerning G_k. In any case, $W = G_k$. That is, G_k is the unique minimal subtree of eccentricity at most k. □

Throughout this section, we will use X_k to denote $V(G_k)$, where G_k is the unique minimal subtree of tree G with eccentricity at most k, $0 \leq k \leq e(C(G))$, given in the previous proof.

Remark 6.1. Observe that the "pruning" process given in the proof gives rise to a linear time algorithm for finding X_k for all k, $0 \leq k \leq e(C(G))$, as described below in Theorem 6.5.

Next we develop a theorem that applies to a much more general class of central structures than merely $\mathcal{T}_D$-centers. For this we introduce a couple of definitions.

Definition 6.2. $\mathcal{T}$ is a *hereditary class of trees* if $\mathcal{T}$ is a nonempty set of trees such that for each $T \in \mathcal{T}$, every subtree of T is in $\mathcal{T}$.

Observe that this resembles the definition of a hereditary class of graphs except that "set of trees" and "subtree" replaces "set of graphs" and "subgraph". However, it is not a special type of hereditary class of graphs because the subtrees of a tree are its *connected* subgraphs.

We first note a few easily checked facts.

Theorem 6.2. *Let $\mathcal{T}$ be a hereditary class of trees. Then*

(1) $K_1 \in \mathcal{T}$,
(2) $K_2 \in \mathcal{T}$ *unless* $\mathcal{T} = \{K_1\}$,
(3) all subtrees of a fixed tree form a hereditary class of trees,
(4) unions and intersections of hereditary classes of trees yield new hereditary classes of trees, and
(5) if $\mathcal{T}$ *is finite and* $\mathcal{M} = \{T \in \mathcal{T}\colon T \text{ maximal in } \mathcal{T}\}$, *then* $\mathcal{T} = \{W\colon W \text{ is a subtree of some } T \in \mathcal{M}\}$.

Many other observations can be easily generated. For example: If $|\mathcal{T}| \geq 2$, then $K_2 \in \mathcal{T}$. If $|\mathcal{T}| \geq 3$, then T also contains $K_{1,2}$. If $|\mathcal{T}| = 4$, then $\mathcal{T}$ contains $K_1, K_2, K_{1,2}$, and either $K_{1,3}$ or P_4 (the path of order 4).

Definition 6.3. Let $\mathcal{T}$ be a hereditary class of trees, let G be a tree, and let $\mathcal{T}'$ be the subtrees of G that are in $\mathcal{T}$. A $\mathcal{T}$*-center* of G is an element of smallest order in the set $\{\, T \in \mathcal{T}'\colon e(T) \leq e(T'),\ T' \in \mathcal{T}'\}$.

Note that this directly generalizes the $\mathcal{T}_D$-center from Definition 6.1, since $\mathcal{T}_D$ is the family of trees of maximum degree at most D, and $\mathcal{T}_D$ is clearly a hereditary class of trees.

Other examples include trees of order at most k, trees of diameter at most d (for $d = 2$ these are stars), trees with at most ℓ leaves, caterpillars (including all paths), lobsters (a lobster is a tree that contains a path of eccentricity at most 2), subdivisions of stars, and all the subtrees of any fixed set of trees. The following theorem applies to each of these classes.

Theorem 6.3. *For a tree G and a hereditary class of trees $\mathcal{T}$, the $\mathcal{T}$-center of G is unique unless both $\mathcal{T} = \{K_1\}$ and $|C(G)| = 2$.*

Proof. Clearly there is a $\mathcal{T}$-center of G. Let T be a $\mathcal{T}$-center of G. If $e(C(G)) < e(T)$, then the subtree $G[C(G)]$ is not in $\mathcal{T}$, and by Theorem 6.2(1), $|C(G)| \neq 1$. Hence $|C(G)| = 2$ and $G[C(G)] = K_2$, so by Theorem 6.2(2), $\mathcal{T} = \{K_1\}$.

On the other hand, if $e(C(G)) \geq e(T)$, then using $k = e(T)$ in Theorem 6.1, we see that $G[X_k]$ is the unique minimal subtree of G with eccentricity at most $e(T)$. Given any subtree W of G with $e(W) \leq k = e(T)$, W contains a minimal subtree W' of eccentricity at most k. But, by the uniqueness of $G[X_k]$, $G[X_k] = W'$, so we deduce that $G[X_k]$ is a subtree of W. Using T in place of W, $G[X_k]$ is a subtree of T; in particular, $|X_k| \leq |V(T)|$. Since $\mathcal{T}$ is a hereditary class of trees, $G[X_k] \in \mathcal{T}$. For every $T' \in \mathcal{T}'$, $e(G[X_k]) \leq k = e(T) \leq e(T')$. So, $G[X_k] \in \mathcal{T}'$. As T is an element of smallest order in the set $\{T \in \mathcal{T}' \colon e(T) \leq e(T'),\ T' \in \mathcal{T}'\}$, $T = G[X_k]$. That is, T is unique unless $\mathcal{T} = \{K_1\}$ and $|C(G)| = 2$. □

Theorem 6.4. *Let G be a tree, and let D be a positive integer. The $\mathcal{T}_D$-center of G is unique, and, for each integer D', $1 \leq D' \leq D$, the $\mathcal{T}_D$-center of G contains the $\mathcal{T}_{D'}$-center of G.*

Proof. Let D and D' be integers, $1 \leq D' \leq D$. By Theorem 6.3, the $\mathcal{T}_D$-center T is unique and the $\mathcal{T}_{D'}$-center T' is unique. As $\mathcal{T}_{D'} \subseteq \mathcal{T}_D$, it follows that $e(T) \leq e(T')$. The proof of Theorem 6.3 shows that T' is the unique minimal subtree with eccentricity at most $e(T')$. As every subtree of G with eccentricity at most $e(T')$ contains a minimal subtree with eccentricity at most $e(T')$ and T is a subtree of G of eccentricity at most $e(T')$, we deduce that T contains T'. □

Similar results can be obtained for other hereditary classes of trees. For example, let $\mathcal{C}$ denote the set of caterpillars (including all paths), let G be a tree, and let $\mathcal{C}'$ be the subtrees of G that are in $\mathcal{C}$. A *caterpillar center* of G is an element of smallest order in $\{T \in \mathcal{C}' \colon e(T) \leq e(T'),\ T' \in \mathcal{C}'\}$. Since $\mathcal{C}$ is a hereditary class of trees, each tree G has a unique caterpillar center. Of course, if G is a caterpillar then the caterpillar center of G is G itself.

Remark 6.2. The caterpillar center of the tree G is not a path unless G is a path. To see this, again note that the proof of Theorem 6.3 shows that the unique caterpillar center is $G[X_k]$, for some integer k, $0 \leq k \leq e(C(G))$. So, $G[X_k]$ is a caterpillar contained in G of smallest order among all caterpillars of G with smallest eccentricity. As $e(C[X_k]) = k$, any caterpillar contained in G of smallest eccentricity has eccentricity k. If $k > 0$ and $G[X_k]$ is a path, then by definition of X_{k-1} in the proof of Theorem 6.1, $G[X_{k-1}]$ is a path or a caterpillar contained in G and $e(G[X_{k-1}]) = k - 1 < k$, a contradiction. So, if $k > 0$, then $G[X_k]$ is not a path. If $k = 0$, $G[X_0] = G$, so the caterpillar center is not a path unless G is a path.

Similar results hold for other hereditary classes of trees, such as those mentioned above following Definition 6.3. Moreover, recall that McMorris and Reid [7] defined a central k-tree of a tree G to be an element of the set $\{T \in A_k \colon e(T) \leq e(T'),\ T' \in A_k\}$, where $A_k = \{T \colon T \text{ a subtree of } G \text{ order } k\}$. Subtrees of order k do not form a hereditary class of subtrees; indeed, a tree need not have a unique central k-tree. However, the class $\mathcal{T}$ of trees of order at most k is a hereditary class. Then the treatment by McMorris and Reid for finding central k-trees amounts to adding arbitrary, new vertices adjacent to the $\mathcal{T}$-center until trees of order exactly k are obtained.

Remark 6.3. In general, the $\mathcal{T}$-center of a tree G can be found by finding each X_k as k decreases from $e(C(G))$ to 0 and stopping at the largest index k so that $G[X_k]$ is not in $\mathcal{T}$. Then $G[X_{k+1}]$ is the $\mathcal{T}$-center. As indicated in Remark 6.1, finding the sets X_k can be accomplished in linear time. In certain instances we can also test quickly whether $G[X_k]$ is in $\mathcal{T}$ as we see next.

Theorem 6.5. *Let G be a tree and let $G[X_k]$ be its unique minimal subtree with eccentricity at most k, for $0 \leq k \leq e(C(G))$. There is a linear time algorithm that finds all X_k for $0 \leq k \leq e(C(G))$.*

Proof. First, $C(G)$ can be found in linear time (successively remove all leaves until K_1 or K_2 remains). If $C(G) = \{x\}$, we let x be the root of G, and if $C(G) = \{x, y\}$ we contract y to x and let x be the root of the "adjusted" tree (and we still refer to the adjusted tree as G). In the resulting rooted tree each non-root vertex v is itself the root of a unique maximal tree G_v induced by all vertices reachable from x via v. Let $e'(v)$ be the eccentricity of v in G_v. We determine $e'(v)$ for each v in V by a depth-first search (DFS) from the root x as follows. Begin with $e'(v)$ set to 0 for all v in V. When we arrive at a vertex v from its child u, update $e'(v)$ to be the maximum of $e'(u) + 1$ and the current value of $e'(v)$. Observe that when the DFS is done, all $e'(v)$ are correctly computed in linear time. Note that $e'(x) = e(C(G))$. For each $0 \leq k \leq e(C(G))$ we create a list, and place a vertex $v \in V$ in the list with $k = e'(v)$ (in linear time). Now let $X_{e(C(G))} = C(G)$, and for each $k < e(C(G))$, let $X_k = X_{k+1} \cup \{v \colon e'(v) = k\}$. This can be done in linear time using the lists we set up. Thus all X_k are found in linear time, as desired. □

Theorem 6.6. *Let G be a tree and let D be a positive integer. There is a linear time algorithm for finding the $\mathcal{T}_D$-center of G. There are similar linear time algorithms for finding the $\mathcal{T}$-center if $\mathcal{T}$ is the family of all of any of the following: any finite hereditary class of trees (such as trees of order at most n), stars (including K_1 and K_2), spiders (subdivisions of stars, including all stars), trees of diameter at most D, trees with at most ℓ leaves, caterpillars (and paths), and lobsters (and caterpillars and paths).*

Proof. To find the $\mathcal{T}_D$-center in linear time, we modify the last step of Theorem 6.5 so that each time we add a vertex of $X_k - X_{k+1}$ to X_k, we check to see whether a vertex of degree greater than D has been created in $G[X_k]$. When that happens, we stop and $G[X_{k+1}]$ is the $\mathcal{T}_D$-center. If $\mathcal{T}$ is the family of trees with at most ℓ leaves, and there are more than ℓ leaves in G, then we keep track of the number of leaves in the current subtree $G[X_k]$ by incrementing a counter (initialized appropriately) each time a leaf is added which is incident to a current non-leaf. When the counter exceeds ℓ as a leaf is added to $G[X_k]$, we stop and $G[X_{k+1}]$ is the $\mathcal{T}$-center. When $\mathcal{T}$ is the family of trees of order at most n, for some $n \geq 1$, the $\mathcal{T}$-center can be found in linear time (much like the algorithm in [7]), since it is trivial to recognize the minimum n such that $|X_{k+1}| > n$. (If $n = 1$, then $C(G)$

is the set of $\mathcal{T}$-centers, which can also be found in linear time.) Moreover, since it takes constant time to check whether a tree is isomorphic to a given fixed tree, whenever $\mathcal{T}$ is a finite hereditary class of trees, the $\mathcal{T}$-center can be found in linear time. (As above, $\mathcal{T} = \{K_1\}$ is a special case.) If $\mathcal{T}$ represents stars (or spiders), one simply chooses the minimum k such that $G[X_{k+1}]$ has more than one vertex of degree greater than 1 (degree greater than 2 for spiders). Thus the $\mathcal{T}$-center is found in linear time. Note that the diameter of $G[X_k]$ is 2 plus the diameter of $G[X_{k+1}]$ if $k \geq 0$, and the diameter of $G[X_{e(C(G))}]$ is 0 or 1 when $|C(G)|$ is 1 or 2, respectively. Then it is easy to see that if $\mathcal{T}$ is the family of trees of diameter at most D (with $D \geq 1$), then, since $dia(G[X_k]) = 2(e(C(G)) - k) + \lfloor \frac{|C(G)|}{2} \rfloor$, $G[X_k]$ is the $\mathcal{T}$-center for $k = \max\{0, e(C(G)) - \lceil (D - |C(G)|)/2 \rceil\}$.

To find the central caterpillar in linear time, we begin by finding k such that $G[X_k]$ is the $\mathcal{T}_2$-center (i.e., path center), as above. Note that this is very similar to the algorithm for finding a path center given in [6]. As noted earlier, unless G is itself a path, $G[X_{k-1}]$ is not a path, in which case it is a caterpillar. Note that no vertex of $X_{k-1} - X_k$ is a leaf in $G[X_{k-2}]$ unless $G[X_{k-1}] = G$ (and $k = 1$). Therefore, $G[X_{k-2}]$ is not a caterpillar when $k \geq 2$, so $G[X_{k-1}]$ is the central caterpillar unless G is a path. Similarly, $G[X_{k-2}]$ is the lobster center unless G is a caterpillar or a path, and thus the lobster center can be found in linear time. □

For an arbitrary hereditary class of trees $\mathcal{T}$, the running time for finding the $\mathcal{T}$-center ultimately depends on how easy it is to recognize whether each $G[X_k]$ is in $\mathcal{T}$.

Similar results can be obtained when the branch weight function $bw(\cdot)$ is used in place of eccentricity $e(\cdot)$. The analogous, fundamental result is given next.

Theorem 6.7. *If G is a tree and k is an integer, $0 \leq k \leq bw(B(G))$, then G contains a unique minimal subtree with branch weight at most k.*

Proof. Let G be a tree, and let k be an integer, $0 \leq k \leq bw(B(G))$. Clearly the conclusion follows if $k = 0$. So, assume $k > 0$ and $|V(G)| \geq 3$. For simplicity, let $b = bw(B(G))$. For a tree H, again let $L(H) = \{x \in V(H)\colon d(x) = 1\}$, the set of leaves of H. For each $v \in V(G) - B(G)$, define G to be the unique maximal subtree of G with vertex set consisting of all vertices reachable from $B(G)$ via v, i.e., $\{u \in V(G)\colon d(B(G), u) = d(B(G), v) + d(v, u)\}$. We determine G_v for all $v \in V(G) - B(G)$. If $v \in L(G)$, G_v is the subtree consisting of v alone. Note that for $v \in V(G) - (L(G) \cup B(G))$, vu is an edge of G_v (so, $u \in V(G_v)$) if and only if $d(B(G), u) = d(B(G), v) + 1$. Recursively, we see that for $v \in V(G) - (L(G) \cup B(G))$, subtree G_v is the subtree of G induced by v and $\bigcup\{V(G_v)\colon u \in V(G),\ d(B(G), u) = d(B(G), v) + 1\}$. So, $|V(G_v)|$ is determined for all $v \in V(G) - B(G)$. Define the nested family of sets Y_k, $0 \leq k \leq b$, as follows: $Y_b = B(G)$, and, for each k, $0 \leq k < b$, let $Y_k = Y_{k+1} \cup \{v\colon |V(G_v)| = k+1\}$. So, if $V_j = \{v \in V\colon |V(G_v)| = j\}$, then $Y_b = B(G)$ and $Y_k = B(G) \cup \bigcup_{j=k+1}^{b} V_j$, for $0 \leq k < b$. In particular, for

$0 \le k \le b$, any vertex $v \in V$ with $|V(G_v)| = k$ is not in Y_k (and, hence, not in any of $Y_b, Y_{b-1}, \ldots, Y_{k+1}$), but is in Y_{k-1} (and, hence also in $Y_{k-2}, Y_{k-3}, \ldots, Y_0$). And, Y_k is exactly the set of vertices of a subtree of G.

First, note that $bw(Y_k) \le k$. For otherwise, there is a connected component W of order at least $k+1$ in $G - Y_k$. Let uv denote the edge of G with $u \in Y_k$ and $v \in V(W)$. As $V(W) \subseteq V(G_v)$, $|V(G_v)| \ge |W| \ge k+1$. By the comments above, this implies that $v \in Y_k$, a contradiction to $Y_k \cap V(W) = \emptyset$.

Second, no proper subtree of $G[Y_k]$ has branch weight at most k. For, suppose W is a proper subtree of $G[Y_k]$ so that $bw(W) \le k$. Let uv be any edge of $G[Y_k]$ so that $u \in V(W)$ and $v \in Y_k - V(W)$. Since $B(G) \subseteq V(W)$, $V(G_v) \cap V(W) = \emptyset$. By the comments above, every vertex $x \in Y_k - B(G)$ has $|V(G_k)| \ge k+1$, so $|V(G_v)| \ge k+1$. Then G_v is a connected component of $G - W$ of order at least $k+1$. This implies $bw(W) \ge k+1$, a contradiction. Hence, $G[Y_k]$ is a minimal subtree of G of branch weight at most k.

Finally, $G[Y_k]$ is the unique minimal subtree of G of branch weight at most k. For, suppose not, i.e., suppose that W is a subtree of G so that (1) $bw(W) \le k$ and (ii) no proper subtree of W has branch weight at most k. We claim that either W is a subtree of $G[Y_k]$, or $G[Y_k]$ is a subtree of W. Suppose not, i.e., suppose that there is a vertex $w \in V(W)$, $w \notin Y_k$ and Y_k is not a subset of $V(W)$. At this point we invoke the following lemma, the proof of which is given below:

Lemma 6.1. *Let G be a tree and let $b = bw(B(G))$. For each integer k, $0 \le k \le b$, and for each subtree W of G with $bw(W) \le k$, $B(G) \subseteq V(W)$.*

By Lemma 6.1, $Y_b = B(G) \subseteq V(W)$. Let h be the index so that $Y_b \subseteq Y_{b-1} \subseteq \ldots \subseteq Y_h \subseteq V(W)$, but $Y_{h-1} \nsubseteq V(W)$. Such an index exists since $Y_k \nsubseteq V(W)$, and, thus, $k \le h-1$. These set inclusions imply that there is a vertex $v \in Y_{h-1}$, $v \notin V(W)$, and as $Y_h \subseteq Y_k$, $W \notin Y_h$. Since $v \in Y_{h-1}$, $|G_v| \ge h$. If any vertex, say z, of $V(G_v) - \{v\}$ is in $V(W)$, then the path in W from z to w must include v, and, thus, $v \in V(W)$, a contradiction. So, $V(W) \cap V(G_v) = \emptyset$. That is, G_v is a subtree of a connected component of $G - W$. This implies, $bw(W) \ge |G_v| \ge h \ge k+1 > k$, contrary to the choice of W. That is, either W is a subtree of $G[Y_k]$, or $G[Y_k]$ is a subtree of W. In the former case, by the first part of this proof and assumption (1) above, we deduce $W = G[Y_k]$. In the latter case, as $bw(G[Y_k]) \le k$ and assumption (2) above, we deduce $G[Y_k] = W$. Consequently, $G[Y_k]$ is the unique minimal subtree of G of branch weight at most k. □

Proof. [Proof of Lemma 6.1] Assume G is a tree, $b = bw(B(G))$, k is an integer, $0 \le k \le b$, and W is a subtree of G with $bw(W) \le k$. Suppose $B(G)$ is not a subset of $V(W)$. If $|B(G)| = 1$, say $B(G) = \{x\}$, let $x = x_0, x_1, \ldots, x_d = w$ be the shortest path from x to a vertex $w \in V(W)$. Since $bw(W) \le k$, the connected component of $G - W$ containing x (and $x_1, \ldots, x_{d-1}$) has order at most k. And, since $bw(x) = bw(B(G)) = b$, this implies that the connected component of $G - x$ containing W (and $x_1, x_2, \ldots, x_{d-1}$) is exactly b. Consequently if $k < b$, then each connected component of $G - w$ has order less than b, i.e., $bw(w) < b = bw(B(G)) = \min\{b(v) : v \in V(G)\}$, a contradiction. Suppose $k = b$. If $d > 1$, then $bw(x_1) \leq b - 1 < b$, again a contradiction. And, if $d = 1$, then $bw(w) = b$, so that $w \in B(G)$, a contradiction. We conclude $|B(G)| \neq 1$. If $|B(G)| = 2$, say $B(G) = \{x, y\}$, either x or y is not in W. Without loss of generality, we may assume that $x \notin V(W)$ and both y and W are in the same connected component of $G - x$. Since $bw(x) = bw(y)$, it is easy to see that y is in a connected component Y in $G - x$ of order $bw(x)$ and x is in a connected component X in $G - y$ of order $bw(y)$. As x and y are adjacent this implies $|V(G)| = |V(X)| + |V(Y)|$ and $|V(X)| = |V(Y)|$. $G - \{x, y\}$ contains a connected component of order b. No such component is adjacent to x, for otherwise, by our conventions for x and y, $bw(W) \geq b + 1 > b \geq k$, contrary to the choice of W. So, all connected components of $G - \{x, y\}$ of order b are adjacent to y. We claim there is exactly one, and it contains W. Suppose that there are at least two such components of $G - \{x, y\}$ of order b. If there are vertices of W in two or more of these components (so, $y \in V(W)$ as well), then connected components of $G - W$ are X and subtrees of the components of $G - y$ that are adjacent to y. The former component, X, has order $|V(X)| = |V(Y)| \geq 2b + 1 > b$, and components of the latter type have order at most b, so $bw(W) = |V(X)| > b \geq k$, contrary to the choice of W. So, there are vertices of W in at most one component in $G - \{x, y\}$ of order b, and there is at least one other component, say D, in $G - \{x, y\}$ of order b adjacent to y. If $y \in V(W)$, then as before, $bw(W) \geq |V(X)| > b \geq k$, contrary to the choice of W. If $y \notin V(W)$, then $G[V(D) \cup \{y\}]$ is a connected subtree of order $b + 1$ contained in $G - W$, so $bw(W) > b \geq k$, again a contradiction. So, there is no such other component D; that is, there is only one component of $G - \{x, y\}$ of order b. Suppose no vertices of W are in the unique component of $G - \{x, y\}$ of order b. Then $bw(W) \geq |V(X)| = |V(Y)| \geq b + 1 \geq k + 1 > k$, again a contradiction. In summary, the unique component F of $G - \{x, y\}$ of order b contains W. Consequently, $G - V(F)$ is a subtree of $G - V(W)$, so that $b \geq k \geq bw(W) \geq |V(G)| - |V(F)| = |V(X)| + |V(Y)| + b = 2|V(Y)| - b \geq 2(b+1) - b = b+2$, a contradiction.

In conclusion, it is impossible for $B(G)$ not to be a subset of $V(W)$. □

We now give the promised definition.

Definition 6.4. Let $\mathcal{T}$ be a hereditary class of trees, let G be a tree, and let $\mathcal{T}'$ be the subtrees of G that are in $\mathcal{T}$. A $\mathcal{T}$-*centroid* of G is a tree of smallest order in the set $\{T \in \mathcal{T}' : bw(T) \leq bw(T'),\ T' \in \mathcal{T}'\}$.

When $\mathcal{T}$ is $\mathcal{T}_D$, the hereditary family of trees of maximum degree D, we obtain the $\mathcal{T}_D$-centroid of Definition 6.1. So, Definition 6.4 is a further generalization of the centroid of a tree.

The proofs of the next two results are quite similar to their eccentricity analogues, Theorems 6.3 and 6.4, so we leave the proofs to the reader.

Theorem 6.8. *For a tree G and a hereditary class of trees $\mathcal{T}$, the $\mathcal{T}$-centroid of G is unique unless both $\mathcal{T} = \{K_1\}$ and $|M(G)| = 2$.*

Theorem 6.9. *Let G be a tree, and let D be a positive integer. The $\mathcal{T}_D$-centroid of G is unique and, for each integer D', $1 \leq D' \leq D$, the $\mathcal{T}_D$-centroid of G contains the $T_{D'}$-centroid of G.*

Remark 6.4. For a tree G, $B(G)$ can be found in linear time [19]. The procedure for finding $|V(G_v)|$ for every $v \in V(G)$ given in the proof of Theorem 6.7 can be implemented in linear time. For each $1 \leq k \leq bw(B(G))$ create a list and place each $v \in V(G) - B(G)$ in the list provided $|V(G_v)| = k$. As in the proof of Theorem 6.7, $Y_{bw(B(G))} = B(G)$, and for each $0 \leq k < bw(B(G))$, set $Y_k = Y_{k+1} \cup \{v \in V(G)\colon |V(G_v)| = k+1\}$. This can be done in linear time using the lists. Thus, the sets Y_k, $0 \leq k \leq bw(B(G))$, so that $G[Y_k]$ is the unique minimal subtree with branch weight at most k, can be found in linear time. This has implications for $\mathcal{T}$-centroids. In the remainder of this section we continue to utilize the sets Y_k.

Theorem 6.10. *Let G be a tree and let D be a positive integer. There is a linear time algorithm for finding the $\mathcal{T}_D$-centroid of G. There are similar linear time algorithms for finding the $\mathcal{T}$-centroid if $\mathcal{T}$ is the family of all of any of the following: any finite hereditary class of trees (such as trees of order at most n), stars (including K_1 and K_2), spiders (subdivisions of stars, including all stars), trees with at most ℓ leaves, caterpillars (and paths), and lobsters (and caterpillars and paths).*

Proof. The $\mathcal{T}_D$-centroid can be found in linear time for the same reasons that the $\mathcal{T}_D$-center can be found in linear time, and the $D = 2$ case resembles the algorithm for finding the path centroid given in [6]. Likewise, we can find the $\mathcal{T}$-centroid in linear time if $\mathcal{T}$ represents any finite hereditary class of tree (such as the trees of order at most n), the stars, the spiders, or the trees with at most ℓ leaves.

Let $\mathcal{C}$ be the hereditary class of all caterpillars (including all paths). The caterpillar centroid (i.e., $\mathcal{C}$-centroid) of a tree G is a bit harder to find than the caterpillar center, since a vertex of $Y_k - Y_{k+1}$ may remain a leaf in Y_{k-1} (indeed, $Y_k = Y_{k-1}$ is a possibility). Suppose $0 \leq k < b(B(G))$. For two vertices x and y in $Y_{k-1} - Y_k$, $|V(G_x)| = |V(G_y)| = k$. If x and y are adjacent in G, then, since $B(G) \subseteq Y_k$, $d(x, B(G))$ and $d(y, B(G))$ are positive and $|d(x, B(G)) - d(y, B(G))| = 1$, say $d(x, B(G)) = 1+d(y, B(G))$. Then $V(G_x) \cup \{y\} \subseteq V(G_y)$ so $|V(G_y)| \geq |V(G_x)|+1$, a contradiction. So, no two vertices in $Y_{k-1} - Y_k$ are adjacent, i.e., $Y_{k-1} - Y_k$ is either empty or an independent set. This implies that any vertex in $Y_{k-1} - Y_k$ is a leaf in $G[Y_{k-1}]$. To find the caterpillar centroid, begin with the path centroid $G[Y_k]$. Then, by the comments above, $G[Y_{k-1}]$ is a caterpillar, denoted C. Label the vertices of the path P obtained by deleteing the leaves of C from C by $v_m, v_{m+1}, \ldots, v_{m+p}$ in the order encountered in a traverse of P from one end to the other end, where m is any integer (e.g., $m > |V(G)|$ will insure no negative indices will be produced later in this process) and $p = l(P)$. If any vertex of $Y_{k-2} - Y_{k-1}$ is adjacent to a leaf of

C that, in turn, is adjacent to some v_i, with $m < i < m + p$, then stop: C is the caterpillar centroid of G. Otherwise, any vertex of the independent set $Y_{k-2} - Y_{k-1}$ is either adjacent to one of $v_m, v_{m+1}, \ldots, v_{m+p}$ or adjacent to one of the leaves u of C that, in turn, is adjacent to v_m or v_{m+p}. The set of vertices of $Y_{k-2} - Y_{k-1}$ of the former type is denoted A and the set of vertices of $Y_{k-2} - Y_{k-1}$ of the latter type is denoted D. If $D \neq 0$, label the vertices in D by $z_1, z_2, \ldots, z_q$. For $1 \leq i \leq q$, define u_i as follows: z_i is adjacent the leaf u_i of C that, in turn, is adjacent to v_m or v_{m+p}. If there exist distinct vertices u_i and u_j, $i \neq j$, both adjacent to either v_m or v_{m+p}, then stop: C is the caterpillar centroid of G. Otherwise, by the pigeonhole principle, there are at most two distinct u_i's, one adjacent to v_m and/or one adjacent to v_{m+p}. So, adjoining every vertex of $Y_{k-2} - Y_{k-1}$ to C yields the caterpillar $G[Y_{k-2}]$. Now, update C to be $G[Y_{k-2}]$ and repeat the process on this (possibly) new caterpillar. Eventually, we obtain the caterpillar centroid, and it will be of the form $G[Y_k]$ for some k. This process can be carried out in linear time.

The process for the $\mathcal{T}$-centroid, where $\mathcal{T}$ is the set of lobsters, is quite similar with suitable adjustments on the stopping rules and will be left to the reader. □

The family $\mathcal{T}$ of trees with diameter at most D is not mentioned in Theorem 6.10, because the formula for k so that $G[X_k]$ is the $\mathcal{T}$-center given in the proof of Theorem 6.6 does not extend to an analogous result for a formula for k so that $G[Y_k]$ is the $\mathcal{T}$-centroid.

6.3. Generalized path medians

Now we turn to the $\mathcal{T}_D$-medians. Recall that Slater's path median (the $\mathcal{T}_2$-median) need not be unique. The following theorem constructs a tree for each $D \geq 0$ in which the $\mathcal{T}_D$-median is not unique. These examples also show something stronger, with an immediate consequence that for any fixed $D \geq 3$, one cannot hope for a polynomial-time algorithm that finds all $\mathcal{T}_D$-medians of a tree.

The proof of the following theorem motivates and serves as a simpler version of the subsequent proof.

Theorem 6.11. *For any integer $D \geq 0$, there exists a tree G that has more than one $\mathcal{T}_D$-median. If $D \geq 2$, then there is a tree G that has two non-isomorphic $\mathcal{T}_D$-medians. If $D \geq 3$, then there is a tree G with a superpolynomial number of $\mathcal{T}_D$-medians.*

Proof. Recall that in a tree G, a $\mathcal{T}_0$-median is simply a median vertex, and $M(G)$ contains either one vertex or two adjacent vertices. In the latter case, the $\mathcal{T}_0$-median is not unique, but the $\mathcal{T}_1$-median is the unique subgraph with one edge induced by $M(G)$.

If there is only one median vertex x and $|V| > 1$, then a $\mathcal{T}_1$-median of G is induced by any set $\{x, y\}$ where y is a neighbor of x that lies in a largest component of $G - x$. For example, a k-star (a tree of order $k+1$ with k leaves, all adjacent to one vertex of degree k) has k distinct $\mathcal{T}_1$-medians.

A $\mathcal{T}_2$-median is a path median; there are examples in [24] and [6] that show that the path median is not unique (and that the path median of a tree does not necessarily contain the the median). However this does not suffice for the case $D = 2$, as we also seek a tree with non-isomorphic $\mathcal{T}_2$-medians.

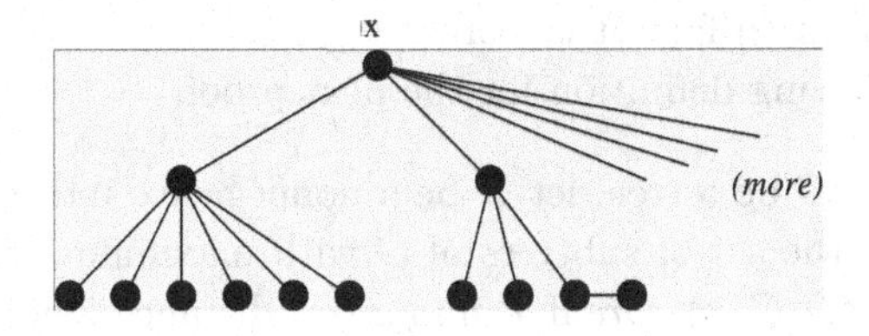

Fig. 6.1. An example with $D = 4$, $k = 6$, and $j = 1$, focused on T_1 and T_2.

Let D, k, j be integers such that $D \geq 2$, $k \geq D+1$, and $1 \leq j \leq k$. For $1 \leq i \leq j$, let T_i be a $(D+3)$-star with a leaf labeled x. For $j+1 \leq i \leq k$, let T_i be a 4-vertex path with a leaf labeled x, and an additional $D-2$ leaves attached to the neighbor of x. Let G be the tree formed by identifying all the vertices labeled x. (See Fig. 6.1 for an example.)

Let T be a $\mathcal{T}_D$-median of G. Then T must be a maximal subgraph of maximum degree at most D. Therefore T cannot be contained in $T_i - x$ for $i \geq j+1$. If T were contained in $T_i - x$ for $i \leq j$, then T would be a D-star and replacing any leaf of T by x would clearly decrease its distance in G; a contradiction. Therefore T must contain x.

T must intersect exactly D components of $G - x$. For $1 \leq i \leq j$, either T intersects $T_i - x$ in a $(D-1)$-star, in which case the vertices of $T_i - x$ contribute 3 to the distance of T, or T does not intersect $T_i - x$, in which case the vertices of $T_i - x$ contribute $2D+5$ to the distance of T. For $j+1 \leq i \leq k$, either T contains $T_i - x$, in which case the vertices of $T_i - x$ contribute 0 to the distance of T, or T does not intersect $T_i - x$, in which case the vertices of $T_i - x$ contribute $2D+2$ to the distance of T. Thus, the "relative cost" of not intersecting any T_i is $2D+2$, for all $1 \leq i \leq k$. Therefore, G has at least one distinct $\mathcal{T}_D$-median for each D-subset of the components of $G - x$, for a total of at least $\binom{k}{D}$ distinct $\mathcal{T}_D$-medians.

The intersection of T and $T_i - x$ is in a different isomorphism class depending on whether $1 \leq i \leq j$ or $j+1 \leq i \leq k$. (For example, with $D = 2$ the intersection can be a 2-vertex path or a 3-vertex path.) When $j \neq 0$ and $j \neq k$, $G - x$ has both types of components. Since $k \geq D+1$, there are $\mathcal{T}_D$-medians that intersect different numbers of the first (and second) type of component of $G - x$. Therefore, there are non-isomorphic $\mathcal{T}_D$-medians of G (when $0 < j < k$, which is always possible since $k - 1 \geq D \geq 2$).

In the case that $k = 2D$ and $D \geq 3$, $\binom{k}{D} \sim 2^{2D}/\sqrt{\pi D}$ (using Stirling's approximation) and $|V(G)| \sim 2D^2$, so the number of $\mathcal{T}_D$-medians is superpolynomial in $|V(G)|$. □

Even though there can be a superpolynomial number of $\mathcal{T}_D$-medians, of different isomorphism classes, the following theorem produces all $\mathcal{T}_D$-medians quickly, relative to the number of $\mathcal{T}_D$-medians sought. The ideas are similar to those in [24] for path medians and in [3; 2] for k-tree cores.

We need the following definition for the next proof.

Definition 6.5. Let G be a tree, let D be a nonnegative integer, let x be a vertex of G, and let $\mathcal{T}_D$ be the set of subtrees of G with maximum degree at most D. A subtree T of G is a (D, G, x)*-core* if T minimizes distance among all subtrees of G in $\mathcal{T}_D$ that contain x. Let $D(D, G, x)$ be the distance of a (D, G, x)-core T, i.e., $D(D, G, x) = \sum\{d(y, T) \colon y \in V(G)\}$.

A $(0, G, x)$-core is the single-vertex tree on $\{x\}$. Thus, $D(0, G, x)$ is the distance of x in G. (Although $D(0, G, x)$ is more simply $D(x)$ when the graph G is clear, this notation allows us to denote distance of x in any subgraph H of G, by $D(0, H, x)$.)

Theorem 6.12. *Let G be a tree, and let D a positive integer, and let $\mathcal{T}_D$ be the set of subtrees of G with maximum degree at most D. There is a recursive algorithm that finds all $\mathcal{T}_D$-medians in time that is linear in the order of G plus the number of $\mathcal{T}_D$-medians. There is an algorithm that finds a single $\mathcal{T}_D$-median in linear time.*

Proof. Let $x_1, \ldots, x_n$ be an ordering of the vertices of G such that x_i is a leaf in $G[\{x_i, \ldots, x_n\}]$ for each $1 \leq i < n$. Considering G as a rooted tree with root x_n, let G_i be the subtree rooted at x_i, let N_i be the set of neighbors of x_i in G_i (its *children*), and let $I_i = \{j \colon x_j \in N_i\}$, for $1 \leq i \leq n$. Each vertex x_i with $1 \leq i < n$ has exactly one neighbor x_j that is not in N_i (its *parent*); let $p(i) = j$ be its index. (Equivalently, for every edge $x_i x_j$ in G with $i < j$: $x_j \in N_i$, $j \in I_i$, and $p(i) = j$.) Note that for $1 \leq i < n$, G_i is the component of $G - x_i x_{p(i)}$ that contains x_i. For $1 \leq i < n$, let G_i' be the subtree of G that contains G_i, $x_{p(i)}$, and $x_i x_{p(i)}$. For $1 \leq i \leq n$, let $G_i'' = G - (V(G_i) - x_i) = G \setminus \bigcup_{j \in I_i} V(G_j)$. For $1 \leq i < n$, G_i'' is the component of $G - x_i x_{p(i)}$ that contains $x_{p(i)}$, with x_i and $x_i x_{p(i)}$ added. If $i \neq n$ then x_i is a leaf in G_i''; x_n is the only vertex of G_n''. For any subtree T of G and any vertex $x_i \in V(T)$, let $N_i^T = N_i \cap V(T)$ and let $I_i^T = \{j \colon x_j \in N_i^T\}$.

For reasons we will see later, for any $1 \leq j \leq n-1$, let $m_j = D(0, G_j', x_{p(j)}) - D(D, G_j', x_{p(j)})$. As part of the solution, we will define four subsets of N_i for all $1 \leq i \leq n$: $R_i^D, S_i^D, R_i^{D-1}, S_i^{D-1}$ with $R_i^D \subseteq S_i^D$, $R_i^{D-1} \subseteq S_i^{D-1}$, $R_i^{D-1} \subseteq R_i^D$, and $S_i^{D-1} \subseteq S_i^D$. This will also be explained later.

Suppose that T is a $\mathcal{T}_D$-median of G. Let i be the largest index such that x_i is in T, or equivalently, let i be smallest such that $T \subseteq G_i$. Since T is a subtree of G_i that contains x_i and T is a $\mathcal{T}_D$-median, T has minimum distance in G among all subtrees of G_i in $\mathcal{T}_D$ that contain x_i. For any subtree T' of G_i in $\mathcal{T}_D$ that contains x_i, the distance of T' in G equals the distance of T' in G_i plus $D(0, G_i'', x_i)$ (the distance of x_i in G_i'') if $1 \leq i \leq n-1$; if $i = n$ then $G_i = G$ so replacing $D(0, G_i'', x_i)$ by zero makes it true. Therefore T has minimum distance in G_i among all subtrees of G_i in $\mathcal{T}_D$ that contain x_i; that is, T is a (D, G_i, x_i)-core.

Suppose that T is a (D, G_i, x_i)-core for any integer i with $1 \leq i \leq n$. Clearly, $|N_i^T| = |I_i^T| = \min(D, |N_i|)$. For each $j \in I_i^T$, let T_j be any subtree of G_j' in $\mathcal{T}_D$ that contains x_i; $T \cap G_j'$ is one such subtree. The union of subtrees T_j taken over all $j \in I_i^T$ forms a subtree of G_i in $\mathcal{T}_D$ that contains x_i; its distance in G_i is equal to the sum of the distances of T_j in G_j', taken over $j \in I_i^T$. Since T is a (D, G_i, x_i)-core, it has minimum distance in G_i among all subtrees of G_i in $\mathcal{T}_D$ that contain x_i, such as the union of subtrees T_j. Therefore, $T \cap G_j'$ must have minimum distance among all subtrees of G_j' in $\mathcal{T}_D$ that contain x_i, for each $j \in I_i^T$. That is, $T \cap G_j'$ is a (D, G_j', x_i)-core for all $j \in I_i^T$.

For any $1 \leq i \leq n$, let T be a subtree of G_i in $\mathcal{T}_D$ that contains x_i, such that $|N_i^T| = \min(D, |N_i|)$, and such that $T \cap G_j'$ is a (D, G_j', x_i)-core for each $j \in I_i^T$; for example, T could be any (D, G_i, x_i)-core. The distance of T in G_i equals the sum of $D(D, G_j', x_i)$ over all $j \in I_i^T$, plus the sum of $D(0, G_j', x_i)$ over all $j \in I_i \setminus I_i^T$. This equals the sum of $D(0, G_j', x_i)$ over all $j \in I_i$, minus the sum of $D(0, G_j', x_i) - D(D, G_j', x_i)$ taken over all $j \in I_i^T$. Let $m_j = D(0, G_j', x_i) - D(D, G_j', x_i)$ represent this *relative benefit* of having x_j in T. The distance of T in G_i is minimized precisely when N_i^T is chosen among size $\min(D, |N_i|)$ subsets of N_i such that $\sum_{j \in I_i^T} m_j$ is maximized.

If $|N_i| > D$, let m_i^* be the minimum m such that $|\{x_j \in N_i \colon m_j > m\}| \leq D$, let $R_i^D = \{x_j \in N_i \colon m_j > m_i^*\}$, and let $S_i^D = \{x_j \in N_i \colon m_j \geq m_i^*\}$. If $|N_i| \leq D$, let $R_i^D = S_i^D = N_i$ (m_i^* is not defined in this case). A subtree T of G_i is a (D, G_i, x_i)-core if and only if $|N_i^T| = \min(D, |N_i|)$, $R_i^D \subseteq N_i^T \subseteq S_i^D$, and $T \cap G_j'$ is a (D, G_j', x_i)-core for all $j \in I_i$.

Note that a (D, G_j', x_i)-core with $x_j \in N_i$ and $1 \leq i \leq n$ is the same as a $(D, G_j', x_{p(j)})$-core with $1 \leq j \leq n-1$. Consider any integer i such that $1 \leq i \leq n-1$. If T is a $(D, G_i', x_{p(i)})$-core, then clearly $|N_i^T| = \min(D-1, |N_i|)$, and the distance of T in G_i' equals the distance of $T - x_{p(i)}$ in G_i. Then a similar argument shows that a subtree T of G_i' is a $(D, G_i', x_{p(i)})$-core if and only if $|N_i^T| = \min(D-1, |N_i|)$, $R_i^{D-1} \subseteq N_i^T \subseteq S_i^{D-1}$, $T \cap G_j'$ is a (D, G_j', x_i)-core for all $j \in I_i$, and T contains $x_{p(i)}$.

Now, we will be able to find desired subtrees and distance values recursively. We describe the algorithm for the computations next.

If x_i is a leaf and $1 \leq i \leq n-1$ then $|V(G_i)| = 1$, otherwise (for all $1 \leq i \leq n$) $|V(G_i)| = 1 + \sum_{j \in I_i} |V(G_j)|$; thusly we compute $|V(G_i)|$ as i increases from 1 to n. For $1 \leq i \leq n-1$, if x_i is a leaf then $D(0, G_i', x_{p(i)}) = 1$, otherwise $D(0, G_i', x_{p(i)}) = |V(G_i)| + D(0, G_i, x_i) = |V(G_i)| + \sum_{j \in I_i} D(0, G_j', x_i)$. Therefore, after having found $|V(G_i)|$ for $1 \leq i \leq n$, we may compute $D(0, G_i', x_{p(i)})$ as i

increases from 1 to $n-1$. For $i=n$, $D(0, G''_i, x_i) = 0$ since $V(G''_n) = \{x_n\}$. For $1 \le i \le n-1$, $D(0, G''_i, x_i) = |V(G) - V(G_i)| + D(0, G - V(G_i), x_{p(i)})$, which equals

$$|V(G)| - |V(G_i)| + D(0, G''_{p(i)}, x_{p(i)}) + \sum_{j \in I_{p(i)}, j \neq i} D(0, G'_j, x_{p(i)}).$$

Therefore, after having found $|V(G_i)|$ for $1 \le i \le n$ and $D(0, G'_i, x_{p(i)})$ for $1 \le i \le n-1$, we can compute $D(0, G''_i, x_i)$ as i decreases from n to 1. Clearly, this can all be done in linear time with respect to $|V(G)|$.

Since $m_i = D(0, G'_i, x_{p(i)}) - D(D, G'_i, x_{p(i)})$ and $D(0, G'_i, x_{p(i)})$ has already been found (for $1 \le i \le n-1$), we can immediately compute m_i as soon as $D(D, G'_i, x_{p(i)})$ is found (for $1 \le i \le n-1$). In the following, we show how to compute $D(D, G'_i, x_{p(i)})$ and m_i $(1 \le i \le n-1)$ and compute $D(D, G_i, x_i)$ and find R_i^D, S_i^D, R_i^{D-1}, and S_i^{D-1} $(1 \le i \le n)$. All values are computed for fixed i, then i is increased by one; thus, we can assume that values have been determined for indices less than the current i.

If $N_i = \emptyset$ (or equivalently, if x_i is a leaf and $1 \le i \le n-1$), then $D(D, G'_i, x_{p(i)}) = 0$, $D(D, G_i, x_i) = 0$, and $R_i^D = S_i^D = R_i^{D-1} = S_i^{D-1} = \emptyset$. Thus we may assume that $N_i \neq \emptyset$.

If $1 \le |N_i| \le D$, let $R_i^D = S_i^D = N_i$. Suppose that $|N_i| > D$. Sort all $j \in I_i$ so that the corresponding m_j are non-increasing, then initialize $R_i^D = S_i^D = \emptyset$, and repeat the following: Let $m^* = \max\{m_j \colon x_j \in N_i \setminus R_i^D\}$ and let $N^* = \{x_j \in N_i \setminus R_i^D \colon m_j = m^*\}$. If $|R_i^D \cup N^*| \le D$, let $R_i^D := R_i^D \cup N^*$ (and repeat). Otherwise (if $|R_i^D \cup N^*| > D$), let $S_i^D = R_i^D \cup N^*$, let $m_i^* = m^*$, and stop repeating.

Note that this correctly computes R_i^D and S_i^D (and m_i^* if $|N_i| > D$), which allows us to compute $D(D, G_i, x_i) = \sum_{x_j \in R_i^D} D(D-1, G'_j, x_i) + (D - |R_i^D|)m_i^*$, for all $1 \le i \le n$. We can similarly compute R_i^{D-1}, S_i^{D-1}, and $D(D, G'_i, x_{p(i)})$ for any $1 \le i \le n-1$. These computations depend only on knowing $D(D-1, G'_j, x_i)$ for $j \in I_i$, so as i increases from 1 to n or $n-1$, the computations can be performed. Since the values are clearly all between 0 and n^2, sorting $|N_i|$ values with radix sort takes $O(|N_i|)$ time. Aside from the sorting, this process takes $O(|N_i|)$ time as well, so overall running time is linear in $|V(G)|$.

Now we can compute the distance of a (D, G_i, x_i)-core T_i with respect to G (for $1 \le i \le n$), since $D(T_i) = D(D, G_i, x_i) + D(0, G''_i, x_i)$ for $1 \le i \le n$. $\mathcal{T}_D$-medians are precisely the (D, G_i, x_i)-cores T_i with values of i for which $D(T_i)$ is smallest, and we can find $\min_i D(T_i)$ and then the set of i that minimize $D(T_i)$ in $O(|V(G)|)$-time.

Finally, we will show how to obtain every $\mathcal{T}_D$-median of G. By the previous remark, there can be a superpolynomial number of $\mathcal{T}_D$-medians, which is why we have thus far not attempted to actually produce $\mathcal{T}_D$-medians. However, we can find any one in linear time, and we can find them all quickly in an appropriate sense:

Begin with any $x_i \in V$ for which $D(T_i) = \min_i D(T_i)$. Add R_i^D and add $\min\{D, |N_i|\} - |R_i^D|$ vertices of S_i^D. Now, as i decreases from $n-1$, if x_i is already selected then add R_j^{D-1} and add $\min\{D-1, |N_i|\} - |R_i^{D-1}|$ vertices of S_i^{D-1}. The choices made determine which of the $\mathcal{T}_D$-medians of G is obtained, and this $\mathcal{T}_D$-median is obtained in linear time. By branching the procedure to follow through with all possible choices, we obtain all $\mathcal{T}_D$-medians. Each $\mathcal{T}_D$-median is found in a unique way, since if x_i is its vertex of maximum index, then it is a (D, G_i, x_i)-core but not a (D, G_j, x_j)-core for all $j \neq i$.

Finally, if we instead add all of S_i^D in the first step, and add all of S_j^{D-1} in later steps, then we obtain the set of all vertices that are are contained in $\mathcal{T}_D$-medians of G. □

Remark 6.5. If we knew in advance that some vertex of G is in every $\mathcal{T}_D$-median, then we could order the vertices so that x_n is that vertex. Then every $\mathcal{T}_D$-median would actually be a (D, G_n, x_n)-core in G, which would somewhat simplify the last procedure. The most obvious candidate would be a median vertex, except that Morgan and Slater [24] showed that sometimes the path median of a tree does not contain a median vertex. Slater [25] has found a different sort of vertex which is contained in every path median, called a *pit vertex*. Michael Lenzen [26] (an undergraduate student at IIT) showed that there is a vertex contained in every $\mathcal{T}_D$-median of a fixed tree T if $D \geq 1$. Unfortunately, the natural algorithm to find such a vertex is no simpler than the algorithm of Theorem 6.12.

6.4. Conclusions

This paper unifies many types of central substructures of trees under the definitions of $\mathcal{T}$-center and $\mathcal{T}$-centroid, subsuming previous definitions and algorithms. It also deals with many other potential generalizations, since our work immediately applies whenever $\mathcal{T}$ is a hereditary class of trees. For many choices of $\mathcal{T}$, one can follow our model and show how to find the $\mathcal{T}$-center and $\mathcal{T}$-centroid in linear time. This leads to the question: are there linear time algorithms for finding the $\mathcal{T}$-center and $\mathcal{T}$-centroid for *any* hereditary class of trees $\mathcal{T}$, and, if so, can the algorithms be described in a unified manner? The answer to the first part of the question would be 'Yes' if, for every hereditary class of trees $\mathcal{T}$, there is a sufficiently fast recognition algorithm to test whether a subtree T of an arbitrary tree G is in $\mathcal{T}$. It might help to have a nice alternative characterization of a hereditary class of trees. Another direction to pursue would be to study $\mathcal{T}$-medians: for any hereditary class of trees $\mathcal{T}$ and any graph G, define a $\mathcal{T}$*-median* of G to be an element in the set $\{ T \in \mathcal{T}' \colon D(T) \leq D(T'),\ T' \in \mathcal{T}' \}$, where $\mathcal{T}'$ is the set of subtrees of G that are in $\mathcal{T}$. Then one might seek fast algorithms for finding one or all $\mathcal{T}$-medians in a tree G, for any hereditary class of trees $\mathcal{T}$, or merely for special hereditary classes of trees $\mathcal{T}$.

Yet another possibility is to explore the relationship between our definitions and *disconnected* central substructures. For example, a p-center [27; 28] and p-median [29; 27; 30] are sets of p vertices that minimize maximum distance and distance, respectively, among all sets of p vertices in a given tree G, and a p-

core [31; 13] is a set of p paths with minimum distance among such sets. Is there a function $f(p)$ such that a p-center (or p-median or p-core) must always be contained in a $\mathcal{T}_D$-center (or $\mathcal{T}_D$-median) for $D \geq f(p)$, and if so then can the disconnected substructure be found quickly within a given central subtree? Lenzen, Pelsmajer, and Pierce [32] have carried this out for p-centers and $\mathcal{T}_D$-centers, which yields a relatively simple linear time algorithm for finding a p-center of a tree G, when p is a constant. (Fredrickson [28] already gave a linear time algorithm for finding the p-center of a tree.) The same strategy has also yielded a linear time algorithm (for fixed p) that finds a *p-path center* in a tree [32]: a forest consisting of at most p disjoint paths, with smallest maximum distance. One might hope for analogous results with $\mathcal{T}_D$-medians, p-medians, and p-cores. (See [13] for a quick survey on the best-known algorithms for finding a p-median and p-core.) It seems to be harder in general to work with disconnected central substructures of trees than with the connected varieties, and the strategy of approaching them via the definitions in this paper has not been fully explored.

Acknowledgements

We thank Buck McMorris for organizing the 2003 DIMACS Reconnect Workshop at Illinois Institute of Technology that initiated this study, and for his suggestion to investigate caterpillar centers, which eventually led us to hereditary classes of trees. And, we thank Martyn Mulder for participating in many of our discussions of these topics.

References

1. C. Jordan, Sur les assemblages de lignes. *J. Reine Agnew. Math.* **70** (1869) 185–190.
2. K.B. Reid, *Centrality Measures in Trees*, this volume.
3. O. Ore, *Theory of Graphs*, American Mathematical Society, Colloquium Publications XXXVIII, Providence, R.I., 1962.
4. P.J. Slater, Maximum facility location, *J. Res. Natl. Bur. Stand B* **79** (1975) 107–115.
5. P.J. Slater, Accretion centers: a generaliztion of branch weight centroids, *Discrete Appl. Math.* **3** (1981) 187–192.
6. P.J. Slater, Locating central paths in a graph, *Transportation Sci.* **16** (1982) 1–18.
7. F.R. McMorris, K.B. Reid, Central k-trees in Trees, *Congress. Numer.* **124**, 139–143, (1997).
8. E. Minieka, The optimal location of a path or tree in a tree network, *Networks* **15**, 309–321, (1985).
9. F.R. McMorris, Personal Communication, 2003.
10. S. Peng, A.B. Stephens, Y. Yesha, Algorithms for a core and k-tree core of a tree, *J. Algorithms* **15** (1993) 143–159.
11. A. Shioura and T. Uno, A linear time algorithm for finding a k-tree core, *J. Algorithms* **23** (1997) 281–290.
12. S. Srivastava , R.K. Ghosh, Distributed algorithms for finding and maintaining a k-tree core in a dynamic network, *Inform. Process. Lett.* **88** (2003) 187–194.

13. B.-F. Wang, Finding a 2-core of a tree in linear time, *SIAM J. Discrete Math.* **15** (2002) 193–210.
14. R. I. Becker, Y.I. Chang, I. Lari, A. Scozzari, G. Storchi, Finding the l-core of a tree, *Third ALIO-EURO Meeting on Applied Combinatorial Optimization (Erice, 1999), Discrete Appl. Math.* **118** (2002) 25–42.
15. B. Mirchandani and R.L. Francis (Eds.), *Discrete Location Theory*, Wiley Interscience, New York, 1990.
16. B. Bhattacharya, Y. Hu, Q. Shi, A. Tamir, Optimal algorithms for the path/tree shaped facility location problems in trees, *Proc. ISAAC 2006, LNCS* **4288** (2006) 379–388.
17. N. Graham, R.C. Entringer, A. Székely, New tricks for old trees: Maps and the pigeonhole principle, *Amer. Math. Monthly* **101** (1994) 664–667.
18. S.M. Hedetniemi, E.J. Cockayne, S.T. Hedetniemi, Linear Algorithms for Finding the Jordan Center and Path Center of a Tree, *Trans. Sci.* **15** (1981) 98–114.
19. A.J. Goldman, Optimal center location in simple networks, *Trans. Sci.* **5** (1971) 212–221.
20. A.J. Goldman, Minimax location of a facility on a network, *Trans. Sci.* **6** (1972) 407–418.
21. G.Y. Handler, Minimax Location of a Facility in an Undirected Tree Graph, *Trans. Sci.* **7** (1973) 287–293.
22. B. Zelinka, Medians and Peripherians of Trees, *Arch. Math. (Brno)* **4** (1968) 87–95.
23. K.S. Holbert, *A note on graphs with distant center and median*, In: ed. V. R. Kulli, *Recent Studies in Graph Theory*, Vishna, Gulbarza, India, 1989, pp. 155–158.
24. C.A. Morgan, P.J. Slater, A Linear Algorithm for the Core of a Tree, *J. Algorithms* **1** (1980) 247–258.
25. P. J. Slater, Centrality of paths and vertices in a graph: cores and pits. In: *The theory and applications of graphs (Kalamazoo, Mich., 1980)*, Wiley, New York, 1981, pp. 529–542.
26. M. Lenzen, manuscript, 2006.
27. S.L. Hakimi, Optimum locations of switching centers and absolute centers in a communication network and some related graph-theoretic problems, *Operations Res.* **13** (1965) 462–475.
28. G.N. Frederickson, Parametric Search and Locating Supply Centers in Trees, *(Algorithms and Data Structures, 2nd Workshop, WADS '91, Ottawa, Canada, August 1991), LNCS* **519**, Springer-Verlag, pp. 299–319.
29. S.L. Hakimi, Optimum locations of switching centers and absolute centers and medians of a graph, *Operations Res.* **12** (1964) 450–459.
30. S.L. Hakimi, O. Kariv, An algorithmic approach to network location problems. II: The p-medians, *Siam. J. Appl. Math.* **37** (3) (1979) 539–560.
31. R.I. Becker, Y. Perl, Finding the two-core of a tree, *Discrete Appl. Math.* **11** (2) (1985) 103–113.
32. M. Lenzen, M.J. Pelsmajer, J.J. Pierce, manuscript, 2006.

Chapter 7

Consensus Centered at Majority Rule

Robert C. Powers
Department of Mathematics, University of Louisville
Louisville, KY 40292 USA
rcpowe01@louisville.edu

Consider an election where there are two alternatives and $n \geq 3$ voters. For any integer q such that $\frac{n}{2} < q \leq n+1$, the winner of the election with respect to absolute q-majority rule is the alternative that receives at least q votes. In this chapter, various generalizations of absolute q-majority rule are presented from an axiomatic point of view.

Introduction

One of the most influential results in the theory of social choice is Kenneth May's characterization of simple majority rule [13]. May models the voting situation where there are two alternatives and a fixed number of voters. Each individual votes for one of the alternatives or abstains and the winner of the election under simple majority rule is the alternative with the most votes. If both alternatives receive the same number of votes, then simple majority rule declares a tie. Notice that this version of majority is based on a *relative* notion of majority where it is possible for the winner to receive less than half of the total number of votes. For example, out of a voting population of 100 individuals, alternative x with two votes would beat alternative y with one vote where 97 voters decided to abstain. If this is bothersome, then it may seem more reasonable to require that the winner of an election receive more than half the total number of votes. This version of majority rule is based on an *absolute* notion of majority and it plays a central role in this chapter.

A well cited paper in the area of systematic biology is Margush and McMorris' paper on the majority consensus rule for hierarchies [11]. A *hierarchy* on a finite set S is a collection H of nonempty subsets of S such that $\{\{x\}, S : x \in S\} \subseteq H$ and $A \cap B \in \{A, B, \emptyset\}$ for all $A, B \in H$. A hierarchy arises when a hierarchical clustering method is applied to the data set S. Suppose several hierarchical clustering methods are applied to S, then we get several hierarchies on S. A consensus method would take these hierarchies as input and would output a single consensus hierarchy. In particular, the output for majority rule consists of all the subsets of S that appear in more than half of the input hierarchies. A simple generalization of this idea is to output subsets of S which appear in more than $t > \frac{1}{2}$ of the input heirarchies. This gives the class of *counting rules.* A nice characterization of the counting rules, similar in spirit to May's characterization of simple majority rule, was given by McMorris and Neumann in [15].

McMorris and Neumann's characterization of the counting consensus rules for hierarchies is a consequence of a very general result proved by Bernard Monjardet in [12]. Monjardet gives an order theoretic model of consensus where join irreducible elements are the basic units of information and the output of a consensus function is an element of a meet semilattice. Details on the Monjardet model of consensus are given in the next section.

For background reading on consensus the reader is referred to Day and McMorris [8].

7.1. The Monjardet Model of Consensus

A good portion of the notation and terminology given in this section follows from Monjardet's paper entitled *Arrowian Characterizations of Latticial Federation Consensus Functions* [12].

Let X be a finite partially ordered set (*poset*), i.e., X is equipped with a reflexive, antisymmetric, transemilatticetion $\leq$. The *join* of a subset A of X, denoted by $\bigvee A$, is the least upper bound of A when it exists. Dually, the *meet* of A is denoted by $\bigwedge A$ and it is the greatest lower bound of A when it exists. If $A = \{x_1, x_2, \ldots, x_n\}$, then $\bigvee A$ and $\bigwedge A$ are written as $x_1 \vee x_2 \vee \ldots \vee x_n$ and $x_1 \wedge x_2 \wedge \ldots \wedge x_n$, respectively. The poset X is a *meet semilattice* if $x \wedge y$ exists for all $x, y \in X$. In this case, $\bigwedge X = \bigvee \emptyset$ is the least element of X and is denoted by 0. Dually, the poset X is a *join semilattice* if $x \vee y$ exists for all $x, y \in X$ and $\bigvee X = \bigwedge \emptyset$ is the greatest element of X and it is denoted by 1. The poset X is a *lattice* if it is both a meet and join semilattice. A lattice X is *distributive* if, for all $x, y, z \in X$, $x \wedge (y \vee z) = (x \wedge y) \vee (x \wedge z)$ and $x \vee (y \wedge z) = (x \vee y) \wedge (x \vee z)$. More generally, a meet semilattice X is *distributive* if, for all x in X, the set $\{y \in X | y \leq x\}$ is a distributive lattice.

A meet semilattice X satisfies the *join-Helly property* if, for all $x, y, z \in X$, $x \vee y$, $x \vee z$, and $y \vee z$ exist, then $x \vee y \vee z$ exists. By an induction argument, for any subset A of X, if $x \vee y$ exists for all $x, y \in A$, then $\bigvee A$ exists. A meet semilattice X is a *median semilattice* if it is distributive and satisfies the join-Helly property. For the remainder of this section, X is assumed to be a finite median semilattice containing at least three elements. A simple example is shown below where $X = \{0, x_{-1}, x_1\}$ and $x_{-1} \vee x_1$ does not exist.

An element s in X is *join irreducible* if $s = \bigvee A$ implies $s \in A$. In other words, a join irreducible element is not equal to the join of the elements strictly below it. Let J be the set of all join irreducible elements of X. Notice that $0 \notin J$ and $x = \bigvee\{s \in J | s \leq x\}$ for all $x \in X$. Since X is a distributive semilattice it can be

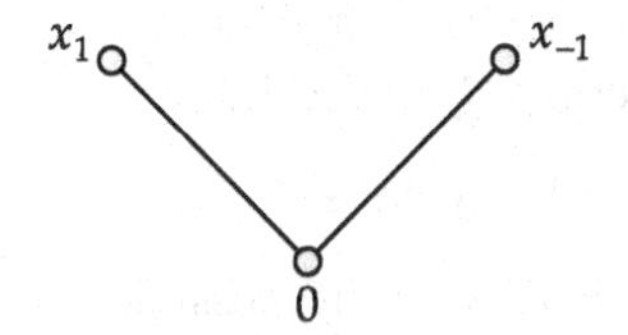

shown that for any $s \in J$ and for any nonempty subset A of X, $s \leq \bigvee A$ implies that there exists $a \in A$ such that $s \leq a$.

The notation $N = \{1, 2, \ldots n\}$ with $n \geq 2$ represents the set of individuals or voters. A function of the form $C : X^n \to X$ is called a *consensus function.* An element $\pi = (x_1, \ldots, x_n) \in X^n$ is called a *profile.* For any profile π and any element s in X, let

$$N_s(\pi) = \{i \in N : s \leq x_i\}.$$

For any integer q such that $\frac{n}{2} < q \leq n + 1$ and for any $\pi \in X^n$, we define the consensus rule C_q on X by

$$C_q(\pi) = \bigvee \{s \in J : |N_s(\pi)| \geq q\}.$$

In particular, if q^* is the least integer strictly greater than $\frac{n}{2}$, then we get the *majority consensus rule* C_{q^*}.

An attractive feature of median semilattices is the close connection between the majority consensus rule C_{q^*} and the median function M. We now briefly explain this connection and refer the reader to [16] for more details.

Let X be a finite median semilattice and let x and y be distinct elements of X. Then y *covers* x if $x \leq y$ and for all $z \in X$ such that $x \leq z \leq y$, either $z = x$ or $z = y$. So if y covers x, then there is no element of X that lies strictly between x and y. The *covering graph* G of X has the elements of X as vertices and xy is an edge of G if and only if either y covers x or x covers y. (The covering graph of a median semilattice is a well known type of graph called a *median graph.*) Let d be the minimum path length metric on the covering graph of X. The *median function* on X is the function $M : \bigcup_{k \geq 1} X^k \to 2^X \setminus \{\emptyset\}$ where, for any $(x_1, \ldots, x_k) \in X^k$, $M((x_1, \ldots, x_k))$ is the set of all elements $x \in X$ such that

$$\sum_{i=1}^{k} d(x, x_i) \leq \sum_{i=1}^{k} d(y, x_i)$$

for all $y \in X$. It turns out that $C_{q^*}(\pi) \in M(\pi)$ for any profile π. Moreover, if n is odd, then $M(\pi) = \{C_{q^*}(\pi)\}$. Finally, if n is even and $\pi = (x_1, \ldots, x_n)$, then $M((x_1, \ldots, x_n, 0)) = \{C_{q^*}(\pi)\}$.

A subset $\mathcal{F}$ of the power set 2^N is called a *federation* if for all $I \in \mathcal{F}$ and for all $J \subseteq N$, $I \subseteq J$ implies that $J \in \mathcal{F}$. A subset $\mathcal{F}$ of the power set 2^N is called *transversal* if for all $I, J \in \mathcal{F}$, $I \cap J \neq \emptyset$. For any transversal federation $\mathcal{F}$ of 2^N, we define the consensus function $C_\mathcal{F}$ on X by

$$C_\mathcal{F}(\pi) = \bigvee\{s \in J : N_s(\pi) \in \mathcal{F}\}.$$

We will call $C_\mathcal{F}$ a *federation rule.* It is possible to have $\mathcal{F} = \emptyset$ in which case $C_\mathcal{F}(\pi) = \bigvee \emptyset = 0$ for all π.

Let $C : X^n \to X$ be a consensus function. We will say that C satisfies

DN: Decisive Neutrality if for all $s, s\prime \in J$ and for all profiles $\pi, \pi' \in X^n$,

$$N_s(\pi) = N_{s'}(\pi'), s \leq C(\pi) \;\Rightarrow\; s' \leq C(\pi').$$

The statement defining DN can also be written as

$$s \leq C(\pi) \;\Leftrightarrow\; s' \leq C(\pi') \text{ whenever } N_s(\pi) = N_{s'}(\pi').$$

The intent behind decisive neutrality is that all the join irreducible elements should be treated equally when determining the consensus output.

We will say that C satisfies

DM: Decisive Monotonicity if for all $s \in J$ and for all profiles $\pi, \pi' \in X^n$,

$$N_s(\pi) \subseteq N_s(\pi'), s \leq C(\pi) \;\Rightarrow\; s \leq C(\pi').$$

Decisive monotonicity implies that if a consensus output is above a join irreducible s and one or more individuals change their vote favorable to s, then the updated consensus output should still be above s.

A consensus function C satisfies

MN: Monotonic Neutrality if for all $s, s\prime \in J$ and for all profiles $\pi, \pi' \in X^n$,

$$N_s(\pi) \subseteq N_{s'}(\pi'), s \leq C(\pi) \;\Rightarrow\; s' \leq C(\pi').$$

It is not hard to show that C satisfies DN and DM if and only if C satisfies MN.

Let $C : X^n \to X$ be a consensus function. For any $s \in J$, we will say that a subset A of N is *s-decisive* if there exists a profile π such that $N_s(\pi) = A$ and $s \leq C(\pi)$. Let $\mathcal{F}_s$ be the set of all s-decisive sets. If C satisfies decisive neutrality, then

$$s \leq C(\pi) \;\Leftrightarrow\; N_s(\pi) \in \mathcal{F}_s$$

for any $s \in J$ and $\pi \in X^n$. Moreover, decisive neutrality implies that $\mathcal{F}_s = \mathcal{F}_{s'}$ for all $s, s' \in J$. In this case, we let $\mathcal{F} = \mathcal{F}_s$ and call the elements of $\mathcal{F}$ *decisive sets.* Observe that

$$C(\pi) = \bigvee\{s \in J : N_s(\pi) \in \mathcal{F}\}.$$

Finally, notice that the family of decisive sets $\mathcal{F}$ is a federation if and only if C satisfies decisive monotonicity.

We will continue to assume that $C : X^n \to X$ satisfies DN and DM. Suppose there exists $x, y \in X$ such that $x \vee y$ does not exist in X. Then the family of decisive sets has the property: $A \in \mathcal{F} \Rightarrow A^c \notin \mathcal{F}$ where A^c is the complement of A with respect to the set N. To see this, assume $A, A^c \in \mathcal{F}$. Define $\pi = (x_1, \ldots, x_n)$ by $x_i = x$ for all $i \in A$ and $x_i = y$ for all $i \in A^c$. Then $\pi \in X^n$ and for every join irreducible $s \leq x$, $N_s(\pi) \supseteq A$. Since $\mathcal{F}$ is a federation it follows that $N_s(\pi) \in \mathcal{F}$ and so $s \leq C(\pi)$. Since $s \leq x$ was arbitrary we get that $x \leq C(\pi)$. A similar argument shows that $y \leq C(\pi)$. Since $x \vee y$ does not exist in X, $C(\pi)$ cannot be an upper bound for x and y. This contradiction shows that $A \in \mathcal{F} \Rightarrow A^c \notin \mathcal{F}$. Since $\mathcal{F}$ is a federation it follows that $A \cap B \neq \emptyset$ for all $A, B \in \mathcal{F}$. This means that $\mathcal{F}$ is a transversal federation.

It is straightforward to verify that the consensus function $C_{\mathcal{F}}$, where $\mathcal{F}$ is a transversal federation, satisfies decisive neutrality and decisive monotonicity.

The preceding remarks lead to the following characterization of federation rules (see Proposition 2.4 in [12]).

Theorem 7.1. *(Monjardet 1990) Assume X is a finite median semilattice such that X is not a lattice. A consensus function $C : X^n \to X$ satisfies DN and DM if and only if there exists a transversal federation $\mathcal{F}$ of 2^N such that $C = C_{\mathcal{F}}$.*

To see Theorem 7.1 in action, let $i \in N$ and let $\mathcal{F} = \{A \subseteq N : i \in A\}$. Then $C(\pi) = x_i$ for all $\pi = (x_1, \ldots, x_n) \in X^n$. In this case, C is called a *dictatorial consensus function.* To avoid such functions, we introduce another axiom. A consensus function C on X satisfies **A: Anonymity** if for any profile $\pi = (x_1, \ldots, x_n) \in X^n$ and for any permutation σ of N,

$$C(x_1, \ldots, x_n) = C(x_{\sigma(1)}, \ldots, x_{\sigma(n)}).$$

If $\mathcal{F}$ is a transversal federation and $C_{\mathcal{F}}$ satisfies anonymity, then $B \in \mathcal{F} \leftrightarrow |B| \geq q$ where $q = \min\{|A| : A \in \mathcal{F}\}$. The fact that $\mathcal{F}$ is transversal implies that $q > \frac{n}{2}$.

The next result follows from the previous remark and Theorem 7.1.

Corollary 7.1. *Assume X is a finite median semilattice such that X is not a lattice. A consensus function $C : X^n \to X$ satisfies DN, DM, and A if and only if there exists an integer q such that $\frac{n}{2} < q \leq n+1$ and $C = C_q$.*

If L is a lattice with largest element 1, then the constant function $C(\pi) = 1$ for all π satisfies DN, DM, and A. So the assumption that X is not a lattice is a necessary assumption in Corollary 7.1.

It is possible to weaken DN in the statement of Corollary 7.1 and still conclude that $C = C_q$. We will say that a consensus function C on X satisfies

0DN: Zero Decisive Neutrality if for all $s, s' \in J$ and for all profiles $\pi, \pi' \in X^n$,

$$N_0(\pi) = N_0(\pi'), N_s(\pi) = N_{s'}(\pi'), \text{and } s \leq C(\pi) \;\Rightarrow\; s' \leq C(\pi').$$

The idea behind 0DN is to apply decisive neutrality under the restriction that $x_i = 0$ in profile π exactly when $x'_i = 0$ in profile π'.

We now come to an improvement of Corollary 7.1.

Theorem 7.2. *Assume X is a finite median semilattice such that X is not a lattice. A consensus function $C : X^n \to X$ satisfies 0DN, DM, and A if and only if there exists an integer q such that $\frac{n}{2} < q \leq n+1$ and $C = C_q$.*

Proof. Assume that $C : X^n \to X$ satisfies 0DN and DM. We will show that C satisfies DN.

Suppose $N_s(\pi) = N_{s'}(\pi')$ where $s, s' \in J$ and $\pi, \pi' \in X^n$. Choose a profile π'' such that $N_0(\pi'') = N_0(\pi')$ and $N_s(\pi'') = N_{s'}(\pi')$. By 0DN, $s \leq C(\pi'') \Leftrightarrow s' \leq C(\pi')$. Since $N_s(\pi'') = N_s(\pi)$ it follows from DM that $s \leq C(\pi'') \Leftrightarrow s \leq C(\pi)$. Thus, $s \leq C(\pi) \Leftrightarrow s' \leq C(\pi')$ and so C satisfies DN.

Our result now follows from Corollary 7.1. □

Let $\mathcal{H}(S)$ be the set of all hierarchies on the set S and assume $|S| \geq 4$. Then $\mathcal{H}(S)$ is a median semilattice with set containment as the partial order and set intersection as the meet operation. The *zero* element of $\mathcal{H}(S)$ is the trivial hierarchy $H_\emptyset = \{\{x\}, S : x \in S\}$. The set of join irreducible elements of $\mathcal{H}(S)$ is given by

$$J = \{H_\emptyset \cup \{A\} : A \subseteq S, A \neq \emptyset, \text{and } A \neq S\}.$$

See Figure 5.4 in [8] for a picture of the median semilattice $\mathcal{H}(S)$ with $|S| = 4$.

Since the median semilattice $\mathcal{H}(S)$ is not a lattice we have the following consequence of Theorem 7.2.

Corollary 7.2. *A consensus function $C : \mathcal{H}(S)^n \to \mathcal{H}(S)$ satisfies 0DN, DM, and A if and only if there exists an integer q such that $\frac{n}{2} < q \leq n+1$ and $C = C_q$.*

The previous result is the McMorris and Neumann characterization of the counting consensus rules for hierarchies with decisive neutrality replaced by zero decisive neutrality.

In the next section we will apply Theorem 7.2 to the case of voting and obtain a characterization of absolute majority voting rules.

7.2. Absolute Majority Rules

In this section we consider the class of voting rules where there are two alternatives and the winner of an election receives more than half of the total number of votes. To distinguish this voting situation with the Monjardet model of consensus, we will use some notation and terminology from [2].

As above, the set of individuals or voters is given by $N = \{1, 2, \ldots n\}$ with $n \geq 2$. Each individual votes for one out of two alternatives or they abstain. The two alternatives can be identified with 1 and -1. The abstention vote is denoted by 0. In this setting, a function of the form

$$F : \{-1,0,1\}^n \to \{-1,0,1\}$$

is called an *aggregation rule* and an n-tuple $R = (R_1,\ldots,R_n) \in \{-1,0,1\}^n$ in the domain of F is called a *profile*. The output $F(R) = 0$ represents a tie, i.e., neither alternative is chosen. For any profile $R = (R_1,\ldots,R_n)$, we let

$$n_+(R) = |\{i \in N : R_i = 1\}| \text{ and } n_-(R) = |\{i \in N : R_i = -1\}|$$

$n_0(R) = |\{i \in N : R_i = 0\}|$. So $n_+(R)$ is the number of individuals who voted for 1, $n_-(R)$ is the number of individuals who voted for -1, and $n_0(R)$ is the number of individuals who abstained.

Using the notation from the previous paragraph, we can now give a key definition.

Definition 7.1. An aggregation rule $F : \{-1,0,1\}^n \to \{-1,0,1\}$ is called *absolute q-majority rule* if there exists an integer q such that $\frac{n}{2} < q \leq n+1$ such that $F = F_q$ where

$$F_q(R) = 1 \text{ if } n_+(R) \geq q \text{ and } F_q(R) = -1 \text{ if } n_-(R) \geq q$$

for any $R \in \{-1,0,1\}^n$. We will say that F is an *absolute majority rule* if it is an absolute q-majority rule for some integer q such that $\frac{n}{2} < q \leq n+1$.

Observe that $F_{n+1}(R) = 0$ for all R since $\max\{n_+(R), n_-(R)\} \leq n$ for all R. So the rule F_{n+1} always declares a tie. Even though the trivial rule F_{n+1} is technically an absolute majority rule, the more interesting and important case is when $q = q^*$ where q^* is the least integer strictly greater than $\frac{n}{2}$.

Let $F : \{-1,0,1\}^n \to \{-1,0,1\}$ be a fixed aggregation rule. Then F may or may not satisfy the following conditions.

Anonymity (A) Given any $R \in \{-1,0,1\}^n$ and any permutation $\Pi : N \to N$, we have $F(R_1,\ldots,R_n) = F(R_{\Pi(1)},\ldots,R_{\Pi(n)})$.

Neutrality (N) $F(-R) = -F(R)$ for all $R \in \{-1,0,1\}^n$.

Anonymity implies that the identities of individual voters are not used in determining the social outcome and neutrality is the requirement that both alternatives should be treated equally. There are many different types of aggregation rules satisfying anonymity and neutrality (see [21]).

Asan and Sanver [2] characterized absolute majority aggregation rules using (A), (N), and the following monotonicity condition.

Maskin Monotonicity (MM) For any $R, R' \in \{-1,0,1\}^n$ such that $R_i \geq 0 \Rightarrow R'_i \geq 0$ for each $i \in N$, we have $F(R) \geq 0 \Rightarrow F(R') \geq 0$. Similarly, for any

$R, R' \in \{-1,0,1\}^n$ such that $R_i \leq 0 \Rightarrow R'_i \leq 0$ for each $i \in N$, we have $F(R) \leq 0 \Rightarrow F(R') \leq 0$.

It is not hard to show that (MM) is equivalent to the following statement.

(MM) For any $R, R' \in \{-1,0,1\}^n$ and for any $s \in \{-1,1\}$, if $\{i : R_i = s\} \subseteq \{i : R'_i = s\}$, then $F(R) = s \Rightarrow F(R') = s$.

We now show how to derive Asan and Sanver's result as a corollary of Theorem 7.2. The set $\{-1,0,1\}$ can be thought of as a median semilattice $X = \{-1,0,1\}$ where the number 0 is the smallest element of X, the set of join irreducibles is $J = \{-1,1\}$, and $1 \wedge -1 = 0$. This is a bit confusing since the ordering on X does not correspond to the usual ordering of the integers.

Using the fact that $X = \{-1,0,1\}$ is a median semilattice we have the following interesting observation.

Proposition 7.1. *The aggregation rule $F : X^n \to X$ satisfies (N) if and only if F satisfies 0DN.*

Proof. Keeping in mind that $J = \{-1,1\}$, observe that for any $s \in J$ and for any $\pi \in X^n$, $s \leq F(\pi) \Leftrightarrow F(\pi) = s$.

Suppose F satisfies 0DN and $\pi \in X^n$. Let $\pi' = -\pi$. Then $N_1(\pi) = N_{-1}(\pi')$ and $N_0(\pi) = N_0(\pi')$. By 0DN, $F(\pi) = 1 \Leftrightarrow F(\pi') = -1$. A similar argument shows that $F(\pi) = -1 \Leftrightarrow F(\pi') = 1$. Hence $F(-\pi) = -F(\pi)$ for all π.

Conversely, assume F satisfies (N). Suppose $N_0(\pi) = N_0(\pi')$ and $N_s(\pi) = N_{s'}(\pi')$ where $s, s' \in J$ and $\pi, \pi' \in X^n$. Then either $s = s'$ and $\pi = \pi'$ or $s = -s'$ and $\pi = -\pi'$. In either case, it follows that $s \leq F(\pi) \Leftrightarrow s' \leq F(\pi')$. $\square$

The previous proposition is not true if we replace 0DN with DN. For example, define $F : \{-1,0,1\}^n \to \{-1,0,1\}$ as follows: $F(\pi) = F_{q^*}(\pi)$ for any profile π such that $n_0(\pi) = 0$ and $F(\pi) = 0$ for any profile π such that $n_0(\pi) \neq 0$. Then F does not satisfies DN but it does satisfy (N).

Observe that for any $R \in \{-1,0,1\}^n$ and for any $s \in \{-1,1\}$, $N_s(R) = \{i : R_i = s\}$. Therefore, DM is equivalent to (MM). Since A and (A) are easily seen to be identical we now have the following consequence of Theorem 7.2.

Theorem 7.3. *(Asan and Sanver 2006) An aggregation rule F satisfies (MM), (N), and (A) if and only if $F = F_q$ where q is an integer satisfying the inequality $\frac{n}{2} < q \leq n+1$.*

We should point out that there are other characterizations of absolute majority aggregation rules. For example, see [3] and [10].

What happens if we drop one of the conditions in Theorem 7.3? For example, if (MM) is dropped, then we are lead to the class of aggregation rules satisfying (A) and (N). One of the most important aggregation rules in this class is the following.

Definition 7.2. The aggregation rule $F_s : \{-1,0,1\}^n \to \{-1,0,1\}$ is called *simple majority rule* if

$$F_s(R) = 1 \text{ iff } n_+(R) > n_-(R) \text{ and } F_s(R) = -1 \text{ iff } n_-(R) > n_+(R)$$

for all $R \in \{-1,0,1\}^n$.

Observe that F_s requires only a relative majority to determine the winning alternative. Consequently, if many voters abstain, then it is possible for the winner to receive less than half of the total number votes.

Kenneth May in 1952 proved a theorem characterizing simple majority rule using (A), (N), and the following condition.

Positive Responsiveness (PR) An aggregation function F satisfies *positive responsiveness* if $R \leq R'$ implies $F(R) \leq F(R')$ for all $R, R' \in \{-1,0,1\}^n$ and, for all $R > R'$ in $\{-1,0,1\}^n$, $F(R') = 0$ implies $F(R) = 1$ and $F(R) = 0$ implies $F(R') = -1$.

Theorem 7.4. *(May 1952) An aggregation rule $F : \{-1,0,1\}^n \to \{-1,0,1\}$ satisfies (PR), (N), and (A) if and only if $F = F_S$.*

As mentioned in the introduction to this chapter, May's theorem is a fundamental result in the area of social choice and it has inspired many extensions. See [2], [4], [5], [7], [12], [18], [19], [28], [29], and [30] for a sample of these results.

Returning to Theorem 7.3, if we drop (A), then we get the class of aggregation rules satisfying (MM) and (N).

Theorem 7.5. *An aggregation rule $F : \{-1,0,1\}^n \to \{-1,0,1\}$ satisfies (MM) and (N) if and only if there exists a transversal federation $\mathcal{F}$ such that for any $R \in \{-1,0,1\}^n$ and for any $s \in \{-1,1\}$, $F(R) = s$ if and only if $N_s(R) \in \mathcal{F}$.*

If we drop (N), then we get the class of aggregation rules satisfying (A) and (MM).

Theorem 7.6. *An aggregation rule $F : \{-1,0,1\}^n \to \{-1,0,1\}$ satisfies (MM) and (A) if and only if there exists nonnegative integers q and ℓ such that $q + \ell \geq n + 1$ and for any $R \in \{-1,0,1\}^n$,*

$$F(R) = 1 \text{ iff } n_+(R) \geq q \text{ and } F(R) = -1 \text{ iff } n_-(R) \geq \ell.$$

We now explain why the previous theorems are true. Suppose $F : \{-1,0,1\}^n \to \{-1,0,1\}$ satisfies (MM). Then, as a consensus rule, F satisfies decisive monotonicity. Therefore, by Propositions 1.4 and 2.3(1) in [12], for each $s \in \{-1,0,1\}$ there exists a federation $\mathcal{F}_s$ such that $F(R) = s$ if and only if $N_s(R) \in \mathcal{F}_s$ for any profile R. Moreover, since the median semilattice $X = \{-1,0,1\}$ is not a lattice, $A \cap B \neq \emptyset$ for all $A \in \mathcal{F}_1$ and all $B \in \mathcal{F}_{-1}$. Since $N_1(R) = N_{-1}(-R)$ for any profile R it follows that F satisfies (N) if and only if $\mathcal{F}_1 = \mathcal{F}_{-1}$. In this case, if we let $\mathcal{F} = \mathcal{F}_1 = \mathcal{F}_{-1}$, then we get Theorem 7.5. On the other hand, if $q = \min\{|N_1(R)| : R \in \{-1,0,1\}$ and $F(R) = 1\}$ and $\ell = \min\{|N_{-1}(R)| : R \in \{-1,0,1\}$ and $F(R) = -1\}$, then F satisfies anonymity exactly when the following conditions hold: $A \in \mathcal{F}_1$ if and only if $|A| \geq q$ and $B \in \mathcal{F}_{-1}$ if and only if $|B| \geq \ell$. Since the median semilattice $X = \{-1,0,1\}$ is not a lattice, we must have $q+\ell > n$. Thus, Theorem 7.6 is true.

Instead of dropping one of (MM), (N), and (A) in Theorem 7.3, we consider cases where one of the axioms is weakened. To illustrate this point, we start with the following weak version of anonymity.

Partial Anonymity (PA) For any $R \in \{-1,0,1\}^n$ and any permutation $\Pi : N \to N$ satisfying $\Pi(1) = 1$, $F(R_1,\ldots,R_n) = F(R_{\Pi(1)},\ldots,R_{\Pi(n)})$.

The idea behind partial anonymity is to incorporate some degree of anonymity and still allow voter 1 not to be anonymous. To make this axiom nontrivial we will assume that $n \geq 3$. The proof of the next result is given in [22].

Theorem 7.7. *If $n \geq 3$ and $F : \{-1,0,1\}^n \to \{-1,0,1\}$ satisfies (MM), (N), and (PA), then there exist nonnegative integers q_0 and q satisfying the inequalities $q_0 + q \geq n$ and $\max\{\frac{n-1}{2}, q-1\} < q_0 \leq n$ such that*

$$F(R) = 1 \Leftrightarrow n_+(R) \geq q_0 + \frac{R_1(R_1+1)}{2}(q+1-q_0)$$

for all $R = (R_1,\ldots,R_n) \in \{-1,0,1\}^n$. Conversely, if q_0 and q are nonnegative integers satisfying the above inequalities and $F : \{-1,0,1\}^n \to \{-1,0,1\}$ is defined by

$$F(R) = 1 \Leftrightarrow n_+(R) \geq q_0 + \frac{R_1(R_1+1)}{2}(q+1-q_0)$$

and

$$F(R) = -1 \Leftrightarrow n_-(R) \geq q_0 + \frac{R_1(R_1-1)}{2}(q+1-q_0)$$

for all $R = (R_1,\ldots,R_n) \in \{-1,0,1\}^n$, then F satisfies (MM), (N), and (PA).

It should be pointed out that the bound $q_0 + \frac{R_1(R_1+1)}{2}(q+1-q_0)$ for $n_+(R)$ shows a close connection to absolute qualified majority rules. In fact, the previous result incorporates absolute qualified majority rules when $q_0 = q+1$. On the other hand, if $q_0 = n$ and $q = 0$, then we get the dictatorship $F(R) = R_1$ for all $R = (R_1,\ldots,R_n) \in \{-1,0,1\}^n$

The next step is to weaken neutrality. We will say that an aggregation rule F is *balanced* if

$$|\{R \in \{-1,0,1\}^n : F(R) = 1\}| = |\{R \in \{-1,0,1\}^n : F(R) = -1\}|.$$

If f satisfies neutrality, then it is clear that f is balanced. The converse is not true. Let Y and Z be two disjoint subsets of $\{-1,0,1\}^n \setminus \{(0,\ldots,0)\}$ such that

$\{(1,0,\ldots,0),(-1,0,\ldots,0)\} \subset Y$ and $|Y| = |Z| = \frac{3^n-1}{2}$. Notice that $X \cup Y = \{-1,0,1\}^n \setminus \{(0,\ldots,0)\}$. Define an aggregation rule G by $G(R) = 1$ if and only if $R \in Y$ and $G(R) = -1$ if and only if $R \in Z$. So $G(R) = 0$ if and only if $R = (0,\ldots,0)$. Notice that G is balanced and that G does not satisfy neutrality.

Even though balanced is less restrictive than neutrality it is still strong enough to give the following result.

Theorem 7.8. *An aggregation rule $F : \{-1,0,1\}^n \to \{-1,0,1\}$ is balanced and satisfies (MM) and (A) if and only if F is a q majority rule where the integer q satisfies the inequality $\frac{n}{2} < q \leq n+1$.*

Proof. Assume $F : \{-1,0,1\}^n \to \{-1,0,1\}$ is balanced, satisfies (MM) and (A). By Theorem 7.6, there exists nonnegative integers q and ℓ such that $q+\ell \geq n+1$ and for any $R \in \{-1,0,1\}^n$,

$$F(R) = 1 \text{ iff } n_+(R) \geq q \text{ and } F(R) = -1 \text{ iff } n_-(R) \geq \ell.$$

Assume without loss of generality that $q \leq \ell$ and note that $\ell > \frac{n}{2}$. Let $Y = \{R : n_-(R) \geq \ell\}$ and let $Z = \{R : n_+(R) \geq q\}$. Observe that $R \in Y$ implies that $-R \in Z$. Thus the mapping $\beta : Y \to Z$ defined by $\beta(R) = -R$ for all $R \in Y$ is well defined. The balanced condition implies that $|Y| = |Z|$. It follows that $q = \ell$ and we're done. □

Finally, consider a less restrictive version of Maskin Monotonicity. Here is a simple example.

Zero Maskin Monotonicity (0MM) For any $R, R' \in \{-1,0,1\}^n$ and for any $s \in \{-1,1\}$, if $N_0(R) = N_0(R')$, $\{i : R_i = s\} \subseteq \{i : R'_i = s\}$, and $F(R) = s$, then $F(R') = s$.

Suppose $F : \{-1,0,1\}^n \to \{-1,0,1\}$ is an aggregation rule satisfying (0MM), (A), and (N). For each integer $k \in \{0,\ldots,n-1\}$, let

$$q_k = \min\{n_+(R) : R \in \{-1,0,1\}^n, F(R) = 1, \text{and } n_0(R) = k\}$$

with the convention that $q_k = n-k+1$ if $\{n_+(R) : R \in \{-1,0,1\}^n, F(R) = 1, \text{and } n_0(R) = k\} = \emptyset$. It follows from (0MM) and (A) that $F(R) = 1$ if and only if $n_+(R) \geq q_k$ where $k = n_0(R)$. Therefore, $\frac{n-k}{2} < q_k \leq n-k+1$. It follows from neutrality that $F(R) = -1$ if and only if $n_-(R) \geq q_k$ where $k = n_0(R)$. If R is the profile where $n_0(R) = n$, then $F(-R) = F(R)$ and so $-F(R) = F(R)$. Thus, $F(0,\ldots,0) = 0$. These comments give us the following result.

Theorem 7.9. *An aggregation rule $F : \{-1,0,1\}^n \to \{-1,0,1\}$ satisfies (0MM), (A), and (N) if and only if there exists a sequence $q_0, q_1, \ldots, q_{n-1}$ of integers such that $\frac{n-k}{2} < q_k \leq n-k+1$ for all k and $F(R) = s$ iff $n_s(R) \geq q_{n_0(R)}$ for all $R \in \{-1,0,1\}^n$ and $s \in \{-1,1\}$.*

It is interesting to compare Theorem 7.9 with Theorem 7.7. In both theorems, $F(R) = 1$ exactly when $n_+(R)$ is greater than or equal to some bound. This bound may depend on the first voter as in Theorem 7.7 or it may depend on the number of zeros in the profile R as in Theorem 7.9. Therefore, the nature of the bound is based on the type of condition we are assuming.

We now introduce another condition.

Bi-idempotent: An aggregation function $F : \{-1,0,1\}^n \to \{-1,0,1\}$ is *bi-idempotent* if $F(R) \neq 0$ for any profile R such that $n_+(R) + n_-(R) = n$ and $n_+(R) \neq n_-(R)$.

Notice that $n_+(R) + n_-(R) = n$ is equivalent to $n_0(R) = 0$. A quick example of an aggregation rule that is bi-idempotent is simple majority rule F_s. In fact, $F_s(R) \neq 0$ whenever $n_+(R) \neq n_-(R)$.

The requirements $n_+(R) + n_-(R) = n$ and $n_+(R) \neq n_-(R)$ imply that either $n_+(R) > \frac{n}{2}$ or $n_-(R) > \frac{n}{2}$. Therefore, if q^* is the smallest integer strictly greater than $\frac{n}{2}$, then the absolute majority aggregation rule F_{q^*} is bi-idempotent. Moreover, F_q is not bi-idempotent for all integers $q > q^*$. The following result now follows directly from Theorem 7.3.

Theorem 7.10. [a] *An aggregation rule $F : \{-1,0,1\}^n \to \{-1,0,1\}$ satisfies (MM), (N), and (A), and is bi-idempotent if and only if it is the absolute majority rule F_{q^*} where q^* is the smallest integer strictly greater than $\frac{n}{2}$.*

The goal of this section was to work with aggregation functions based on a set with two alternatives. In the next section we will work with voting rules where the set of alternatives is not restricted to just two elements. In this context, a voting rule is called a *social welfare function.*

7.3. Social Welfare Functions

Let X be a finite set of alternatives and assume that $|X| \geq 2$. A binary relation ρ on X is *asymmetric* if $(x, y) \in \rho$ implies that $(y, x) \notin \rho$ for all $x, y \in X$. The binary relation ρ is *complete* if $(x, y) \notin \rho$ implies that $(y, x) \in \rho$ for all $x \neq y$ in X. Next, ρ is *transitive* if $(x, y) \in \rho$ and $(y, z) \in \rho$ implies that $(x, z) \in \rho$ for all $x, y, z \in X$. On the other hand, ρ is *negatively transitive* if $(x, y) \notin \rho$ and $(y, z) \notin \rho$ implies that $(x, z) \notin \rho$ for all $x, y, z \in X$. An asymmetric, complete, and transitive binary relation on X is called a *linear order* on X. An asymmetric and negatively transitive binary relation on X is called a *weak order.* Let $\mathcal{L}(X)$ be the set of all linear orders on X and let $\mathcal{W}(X)$ be the set of all weak orders on X. Also, let $\mathcal{A}(X)$ be the set of all asymmetric binary relations on X. Then $\mathcal{L}(X) \subseteq \mathcal{W}(X) \subseteq \mathcal{A}(X)$. In this context, a *social welfare function* is a function of the form $f : D^n \to \mathcal{A}(X)$

[a] A characterization of the majority consensus rule for hierarchies, using a bi-idempotent condition as part of the axiom list, is given in [17].

where $D = \mathcal{W}(X)$ or $D = \mathcal{L}(X)$ and $n \geq 2$ is the number of voters. A profile $R \in D^n$ is either an n-tuple of weak orders on X or an n-tuple of linear orders on X and the social output $f(R)$ is an asymmetric binary relation on X. For any $x, y \in X$ and for any profile $R = (R_1, \ldots, R_n) \in D^n$, let

$$n_R(x,y) = |\{i : (x,y) \in R_i\}|.$$

So $n_R(x,y)$ is the number of voters in the profile R that prefer x over y.

For any integer $q > \frac{n}{2}$, define the social welfare function $f_q : D^n \to \mathcal{A}(X)$ by

$$f_q(R) = \{(x,y) \in X \times X : n_R(x,y) \geq q\}$$

for all $R \in D^n$. The rule f_q is well defined since $q > \frac{n}{2}$ and $f_q(R) = \emptyset$ for all R whenever $q \geq n+1$. In the sequel, the social welfare function f_{q^*} where q^* is the least integer strictly greater than $\frac{n}{2}$ will be called *majority rule.* In fact, for any integer $q > \frac{n}{2}$ the social welfare function f_q with $D = \mathcal{W}(X)$ is a generalization of the absolute q-majority aggregation rule F_q. To see this, assume $X = \{x, y\}$. Then $\mathcal{W}(X) = \mathcal{A}(X) = \{\emptyset, \{(x,y)\}, \{(y,x)\}\}$. Now identify the binary relations $\emptyset$, $\{(x,y)\}$, and $\{(y,x)\}$ with 0, 1, and -1, respectively and notice that $n_R(x,y) = n_+(R)$ and $n_R(y,x) = n_-(R)$ for any profile R. It now follows that the social welfare function f_q is equivalent to the aggregation rule F_q when $|X| = 2$ and $D = \mathcal{W}(X)$.

Observe that $\mathcal{A}(X)$ is a median semilattice with set containment as the partial order and set intersection as the meet operation. The semilattice $\mathcal{A}(X)$ is not a lattice and the empty relation $\emptyset$ is the *zero* of $\mathcal{A}(X)$. The set of join irreducible elements is given by

$$J(X) = \{\{(x,y)\} : x, y \in X \text{ and } x \neq y\}.$$

For convenience we will write $(x,y) \in J(X)$ instead of $\{(x,y)\} \in J(X)$. For any relation $R_i \in \mathcal{A}(X)$ and for any $(x,y) \in J(X)$, $(x,y) \leq R_i$ if and only if $(x,y) \in R_i$. In addition, for any profile $R = (R_1, \ldots, R_n) \in D^n$ and for any $(x,y) \in J(X)$,

$$N_{(x,y)}(R) = \{i \in N : (x,y) \in R_i\} \text{ and } N_\emptyset(R) = \{i \in N : R_i = \emptyset\}.$$

So $n_R(x,y) = |N_{(x,y)}(R)|$. Note that $\emptyset \notin \mathcal{L}(X)$ and so $N_\emptyset(R) = \emptyset$ any profile $R = (R_1, \ldots, R_n) \in \mathcal{L}(X)^n$.

Using the above notation, the axioms DN, DM, and A defined for consensus functions can easily be translated to conditions DN, DM, and A for social welfare functions. For example, $f : D^n \to \mathcal{A}(X)$ satisfies DN if for all $(x,y), (u,v) \in J(X)$ and for all $R, R' \in D^n$, $N_{(x,y)}(R) = N_{(u,v)}(R')$ and $(x,y) \in f(R)$ implies that $(u,v) \in f(R')$. Next, f satisfies DM if for all $(x,y) \in J(X)$ and for all $R, R' \in D^n$, $N_{(x,y)}(R) \subseteq N_{(x,y)}(R')$ and $(x,y) \in f(R)$ implies that $(x,y) \in f(R')$. Finally, f satisfies A if for any profile $R = (R_1, \ldots, R_n) \in D^n$ and for any permutation Π of N, $f(R_1, \ldots, R_n) = f(R_{\Pi(1)}, \ldots, R_{\Pi(n)})$.

The next result is equivalent to Theorem 3.7 in [3].

Theorem 7.11. *Let $f : \mathcal{W}^n \to \mathcal{A}(X)$ be a social welfare function such that $|X| \geq 3$. Then f satisfies DN, DM, and A if and only if there exists an integer q such that $\frac{n}{2} < q \leq n+1$ and $f = f_q$.*

It may seem more reasonable to require that the range of a social welfare function $f : D^n \to \mathcal{A}(X)$ to be D. This restriction does not work well for majority rule f_{q^*}. McGarvey [14] showed that if $D = \mathcal{L}(X)$ and the number n is large compared to the number $|X|$ of alternatives, then the range of f_{q^*} is the entire set $\mathcal{A}(X)$. Stearns [26] improved McGarvey's result by showing that $f_{q^*} : \mathcal{L}(X)^n \to \mathcal{A}(X)$ is onto if $n \geq |X|+1$ when $|X|$ is odd or $n \geq |X|+2$ when $|X|$ is even.

There is a rich literature in the theory of social choice which identifies conditions on a profile $R = (R_1, \ldots, R_n)$ in D^n such that the image $f_{q^*}(R)$ belongs to D. For example, a profile $R = (R_1, \ldots, R_n)$ in $\mathcal{L}(X)^n$ is *value resticted* if for every 3-element subset $\{x, y, z\}$ of X there is one element of $\{x, y, z\}$ that is not below the other two in any R_i, or is not above the other two in any R_i, or is not between the other two in any R_i. Ward [27] and Sen [24] proved that if n is odd and R in $\mathcal{L}(X)$ is value restricted, then $f_{q^*}(R) \in \mathcal{L}(X)$. An even more restrictive condition on a profile R of linear orders is the requirement that for every 3-element subset $\{x, y, z\}$ there is some alternative in $\{x, y, z\}$ that is not ranked below the other two in any R_i. In this case, the profile R is said to be *single-peaked.*[b]

Following Campbell and Kelly [5] we will say that a social welfare function $f : \mathcal{L}(X)^n \to \mathcal{A}(X)$ satisfies

LT: Limited Transitivity if $f(R) \in \mathcal{W}(X)$ whenever $R = (R_1, \ldots, R_n) \in L(X)^n$ is single-peaked and $\{R_i : i \in N\}$ has at most three members.

It is not hard to verify that a weak order on X is a transitive binary relation and that majority rule f_{q^*} satisfies limited transitivity. We now outline Campbell and Kelly's simple and elegant characterization of majority rule. This characterization of majority rule is an extension of a result due to Eric Maskin [12].

First, we need to list some conditions a given social welfare function $f : \mathcal{L}(X)^n \to \mathcal{A}(X)$ may or may not satisfy. We will say that f satisfies

IIA: Independence of Irrelevant Alternatives if for any $(x, y) \in J(X)$ and for any profiles $R, R' \in \mathcal{L}(X)^n$, $(x, y) \in f(R)$ if and only if $(x, y) \in f(R')$ whenever $N_{(x,y)}(R) = N_{(x,y)}(R')$.

Since we are working with linear orders, $N_{(x,y)}(R) = N_{(x,y)}(R')$ implies that $N_{(y,x)}(R) = N_{(y,x)}(R')$ and so IAA implies that $f(R) \cap \{(x, y), (y, x)\} = f(R') \cap \{(x, y), (y, x)\}$. Next, f satisfies

Pareto if, for any $(x, y) \in J(X)$ and for any profile R, $(x, y) \in f(R)$ whenever $N_{(x,y)}(R) = N$.

---

[b]See Chapter 9 in Fishburn [9] for a complete discussion of single-peaked preferences.

We will say that a subset I of N is *decisive* if, for any $(x, y) \in J(X)$ and for any profile R, $(x, y) \in f(R)$ whenever $N_{(x,y)}(R) = I$. The Pareto condition implies that N is decisive.

The next result is due to Campbell and Kelly [5].

Theorem 7.12. *Let $f : \mathcal{L}(X)^n \to \mathcal{A}(X)$ be a social welfare function and assume that $|X| \geq 3$. Then f satisfies Limited Transitivity, IIA, and Pareto if and only if the collection $\mathcal{D}$ of decisive sets of f satisfies:*

$$1) I, J \in \mathcal{D} \text{ implies } I \cap J \neq \emptyset$$

and

$$2) I \subseteq N \text{ implies } I \in \mathcal{D} \text{ or } N \setminus I \in \mathcal{D}.$$

We give a brief and incomplete argument to show that property 2) holds when f satisfies Limited Transitivity, IIA, and Pareto. Let I be a nonempty subset of N. Choose a profile $R = (R_1, \ldots, R_n)$ in $\mathcal{L}(X)^n$ such that $|\{R_i : i \in N\}| = 2$ and R_i restricted to the 3-element subset $\{x, y, z\}$ has the following pattern:

$i \in I$	$i \in N \setminus I$
x	y
y	z
z	x

It follows from Pareto that $(y, z) \in f(R)$. Observe that profile R is single-peaked and so, by Limited Transitivity, $f(R)$ belongs to $\mathcal{W}(X)$. Since $f(R)$ is negatively transitive it follows that either $(x, z) \in f(R)$ or $(y, x) \in f(R)$. Thus either I is a decisive set for (x, z) or $N \setminus I$ is a decisive set for (y, x). Using the previous fact and the assumptions on f it can be shown that either I or $N \setminus I$ is decisive for any order pair of distinct alternatives, i.e, either $I \in \mathcal{D}$ or $N \setminus I \in \mathcal{D}$. This establishes property 2).

Now suppose $f : \mathcal{L}(X)^n \to \mathcal{A}(X)$ satisfies properties 1) and 2). If $I \in \mathcal{D}$ and $I \subseteq J \subseteq N$, then $N \setminus J \notin \mathcal{D}$ since $I \cap N \setminus J = \emptyset$. Therefore, by property 2), $J \in \mathcal{D}$. Hence the set $\mathcal{D}$ of decisive sets is a transversal federation. The next result follows from the previous observation and Theorem 7.12.

Theorem 7.13. *(Campbell & Kelly 2000) Let $f : \mathcal{L}(X)^n \to \mathcal{A}(X)$ be a social welfare function such that $|X| \geq 3$ and n is odd. Then f satisfies Limited Transitivity, IIA, Pareto, and Anonymity if and only if $f = f_{q^*}$.*

One of the most interesting aspects of the previous result is that there is no monotonicity assumption. Moreover, we can view IIA as a very weak version of neutrality. If we think of Limited Transitivity and Pareto as a replacement for decisive monotonicity, then we can view Theorem 7.13 in the same light as Corollary 7.1.

The fact that majority rule produces a large selection of asymmetric outcomes does not preclude the possibility that another social welfare function $f : D^n \to \mathcal{A}(X)$ has range a subset of D. There is still a problem. If $|X| \geq 3$ and $f : D^n \to D$ satisfies Pareto and IIA, then K. Arrow [1] proved that f is *dictatorial*, i.e., there exists $i \in N$ such that $R_i \subseteq F(R)$ for all $R = (R_1, \ldots, R_n) \in D^n$. See [6] for a thorough discussion on Arrow's Theorem and its various generalizations.

To state the next result, we need some more terminology. Let Ω be a subset of $\mathcal{L}(X)$. We will say that Ω^n is *without Condorcet triples* if each profile R in Ω^n is value restricted. If we restrict majority rule f_{q^*} to Ω^n and n is odd, then we get a well defined mapping $f_{q^*} : \Omega^n \to \mathcal{L}(X)$. On the other hand, if n is even, then $f_{q^*}(R)$ need not be a complete relation. To bypass this problem, we can work with a more general version of majority rule.

Definition 7.3. Let Ω be a subset of $\mathcal{L}(X)$. A function $f : \Omega^n \to \mathcal{L}(X)$ is called a **generalized majority rule (GMR)** if there exists $n-1$ linear orders ℓ_1, ..., ℓ_{n-1} on X such that

$$f(R) = f_{q^*}(R_1, \ldots, R_n, \ell_1, \ldots, \ell_{n-1})$$

for all $R = (R_1, \ldots, R_n) \in \Omega^n$.

The $n-1$ additional voters having the fixed preference orders ℓ_1, ..., ℓ_{n-1} are called *dummy agents*. Observe that the alternative x ranks above the alternative y in $f(R)$ if and only if the majority of real voters and dummy agents prefer x over y.

To state the next result we need one more piece of terminology. For any linear order q on X, the *inverse* of q is denoted by q^{-1} and is defined as follows: $(x, y) \in q^{-1}$ if and only if $(y, x) \in q$.

Theorem 7.14. *(Sethuraman et al. 2006 [25]) Let Ω be a subset of $\mathcal{L}(X)$ containing an ordering q and its inverse q^{-1} such that Ω^n is without Condorcet triples. If $|X| \geq 3$ and $f : \Omega^n \to \mathcal{L}(X)$ satisfies Pareto, IIA, DM, and anonymity, then f is a generalized majority rule.*

The previous result extends and generalizes an earlier result due to H. Moulin [20] where the domain of f is assumed to be single-peaked.

A comparison between Theorems 7.13 and 7.14 show that if there is an appropriate assumption about the domain or range of the function f and f satisfies some reasonable conditions, then it has to be a majority type rule.

7.4. Conclusion

In this chapter we considered consensus functions on median semilattices, aggregation rules based on a set with two alternatives, and social welfare functions based on a set with two or more alternatives. Roughly, we observed that if one of these functions is anonymous and satisfies certain conditions like neutrality and monotonicity, then it has to be an absolute majority rule. This observation is confirmed by looking at Theorems 7.2, 7.3, 7.11, and 7.13. On the other hand, once we deviate (even slightly) from the conditions of anonymity, neutrality, and monotonicity, then a lot can happen. The work on understanding majority type rules in various situations is of current interest as demonstrated by Theorems 7.13 and 7.14 (see also [23]). It has been close to sixty years since May's paper on simple majority rule appeared in print and we still continue to learn new things about the majority rule consensus function.

References

1. K.J. Arrow, *Social Choice and Individual Values*, Wiley, New York 2nd ed. (1963).
2. G. Asan, M.R. Sanver, Maskin monotonic aggregation rules, *Economic Letters* **75** (2006) 179–183.
3. D. Austen-Smith, J.S. Banks, *Positive Political Theory I Collective Preference*, University of Michigan Press, (1999).
4. D.E. Campbell, A characterization of simple majority rule for restricted domains, *Economic Letters* **28** (1988) 307–310.
5. D.E. Campbell, J.S. Kelly, A simple characterization of majority rule, *Economic Theory* **15** (2000) 689–700.
6. D.E. Campbell, J.S. Kelly, *Impossibility Theorems in the Arrovian Framework, Handbook of Social Choice and Welfare*, Volume 1, Edited by K.J. Arrow, A.K. Sen, K. Suzumura, Elsevier, Amsterdam, (2002) pp. 35–94.
7. E. Cantillon, A. Rangel, A graphical analysis of some basic results in social choice, *Social Choice and Welfare* **19** (2002) 587–611.
8. W.H.E. Day, F.R. McMorris, *Axiomatic consensus theory in group choice and biomathematics.* With a foreword by M.F. Janowitz. Frontiers in Applied Mathematics, 29, Society for Industrial and Applied Mathematics (SIAM), Philadelphia, PA, 2003
9. P.C. Fishburn, *The theory of social choice*, Princeton University Press, Princeton, N.J. (1973).
10. N. Houy, A new characterization of absolute qualified majority voting, *Economic Bulletin* **4** (2007) 1–8.
11. T. Margush, F.R. McMorris, Consensus n-trees, *Bull. Math. Biol* **43** (1981) 239–244.
12. E. Maskin, Majority rule, social welfare functions, and game forms, in: K. Basu, P.K. Pattanaik, K. Suzumura, eds., *Choice, Welfare and Development*, Festschrift for Amartya Sen, Clarendon Press, Oxford (1995).
13. K.O. May, A set of independent necessary and sufficient conditions for simple majority decision, *Econometrica* **20** (1952) 680–684.
14. D.C. McGarvey, A theorem on the construction of voting paradoxes, *Econometrica* **11** (1953) 608–610.
15. F.R. McMorris, D.A. Neumann, Consensus functions defined on trees. Math. Social Sci. 4 (1983), no. 2, 131–136.
16. F.R. McMorris, H.M. Mulder, R.C. Powers, The median function on median graphs and semilattices, *Discrete Applied Math.* 101 (2000) 221–230.
17. F.R. McMorris, R.C. Powers, A Characterization of Majority Rule, *Journal of Classification* **25** (2008) 153–158.

18. F.R. McMorris, R.C. Powers, Mayös Theorem for Trees, *Annals of Operations Research* **163** (2008) 169–175.
19. A. Miroiu, Characterizing majority rule: from profiles to societies, *Economics Letters* **85** (2004) 359–363.
20. H. Moulin, *Axioms of Cooperative Decision Making*, Cambridge University Press, New York, 1988.
21. J. Perry, R.C. Powers, Aggregation rules that satisfy anonymity and neutrality, *Economic Letters* **100** (2008) 108–110.
22. R.C. Powers, Maskin Monotonic Aggregation Rules and Partial Anonymity, *Economic Letters* **106** (2010) 12–14.
23. M.R. Sanver, Characterizations of Majoritarianism -A Unified Approach, *Social Choice and Welfare* **33** (2009) 159–171.
24. A.K. Sen, A possibility theorem on majority decisions, *Econometrica* **34** (1966) 491–499.
25. J. Sethuraman, C.-T. Teo, R.V. Vohra, Anonymous monotonic social welfare functions, *Journal of Economic Theory* **128** (2006) 232–254.
26. R. Stearns, The voting problem, *American Mathematical Monthly* **66** (1959) 761–763.
27. B. Ward, Majority voting and the alternate forms of public enterprise. In: J. Margolis (ed), *The public economy of urban communities*, John Hopkins Press, Baltimore, 1965, pp. 112–125.
28. G. Woeginger, A new characterization of the majority rule, *Economic Letters* **81** (2003) 89–94.
29. G. Woeginger, More on the majority rule: Profiles, societies, and responsiveness, *Economic Letters* **88** (2005) 7–11.
30. J. Yi, A complete characterization of majority rules, *Economics Letters* **87** (2005) 109–112.

Chapter 8

Centrality Measures in Trees

K.B. Reid*

Department of Mathematics, California State University San Marcos
San Marcos, California 92096-0001 USA
breid@csusm.edu

The classical measures of the "middle" of a tree include the center and the median. It is well known that another seemingly different measure, the branch weight centroid of a tree, yields the same middle as the median. Many other descriptions of "middleness" of a tree are not so well known, some of which also yield the same set of vertices as the median. In this survey we gather together many descriptions of middle sets in trees, some of which yield the same set of vertices as the median, and others that are distinct from the center and the median. We discuss, in turn, the center, the median, the branch weight centroid, the absolute p-center, the vertex p-center, the absolute p-median, the vertex p-center, the security center, the accretion center, the telephone center, the weight balance center, the latency center, the pairing center, the processing center, the leaf weight median, the leaf branch weight centroid, the n-th power center of gravity, the distance balance center, the k-nucleus, the k-branch weight centroid, the R-center, the k-centrum, the k-ball branch weight centroid, the k-ball l-path branch weight centroid, the k-processing center, the centdian (or cendian), the R-median, the R-branch weight centroid, the k-broadcast center, the distance balanced edge center, the weight balanced edge center, the Steiner k-center, the Steiner k-median, a central k-tree, the subtree center, the majorization center, the cutting center, the centrix, the security centroid, the harmonic center, the path center, a path median (core), and the path centroid (spine). We also mention some general central subtrees of maximum degree D that generalize some of the previous concepts, and we conclude with a further generalization involving hereditary class centrality. We also reference some work on tree networks.

Introduction

If G is a graph, $V(G)$ denotes its vertex set and $E(G)$ denotes its edge set. The number of edges of G incident with vertex x of G is the *degree* of x, denoted $d(x)$. A vertex x in a tree T with $d(x) = 1$ is called a *leaf* of T. The set of all leaves of a tree T is denoted $Lf(T)$, i.e., $Lf(T) = \{x \in V(T) : d(x) = 1\}$. If $S \subseteq V(G)$, then the subgraph of G induced by S, denoted $G[S]$, is the subgraph of G with

*Dedicated to F.R. "Buck" McMorris on his 65th birthday

vertex set S and edge set given by all edges in G with both ends in S. If G is connected and $S \subseteq V(G)$, the *smallest, connected subgraph of* G with vertex set S is denoted $G\langle S\rangle$; of course, $G[S]$ is a subgraph of $G\langle S\rangle$. If S is either a subgraph of G or a set of vertices of G, then $G - S$ denotes the subgraph of G induced by the vertices of G that are not vertices of S, i.e., $G - S = G[V(G) - S]$ (or, $G[V(G) - V(S)]$, if S is a subgraph). K_n denotes the complete graph on n vertices, and $K_{m,n}$ denotes the complete bipartite graph with partite sets of order m and n. If e is an edge of a graph G with ends x and y, then we will denote e by xy or yx. The *length* P in G, denoted $l(P)$, is the number of edges in P. If x and y are two vertices in a connected graph G, $d_G(x,y)$ (or simply $d(x,y)$, if there is no confusion as to the underlying graph G) denotes the length of a shortest path in G with ends x and y, i.e., $d(x,y) = \min\{l(P) : P$ a path in G with ends x and $y\}$. If $x, y \in V(G)$, then $V(x,y)$ denotes all vertices in G that are closer to x than to y, i.e., $V(x,y) = \{z \in V(G) : d(x,z) < d(y,z)\}$.

We now define some middle sets. If R and S are subsets of vertices of G or subgraphs of G, then the *distance between* R *and* S, denoted $d(R,S)$, is the smallest distance between a vertex of R and a vertex of S, i.e., $d(R,S) = \min\{d(r,s) : r \in R, s \in S\}$. If $|R| = 1$, say $R = \{r\}$, then $d(R,S)$ will be abbreviated $d(r,S)$. The *eccentricity* of S, denoted $e(S)$, is the largest of the distances between S and all vertices of $V(G)$, i.e., $e(S) = \max\{d(v,S) : v \in V(G)\}$. If $|S| = 1$, say $S = \{x\}$, then $e(S)$ is abbreviated $e(x)$. So, $e(x) = max\{d(x,y) : y \in V(G)\}$. The *radius* of G, denoted $r(G)$, is the smallest eccentricity of the vertices of G, i.e., $r(G) = \min\{e(x) : x \in V(G)\}$, and the *diameter* of G, denoted $dia(G)$, is the largest eccentricity of all of the vertices of G, i.e., $dia(G) = \max\{e(x) : x \in V(G)\}$. The *center* of G, denoted $C(G)$, consists of all vertices of G of smallest eccentricity, i.e., $C(G) = \{x \in V(G) : e(x) = r(G)\}$. The *total distance* of S (or simply the *distance* of S) denoted $D(S)$, is the sum of the distances between S and all vertices of $V(G)$, i.e., $D(S) = \sum\{d(v,S) : v \in V(G)\}$. If $|S| = 1$, say $S = \{x\}$, then $D(S)$ is abbreviated $D(x)$. So, $D(x) = \sum\{d(x,y) : y \in V(G)\}$. The *median* of G, denoted $M(G)$, consists of all vertices of G with smallest distance, i.e., $M(G) = \{x \in V(G) : D(x) \leq D(y), y \in V(G)\}$. If G is a tree T, then the *branch weight* of S, denoted $bw(S)$, is the number of vertices in a largest component of $T - S$, i.e., $bw(S) = \max\{|V(C)| : C$ a connected component of $T - S\}$. If $|S| = 1$, say $S = \{x\}$, then $bw(S)$ is abbreviated $bw(x)$. So, $bw(x) = \max\{|V(C)| : C$ a connected component of $T - x\}$. Equivalently, $bw(x)$ is the maximum number of edges in a maximal subtree of T having x as a leaf. The *branch weight centroid* of T (or simply *centroid* of T), denoted $Bw(T)$, is the set of vertices of T with smallest branch weight, i.e., $Bw(T) = \{x \in V(T) : bw(x) \leq bw(y), y \in V(T)\}$. There have been some generalizations of centroids to general, connected graphs (e.g., see Slater [53], [58] and Piotrowski [44]).

In the language of location of facility theory, the central set $C(G)$ is central with respect to minimizing the worst response time for an emergency facility, and the central set $M(G)$ is central with respect to minimizing the total delivery distance for a supply location. We elaborate a bit on this at the end of this section.

The two central sets $C(G)$ and $M(G)$ might be the same set, as can be seen, for example, by a cycle or by a path of odd length. Or, they may be arbitrarily far apart as can be seen by the tree T with $4m+3$ vertices obtained from $K_{1,2m+2}$, by subdividing one edge with $2m$ new internal vertices. The vertex of degree $2m+2$ is the only vertex in $M(T)$, and the vertex of distance m from the vertex of degree $2m+2$ is the only vertex in $C(T)$. However, if a tree T contains n vertices, $n \geq 3$, then Entringer, Jackson, Snyder [14] showed that the maximum distance between the center and the median is at most $\lfloor \frac{n}{4} \rfloor$.

Jordan [29] introduced the branch weight centroid and the center of a tree in 1869; he showed

Theorem 8.1. *The center of a tree consists of a single vertex or two adjacent vertices, and the branch weight centroid of a tree consists of a single vertex or two adjacent vertices.*

Modern proofs of these two facts can be found in many textbooks, often as exercises (e.g. Bondy, and Murty [4] or Chartrand and Oellermann [7]). A "pruning procedure" is usually employed for the case of the center. A particular attractive proof of both facts was given by Graham, Entringer, Székely [17].

In 1877 Cayley [6] was under the impression that Sylvester had discovered centers, but Sylvester stated in 1873 [65] and again in 1882 [66] that centers were discovered by Jordan. In 1962 Ore [40] presented, without comment, the concept of the median. In 1966 Sabidussi [50] showed, in a discussion of axiomatic considerations of centrality, that the median consists of a single vertex or two adjacent vertices (although he referred to this central set as the center). In 1968 Zelinka [73] proved

Theorem 8.2. *In a tree, the branch weight centroid and the median are the same set of vertices.*

Theorems 8.1 and 8.2 form the motivating results for most of the remainder of this survey. These two measures of centrality are but the tip of the iceberg in the sea of studies done on centrality in graphs, particularly trees, and networks. Many of these studies are motivated by the study of the location of one or more facilities on networks. So, the underlying structure of the network, a graph, is a fundamental structure in which to study centrality. Trees have been the subject of a host of such studies, and provide a rich context for fundamental results. However, there is much more to the issue of location of facilities for real applications, issues that often necessitate cycles in the underlying graph. Even in tree networks complications arise, because edge weights (called lengths), and in some cases vertex weights, are also factored in. Additional complications arise if locations of facilities are to be considered along edges as well as at vertices, so the terms absolute centers and absolute medians are employed. Continuous mathematics enters the picture, and the subject also becomes a branch of combinatorial optimization. Accordingly, algorithmic considerations make up a large part of general facility location theory. Before embarking on this survey, we include, for completeness, the definition of a network, the context for most work in facility location theory, and the definitions of an absolute p-center and an absolute p-median, the centrality concepts most frequently encountered in facility location theory. And, we make some remarks about axiomatics.

Let $\mathcal{R}$, $\mathcal{R}^+$, and $\mathcal{R}^-$ denote the set of real numbers, the set of positive real numbers and the set of negative real numbers, respectively. A *network* is a connected, undirected graph G together with functions $w : V(G) \to \mathcal{R} - \mathcal{R}^-$ and $l : E(G) \to \mathcal{R}^+$. For $v \in V(G)$, the nonnegative real number $w(v)$ is called the *weight* of v, and for $e \in E(G)$, the positive number $l(e)$ is called the *length* of e. Formally, G is embedded into some space, such as $\mathcal{R}^2$ or $\mathcal{R}^3$. A point on G is any point along any edge and may or may not be a vertex. The *length of a path* P connecting two points x and y, denoted $l(P)$, is the sum of the lengths of the edges (and perhaps partial edges) that make up P and the *distance between* x *and* y, denoted $d(x,y)$, is the length of a shortest path between x and y, i.e., $d(x,y) = \min\{l(P) : P \text{ a path connecting } x \text{ and } y\}$. For a set V_p of p points on G and a vertex x, the *distance between* x *and* V_p, denoted $d(x, V_p)$, is the smallest distance between x and the points of V_p, i.e., $d(x, V_p) = \min\{d(x,v) : v \in V_p\}$. The *weighted eccentricity* of V_p, denoted $e(V_p)$, is the maximum value of $w(x) \times d(x, V_p)$ as x ranges over all vertices $x \in V(G)$, i.e., $e(V_p) = \max\{w(x) \times d(x, V_p) : x \in V(G)\}$. A set of p points V_p^* is an ***absolute p-center*** if its weighted eccentricity is the minimum weighted eccentricity among all sets of p points, i.e., $e(V_p^*) \le e(V_p)$, for all sets of p points V_p on G. If the points in V_p are restricted to be vertices of G, then V_p^* is called a *vertex p-center.* The *weighted distance* of V_p, denoted $D(V_p)$, is the sum of all the terms of the form $w(v) \times d(v, V_p)$ as v ranges over all vertices of G, i.e., $D(Vp) = \sum\{w(v) \times d(v, V_p) : v \in V(G)\}$. A set of p points V_p^* is an ***absolute*** *p-median* provided its weighted distance is the minimum weighted distance of all sets of p points on G, i.e., $D(V_p^*) \le D(V_p)$ for all sets of p points V_p on G. If the points in V_p are restricted to be vertices of G, then V_p^* is called a *vertex p-median.* If both the weight function $w(\cdot)$ and the length function $l(\cdot)$ are constant functions, then the definitions of vertex p-center and vertex p-median are basically concepts for ordinary graphs, and, if in addition $p = 1$, then a vertex 1-center is just the ordinary center of the underlying undirected graph G, and the vertex 1-median is just the ordinary median of G. If $p = 1$, the terms absolute 1-center and absolute 1-median are shortened to absolute center and absolute median, respectively. There is a unique point in a tree network that is the absolute center, and it need not be a vertex. There might be many points in a tree network that are absolute medians, but at least one must be at a vertex as was shown in an important early result in 1965 by Hakimi [19].

Theorem 8.3. *In a tree network there is always a vertex p-median that is also an absolute p-median.*

Since Hakimi's seminal work [18], [19], the field of location theory has blossomed. In the early years it was particularly focused on algorithmic work aimed at finding absolute p-centers, absolute p-medians, and various generalizations (some of which serve all the points of the network, not merely the vertices), including combinations of centers and medians. See the 1979 text by Handler and Mirchandani [22] and the 1983 two-part survey by Tansel, Francis, and Lowe [68], [69]. An example of early work on central structures is the paper by Minieka [34]. A 1990 reference book edited by Mirchandani and Francis [33] and a 1995 text by M. Daskin [12] illustrate the growth of the subject, including new directions such as voting and competitive location problem. More recent collections of articles edited by Drezner and Hamacher [13] and recent papers, such as those by Tamir, Puerto, Mesa, and Rodríguez-Chía [67] and Bhattacharya, Hu, Shi, and Tamir [3] illustrate the current depth and breath of the subject.

We make only some quick comments about axiomatics, see also McMorris, Mulder and Vohra (this volume). In 1966 Sabidussi [50] undertook an axiomatic approach of centrality in graphs in order to explain why several previously suggested sociological centrality functions in graphs are not very satisfactory. In 1990 Holzman [28] gave axioms that characterize the centrality measure in a tree that minimizes the sum of the squares of the distances from a vertex to the other vertices. An axiomatic characterization of the median function on a tree was given by Vohra [70] in 1996 and more generally by McMorris, Mulder, and Roberts [31] in 1998. In 1998 Foster and Vohra [15] give an axiomatic characterization of general centrality functions on trees that includes the weighted absolute center, the absolute center, and more. In 2001 McMorris, Roberts and Wang [33] gave an axiomatic characterization of the center function on trees. Subsequently, Mulder, Pelsmajer, and Reid [39] gave a short proof of that characterization in 2008. In 2004 Monsuur and Storcken [37] gave axiomatic characterizations of some centrality concepts on graphs, including examples from sociology, game theory, and business.

In this restricted survey we concentrate on trees; edge and vertex weights will rarely be mentioned or can be taken to be constant. Many of the concepts described were originally defined for all connected graphs, but in most instances we give the definition for trees only. The primary focus will be on the graph theory of centrality in trees (so, any facilities and customers will be assumed to be located at vertices). We will present descriptions of middle vertices in trees given by many different criteria, present the basic results about each concept, and give references for further study. Sections 8.1-8.4 deal with small connected central sets, Section 8.5 treats central sets that can possibly induce large subtrees, Section 8.6 deals with central sets that can possibly induce disconnected subgraphs, and Section 8.7 deals with central subtrees of specific types and concludes with some generalizations. Many of these measures of centrality are related in ways to be revealed, but most of the relations between these measures have yet to be completely determined and await additional research. This survey contains descriptions of many more central sets in trees than earlier surveys (e.g., see [5], [30], [61], [62]) but surely there are more examples of central sets in trees. However, we trust that most readers will become acquainted with at least one new central set.

8.1. The Center

The center is perhaps the most obvious candidate for being the middle of a graph. Ádám [1] introduced five centrality functions for trees, two of which yield the center (one function is essentially the eccentricity function), one that yields the centroid (in essentially the same way as in the definition in the Introduction), and two that yield the same distinct set that is different from the center and the centroid. We describe his alternative description of the center and discuss the later two functions at the end of Section 8.2.

Definition 8.1. If xy is an edge of a tree T, then define $m(x, xy) = \max\{d(x, z) :$ z a vertex in the component of $T - x$ that contains $y\}$. Label the $d = d(x)$ vertices adjacent to x as $y_1, y_2, \ldots, y_d$ so that $m(x, xy_1) \geq m(x, xy_2) \geq \ldots \geq m(x, xy_d)$. Define the function on $V(T)$ as follows: $f(x) = 0$ if x is a leaf, and $f(x) = m(x, xy_2)$ otherwise.

Note that the condition in the set defining $m(x, xy)$ is equivalent to $d(x, z) = 1 + d(y, z)$ which is equivalent to $z \in V(y, x)$ (since $xy \in E(T)$). And, in the definition of $f(x)$, if x is not a leaf of T, then $f(x)$ is the second term in the string of inequalities involving the $m(x, xy_i)$ given above. The vertices that maximize the function $f(\cdot)$ are of interest. Ádám [1] proved the following theorem.

Theorem 8.4. *Suppose that T is a tree. Then $C(T) = \{x \in V(T) : f(x) \geq f(y), y \in V(T)\}$.*

8.2. The Median

We briefly introduce several measures of centrality in trees before stating the main result of this section.

Definition 8.2. For a vertex x in a tree T, the *security number* of x, denoted $sec(x)$, is defined to be the smallest value of $|V(x, v)| - |V(v, x)|$ over all $v \in V(T) - \{x\}$, i.e., $sec(x) = \min\{|V(x, v)| - |V(v, x)| : v \in V(T) - \{x\}\}$. The *security center* of T, denoted $Sec(T)$, consists of all vertices of T with largest security number, i.e., $Sec(T) = \{x \in V(T) : sec(x) \geq sec(y), y \in V(T)\}$.

The security center was introduced and studied by Slater [53] as a model for the competitive location of one facility by two competitors.

Definition 8.3. Suppose that T is a tree of order n. An ordered n-tuple $(x_1, x_2, \ldots, x_n)$ of the n vertices of T is called a *sequential labeling* provided the subgraph $T[\{x_1, x_2, \ldots, x_j\}]$ induced by $\{x_1, x_2, \ldots, x_j\}$ is connected, for all j, $1 \leq j \leq n$. The *sequential number* of a vertex x, denoted $seq(x)$, is the number of sequential labeling of T with x as first entry, i.e., $seq(x) = |\{(x_1, x_2, \ldots, x_n) : (x_1, x_2, \ldots, x_n)$ is a sequential labeling of $V(T), x = x_1\}|$. The *accretion center* of T, denoted $Acc(T)$, is the set of all vertices of T with largest sequential number, i.e., $Acc(T) = \{x \in V(T) : seq(x) \geq seq(y), y \in V(T)\}$.

The accretion center was introduced and studied by Slater [58] as a model of sequencing the establishment of facilities, one at each vertex of a tree network.

Definition 8.4. Define two paths in a tree T to be *distinct* if all four of their ends are distinct. The switchboard number of a vertex $x \in V(T)$, denoted $sb(x)$, is the maximum number of distinct paths having x as an interior vertex. The *telephone center* of T, denoted $Tel(T)$, is the set of all vertices of T with largest switchboard number, i.e. $Tel(T) = \{x \in V(T) : sb(x) \geq sb(y), y \in V(T)\}$.

Mitchell [36] introduced and studied the telephone center in 1978 as a model for certain efficiency in communication networks.

Definition 8.5. The *weight balance* of a vertex x in a tree T, denoted $wb(x)$, is defined to be the integer $\min\{|n_1 - n_2|\}$, where the minimum is taken over all subtrees T_1 and T_2 of T such that $V(T) = V(T_1) \cup V(T_2)$, $V(T_1) \cap V(T_2) = \{x\}$, $|V(T_1)| = n_1$, and $|V(T_2)| = n_2$. The *weight balance center* of T, denoted $Wb(T)$, is the set of all vertices of T with smallest weight balance, i.e., $Wb(T) = \{x \in V(T) : wb(x) \leq wb(y), y \in V(T)\}$.

The weight balance center was introduced and studied by Reid and DePalma [49] as a "best" balance vertex x of a tree so that there are a nearly equal number of vertices on "either side" of x, i.e., nearly equal number of vertices in two subtrees such that x is the only vertex contained in both, but together the two subtrees contain all vertices of T.

Definition 8.6. Let W denote the walk resulting from a depth-first-search through all of the vertices of T starting with x and ending exactly at that vertex that is the first vertex so that all vertices occur in W. For each vertex $y \in V(T)$, let $i_x(y, W)$ denote the length of that part of W from the first occurrence of x to the first occurrence of y.

Lemma 8.1. *If Q and R are any two such depth-first walks in a tree T, each starting at $x \in V(T)$, then $\sum i_x(y, Q) = \sum i_x(y, R)$, where both sums are over all vertices $y \in V(T)$.*

Definition 8.7. The *latency* of x, denoted $l(x)$, is defined to be the unique value $\sum\{i_x(y, W) : y \in V(T)\}$ (by Lemma 8.1), where W is any depth-first walk as described above. The *latency center*, denoted $L(T)$, consists of all vertices of T with largest latency number, i.e., $L(T) = \{x \in V(T) : l(x) \geq l(y), y \in V(T)\}$.

The latency center was suggested by Hedetniemi [27] and studied by Reid [48].

Definition 8.8. A *pairing* of a tree T of order n, denoted $P(T)$, is a partition of $V(T)$ into 2-sets (and one singleton, called a *free vertex*, if n is odd). If $P(T) = \{\{a_i, b_i\} : 1 \leq i \leq \lceil \frac{n}{2} \rceil\}$, where $a_{\lceil \frac{n}{2} \rceil} = b_{\lceil \frac{n}{2} \rceil}$ is the free vertex if n is odd, let P_i denote the unique path in T with ends a_i and b_i. Note that if n is odd, then $l(P_{\lceil \frac{n}{2} \rceil})$ is zero. Let $G(P(T)) = \sum\{l(P_i) : 1 \leq i \leq \lceil \frac{n}{2} \rceil\}$, and define $G(T)$ to be the largest possible value of $G(P(T))$ over all pairings $P(T)$ of T, i.e., $G(T) = \max\{G(P(T)) : P(T)$ a pairing of $T\}$. A pairing $P_0(T)$ of T such that $G(T) = G(P_0(T))$ is called a *maximum pairing* of T. The *pairing center* of T, denoted $PC(T)$, is defined separately for n even and n odd. If n is even, $PC(T)$ consists of all vertices x of T so that there is a maximum pairing of T with x on all of the paths P_i, $1 \leq i \leq \frac{n}{2}$, i.e., $PC(T) = \{x \in V(T) :$ there is a maximum pairing of T with x on all $\frac{n}{2}$ paths of the pairing$\}$. If n is odd, then $PC(T)$ consists of all vertices x of T so that there is a maximum pairing $P_0(T)$ of T with x free in $P_0(T)$, i.e., $PC(T) = \{x \in V(T) :$ there is a maximum pairing of T with x free$\}$.

The pairing center was introduced and studied by Gerstel and Zaks [16] to discuss lower bounds on the message complexity of the distributed sorting problem.

Definition 8.9. A *processing sequence* for a tree T of order n is a permutation $x_1, x_2, \ldots, x_n$ of its vertices so that x_1 is a leaf of T, and for each i, $2 \leq i \leq n$, x_i is a leaf of the subtree $T - \{x_1, \ldots, x_{i-1}\}$. The *processing number* of a vertex x, denoted $proc(x)$, is the index of the earliest possible position for x over all processing sequences, i.e., $proc(x) = \min\{i : x = x_i$ in some processing sequence $x_1, x_2, \ldots, x_n\}$. The *processing center* of T, denoted $Proc(T)$, consists of all vertices with largest processing number, i.e., $Proc(T) = \{x \in V(T) : proc(x) \geq proc(y), y \in V(T)\}$.

The processing center was introduced and studied by Cikanek and Slater [10] as a model for a multiprocessor job-scheduling problem. We now give a result pertinent to each of the centrality measures described thus far in this section. For a tree T, let $X(T)$ denote any one of the central sets discussed thus far in this section. Each of the references corresponding to the choice for $X(T)$ essentially contains the following two theorems, among other things. The second of these results can be thought of as a "Zelinka-type" theorem (see Theorem 8.2)

Theorem 8.5. *If T is a tree, then $X(T)$ consists of either a single vertex or two adjacent vertices.*

Theorem 8.6. *If T is a tree, then $X(T) = M(T) = Bw(T)$.*

Consequently, we see that the median can be expressed in many different ways, and we can restate Theorem 8.6 as

Theorem 8.7. *Let T be a tree. Then the branch weight centroid of T = the median of T = the security center of T = the accretion center of T = the telephone center of T = the weight balance center of T = the latency center of T = the pairing center of T = the processing center of T.*

If x is a vertex in a tree T, then $\max\{d(x,y) : y \in V(T)\} = \max\{d(x,y) : y \in Lf(T)\}$. So, taking the maximum over the leaves of T in the definition of $e(x)$ results in no new value. This is not necessarily the case if such a change is made in the definitions of branch weight of a vertex and distance of a vertex. Ádám [1] obtained some possibly different central sets by making this change. These sets are also linked by a Zelinka-type result (as in Theorem 8.2). Recall that for two distinct vertices x and y in a graph, $V(x,y) \cap V(y,x) = \emptyset$, and if xy is an edge of a tree, then $V(T) = V(x,y) \cup V(y,x)$.

Definition 8.10. The *leaf branch weight* of vertex x, denoted $lbw(x)$, is the maximum number of leaves of T in any connected component of $T - x$, i.e., $lbw(x) = \max\{|V(y,x) \cap Lf(T)| : yx \in E(T)\}$. The *leaf branch weight centroid* of T, denoted $lbw(T)$, is the set of all vertices of T with smallest leaf weight, i.e., $lbw(T) = \{x \in V(T) : lbw(x) \leq lbw(y), y \in V(T)\}$. The *leaf distance* of x, denoted $lD(x)$, is the sum of the distances from x to all the leaves of T, i.e., $lD(x) = \sum\{d(x,z) : z \in Lf(T)\}$. The *leaf median* of T, denoted $lM(T)$, is the set of all vertices of T with smallest leaf distance, i.e., $lM(T) = \{x \in V(T) : lD(x) \leq lD(y), y \in V(T)\}$.

Ádam [1] proved the following theorem, part (a) of which is a Zelinka-type theorem (see Theorem 8.2).

Theorem 8.8. *Suppose that T is a tree.*

(a) $lbw(T) = lM(T)$. That is, the leaf branch weight centroid of T is equal to the leaf median of T.
(b) If T is not a path, then the leaf branch weight centroid of T consists either of the vertices of some $K_{1,m}$, for some $m \geq 0$, or of all the vertices of some path P of one or more edges such that P contains no leaf of T, all internal vertices of P are of degree 2 in T, and the ends of P have degree at least 3 in T. (Here, $K_{1,0}$ means K_1.)

Subsequently, Slater [54] considered the possible location for the leaf median and independently proved

Theorem 8.9. *For each positive integer k there is a tree T for which $C(T)$, $M(T)$, and $lM(T)$ are pairwise separated by distances greater than k, and no path of T contains a vertex of each of these sets.*

Other central sets can be obtained by replacing $Lf(T)$ in the above considerations with other subsets of $V(T)$, as was done by Slater [55] in 1978. In fact, instead of merely one subset, Slater [56] subsequently studied the extension of these ideas to a family of subsets, as we shall see in Section 8.5.

8.3. Two more central sets that induce K_1 or K_2

In this section we see some more central sets consisting of a single vertex or two adjacent vertices. The first such concept is a natural generalization of the median.

Definition 8.11. Let n be a positive integer, and let x be a vertex in a tree T. The *n-th power distance* of x, denoted $D_n(x)$, is the sum of the n-th power of the distances to all other vertices, i.e., $D_n(x) = \sum\{(d(x,y))^n : y \in V(T)\}$. The *n-th power center of gravity* of T, denoted $M_n(T)$, is the set of vertices of T of least n-th power-distance, i.e., $M_n(T) = \{x \in V(T) : D_n(x) \leq D_n(y), y \in V(T)\}$. When $n = 1$, $M_1(T)$ is just the median of T. Since $\frac{D_n(x)}{(|V(T)|-1)}$ is the average of the n-th power of the distances from x to the other vertices, a vertex in $M_n(T)$ minimizes this average.

In 1970 Zelinka [74] proved

Theorem 8.10.

(a) If T is a tree and n is a positive integer, then $M_n(T)$ consists of either a single vertex or two adjacent vertices.

(b) For any two positive integers k and n there exists a tree T so that $d(M_n(T), C(T)) > k$. That is, the n-th power center of gravity and the center can be arbitrarily far apart.

The next concept is more analogous to the center of gravity of a physical body than was the previous concept. The pairs of subtrees described next were encountered in Definition 8.5 of the weight balance of a vertex.

Definition 8.12. For each vertex x in a tree T, let $P(x)$ denote the set of all pairs of subtrees of T such that x is the only vertex in common to the pair and the union of the vertex sets of the pair is $V(T)$. Of course, $(T_1, T_2) \in P(x)$ if and only if $(T_2, T_1) \in P(x)$, and in either case, $V(T_1) - \{x\}$ and $V(T_2) - \{x\}$ yield a bipartition of $V(T) - \{x\}$. For $(T_1, T_2) \in P(x)$, define the *distance balance* of x with respect to (T_1, T_2), denoted $dbal(x; (T_1, T_2))$, to be the absolute value of the difference of the sum of the distances between x and all vertices in T_1 and the sum of the distances between x and all vertices in T_2, i.e.,

$$dbal(x; (T_1, T_2)) = |\sum\{d(x,y) : y \in V(T1)\} - \sum\{d(x,z) : z \in V(T_2)\}|.$$

Define the *distance balance (in T)* at the vertex x, denoted $dbal(x)$, to be the integer given by $\min\{bal(x; (T_1, T_2)) : (T_1, T_2) \in P(x)\}$. The *distance balance center* of T, denoted $DB(T)$, consists of all vertices of T with smallest distance balance, i.e., $DB(T) = \{x : x \in V(T), dbal(x) \leq dbal(y), y \in V(T)\}$. Vertices in $DB(T)$ are called *distance balanced vertices.*

The term "distance balance" in these definitions is simply called "balance" in references [47] and [51], but Reid and DePalma [49] adopted the term "distance balance" to distinguish this concept from the weight balanced vertices discussed in Section 3. Distance balanced vertices in trees can be considered as a discrete version of the continuous notion of the "center of gravity" of a physical body. For example, consider a vertex-weighted tree T in which each vertex has the same weight h. Let $(T_1, T_2) \in P(x)$. Then for a vertex z in $V(T_1)$, $h \times (d(x,z))$ can be thought of as the moment of z about x in the direction of T_1, and for a vertex y in $V(T_2)$, $h \times (d(x,y))$ can be thought of as the moment of y about x in the direction of T_2. A vertex x in $DB(T)$ is one for which the absolute value of the differences of the total moment about x in the direction of T_1 and the total moment about x in the direction of T_2 is as small as possible. If there is a vertex x with $dbal(x) = 0$, then x is the "center of gravity" of T; in general, a vertex in $DB(T)$ is a "best approximation" to a vertex center of gravity of T. Reid [47] introduced this concept and obtained the next two results. Subsequently, Shan and Kang [51] gave a shorter proof of the first result, and Reid and DePalma [49] gave, among other things, another shorter, independent proof. The reader may not be surprised with the first result, given the results from the previous section.

Theorem 8.11. *The distance balance center of a tree consists of a single vertex or two adjacent vertices.*

However, there may be a bit of a surprise by the fact that the distance balance center might be far from center vertices or median vertices, as the next theorem states.

Theorem 8.12. *There exist trees in which a median vertex, a balance vertex, and a distance balanced vertex can be arbitrarily far apart.*

8.4. Families of central sets inducing K_1 or K_2

In this section we describe several families of central sets, where each member of the family induces K_1 or K_2.

Definition 8.13. Let k be an integer, $0 \leq k \leq dia(T)$. The *ball of radius k about the vertex x* of a tree T, denoted $B(x;k)$ and abbreviated as the *k-ball about x*, is the set of vertices of distance no more than k from x, i.e. $B(x;k) = \{y \in V(T) : d(x,y) \leq k\}$. The *$k$-ball about a set S of vertices* of a tree T is the union of the k-balls about each vertex in S, i.e., $\cup\{B(x;k) : x \in S\}$. The *k-distance* of x, denoted $D(x,k)$, is equal to the sum of the distances between the k-ball about x

and other vertices of T, i.e., $D(x,k) = \sum\{d(y, B(x;k)) : y \in V(T)\}$. The *k-nucleus* of T, denoted $M(T,k)$, consists of all vertices of T with smallest k-distance, i.e., $M(T,k) = \{x \in V(T) : D(x,k) \leq D(y,k), y \in V(T)\}$.

In 1981 Slater [57] introduced this one-parameter family of central sets and studied some of its properties. Note that $M(T,0)$ is the median of T (and the branch weight centroid) and $M(T, r(T))$ is the center of T. So, this is a family of central sets that moves along T from the one or two adjacent vertices in the median to the one or two adjacent vertices in the center as k increases from 0 to $r(T)$. Slater showed that

Theorem 8.13. *Suppose that T is a tree.*

(a) For $0 \leq k \leq r(T)$, $M(T,k)$ consists of a single vertex or two adjacent vertices. This is no longer true for k in the range $r(T)+1 \leq k \leq dia(T)$; in particular, $M(T, dia(T)) = V(T)$.
(b) $\cup\{M(T,k) : 0 \leq k \leq r(T)\}$ induces a subtree of T that contains the center of T and the median of T.

A short proof of part (a) can be found in Reid [46]. The next family of central sets involves minimizing the largest number of certain vertices outside of a k-ball.

Definition 8.14. For an edge ab of a tree T, let $S(a,b;k)$ denote the vertices of $V(a,b)$ that are outside of the k-ball about b, i.e., $S(a,b;k) = V(a,b) \cap (V(T) - B(b;k))$. Equivalently, $S(a,b;k)$ consists of all vertices of T that are in the component of $T-b$ that contains a, and are at least distance $k+1$ from b. The *k-branch weight* of x, denoted $b(x,k)$, is the integer given by $\max\{|S(y,x;k)| : yx \in E(T)\}$, and the *k-branch weight centroid* of T, denoted $Bw(T,k)$, is the set of all vertices of T with smallest k-branch weight, i.e., $Bw(T,k) = \{x \in V(T) : b(x,k) \leq b(y,k), y \in V(T)\}$.

Note that $Bw(T,0)$ is the ordinary branch weight centroid of T (and the median of T), and $Bw(T, r(T))$ is the center of T. In 1996 Zaw Win [71] introduced this one-parameter family of central sets that moves from the one or two adjacent vertices in the centroid to the one or two adjacent vertices in the center as k increases from 0 to $r(T)$, and he proved

Theorem 8.14. *Suppose that T is a tree. Then for $0 \leq k \leq r(T)$, $Bw(T,k) = M(T,k)$. That is, the k-branch weight centroid of T is the same as the k-nucleus.*

The importance of this result is that it shows that the k-branch weight centroid is the proper way to generalize the ordinary branch weight centroid so that Zelinka's Theorem 8.2 holds when the k-nucleus is taken as the generalization of the median. As a result we deduce

Corollary 8.1. *Suppose that T is a tree. Then for $0 \leq k \leq r(T)$, $Bw(T,k)$ is a single vertex or two adjacent vertices.*

The motivating idea for the p-center of a tree was the placement of p facilities to "best" serve customers at all of the vertices. Now, instead, we consider a fixed set of customers, one customer at each vertex of the subset, and we wish to locate a single facility to "best" serve these customers. If there is a customer at every vertex of T, then we simply obtain the (ordinary) center. So, we now consider other subsets of the vertices, and even families of subtrees, at which customers are located and define the corresponding concept of center.

Definition 8.15. If $R = \{R_1, R_2, \dots, R_m\}$ is a collection m vertex sets of m subtrees of a tree T and $x \in V(T)$, then the *R-eccentricity* of x, denoted $e_R(x)$, is the largest distance to the m subtrees, i.e., $e_R(x) = \max\{d(x, R_i) : 1 \leq i \leq m\}$. The *R-center* of T, denoted $C_R(T)$, consists of all vertices of T of smallest R-eccentricity, i. e., $C_R(T) = \{x \in V(T) : e_R(x) \leq e_R(v), v \in V(T)\}$.

When $m = 1$, $e_R(x) = d(x, R_1)$, so $C_R(T) = R_1$. When $|V(R_i)| = 1$, for all i, $1 \leq i \leq m$, set $S = \cup\{V(R_i) : 1 \leq i \leq m\}$. Then the R-eccentricity of x is just the largest distance between x and all vertices of S, i.e., $e_R(x) = \max\{d(x, s) : s \in S\}$. Note that if either $S = V(T)$ or $S = Lf(T)$, then the R-center of T is just the ordinary center of T. In 1978 [55] Slater gave the versions of Definition 8.15 for other subsets $S = \cup\{V(R_i) : 1 \leq i \leq m\}$ when $|V(R_i)| = 1$, for all i, $1 \leq i \leq m$, and proved

Theorem 8.15. *If $R = \{R_1, R_2, \dots, R_m\}$ is a collection m singleton sets of vertices of a tree T (i.e., $|R_i| = 1$, for all i, $1 \leq i \leq m$) and $S = \cup\{R_i : 1 \leq i \leq m\}$, then the R-center of T is the center of $T\langle S\rangle$. Consequently, the R-center of T consists of either one vertex or two adjacent vertices.*

Now we return to the general case where $m > 1$ and there is no restriction on the numbers $|V(R_i)|$. In 1981, Slater [59] presented the definitions in Definition 8.15 and proved the following extension of Theorem 8.15. Part (b) is an interloper in this Section since, under the conditions given, it is possible for $C_R(T)$ to induce a subtree different from K_1 and K_2.

Theorem 8.16.

(a) Suppose that T is a tree and $R = \{R_1, R_2, \dots, R_m\}$ is a collection of m subtrees of T such that $\cap\{V(R_i) : 1 \leq i \leq m\} = \emptyset$. Then $C_R(T)$ is equal to the center of some subtree of T, and hence $C_R(T)$ consists of either one vertex or two adjacent vertices.

(b) Suppose that T is a tree and $R = \{R_1, R_2, \dots, R_m\}$ is a collection m subtrees of T such that $I = \cap\{V(R_i) : 1 \leq i \leq m\} \neq \emptyset$. Then $C_R(T) = I$.

Central sets analogous to the R-center, but using the distance function $D(\cdot)$ (obtaining the R-median) and the branch weight function $bw(\cdot)$ (obtaining the R-branch weight centroid) are described in Section 8.5. Next we introduce a family of central sets involving minimizing the largest total distance to any k customers.

Definition 8.16. Suppose that T is a tree. For $1 \le k \le |V(T)|$ and $x \in V(T)$, define the integer $r_k(x)$ to be the largest possible sum of k distances from x, i.e., $r_k(x) = \max\{\sum d(x,s) : s \in S, S \subseteq V(T), |S| = k\}$. The *k-centrum* of T, denoted $C(T,k)$, consists of all vertices of T that minimize the function $r_k(\cdot)$, i.e., $C(T,k) = \{x \in V(T) : r_k(x) \le r_k(y), y \in V(T)\}$.

Observe that $C(T,1)$ is the center of T and $C(T,|V(T)|)$ is the median of T, which is also the branch-weight centroid of T. So, as the parameter k increases from 1 to $|V(T)|$, this central set moves from the one or two vertices in the center to the one or two vertices in the branch weight centroid, as suggested in the title of Slater's 1978 paper [55] in which he proved (a) - (d) of the following theorem. Part (e) occurs in his paper on the k-nucleus [57].

Theorem 8.17. *Suppose that T is a tree and k is an integer, $1 \le k \le |V(T)|$.*

(a) $C(T,k)$ consists of either one vertex or two adjacent vertices.
(b) For $2 \le k \le |V(T)|$, $C(T,k) \cap C(T,k-1) \ne \emptyset$.
(c) If $S = \cup\{C(T,k) : 1 \le k \le |V(T)|\}$, u is in the center of T, and v is in the centroid of T, then $T[S]$ is a subtree of T containing the path between u and v.
(d) There exist trees T so that if S is as in (c), then $T[S]$ is not a path.
(e) $\cup\{C(T,k) : 1 \in k \in |V(T)|\} = \cup\{M(T,k) : 0 \le k \le r(T)\}$ (recall, from above, that $M(T,k)$ denotes the k-nucleus of T).

A short proof of part (a), based on the recommended proof of Theorem 8.1, can be found in Reid [46]. Next, we present a family minimizing the largest connected set of customers, all at more than distance k.

Definition 8.17. If x is a vertex of a tree T, and k is a non-negative integer, the *k-ball branch weight* of x, denoted $b(x,k)$, is the order of a largest subtree in the forest of trees obtained by removing from T the k-ball about x, i.e., $b(x,k) = \{|V(C)| : C \text{ a component of } T - B(x;k)\}$. The *k-ball branch weight centroid* of T, denoted $W(T,k)$, is the set of all vertices with smallest k-ball branch weight, i.e., $W(T,k) = \{x \in V(T) : b(x,k) \le b(y,k), y \in V(T)\}$.

Observe that $W(T,0)$ is the ordinary branch weight centroid of T (which is the median of T), and $W(T,r(T))$ is the center of T. In 1991, Reid [45] introduced this one-parameter family of central sets that moves along T from the branch weight centroid to the center as k increases, as suggested in the title of his 1991 paper. These sets are certainly reminiscent of the k-nucleus $M(T,k)$ (which is the k-branch weight centroid $Bw(T,k)$ by Theorem 8.14) or the k-centrum $C(T,k)$. Indeed, $W(T,0) = M(T,0) = Bw(T,0)$ and $W(T,r(T)) = M(T,r(T)) = B(T,r(T))$. However, Reid [45] showed

Theorem 8.18.

(a) There exists trees T so that $\cup\{W(T,k) : 1 \leq k \leq r(T)-1\}$ and $\cup\{M(T,k) : 1 \leq k \leq r(T)-1\}$ are disjoint.

(b) There exists trees T so that each of the two sets $\cup\{W(T,k) : 0 \leq k \leq r(T)\}$ and $\cup\{C(T,k) : 1 \leq k \leq |V(T)|\}(= \cup\{M(T,k) : 0 \leq k \leq r(T)\} = \cup\{B(T,k) : 0 \leq k \leq r(T)\}$, by Theorems 8.14 and 8.17(e)) contains a vertex not in the other.

(c) Let T be a tree. For each k, $0 \leq k \leq r(T)$, $W(T,k)$ consists of a single vertex or two adjacent vertices.

(d) Let T be a tree. If $S = \cup\{W(T,k) : 0 \leq k \leq r(T)\}$, then $T[S]$ is a subtree of T containing the centroid of T and the center of T.

Subsequently, Reid [46] proved

Theorem 8.19. *For each $k \geq 0$, there is a tree T so that the k-ball branch weight centroid and the k-branch weight centroid (= the k-nucleus) are distance k apart.*

Another way to describe $b(x;k)$, the k-ball branch weight of a vertex x in a tree, is that it is the largest number of vertices of T outside of $B(x;k)$, the k-ball about x, that are reachable in $T - B(x;k)$ from exactly one vertex of distance $k+1$ from x. This suggests a second parameter for consideration, namely, path lengths.

Definition 8.18. A vertex y in a tree T is *reachable from a vertex x via a path P* that has one end at x if the unique path from x to y contains the path P. For $0 \leq l \leq dia(T)$, let $\mathcal{P}(x,l)$ denote the set $\{P : P$ a path in T of length l with one end at $x\}$, and for $P \in \mathcal{P}(x,l)$, let $\beta_P(x,k) = |\{y \in V(T) : y$ is reachable from x via $P\} \cap (V(T) - B(x;k))|$. That is, $\beta_P(x,k)$ is the number of vertices that are outside the k-ball about x and that are reachable from x via path P. The *k-ball l-path branch weight* of x, denoted $\beta_l(x,k)$, is the integer $\max\{\beta_P(x,k) : P \in \mathcal{P}(x,l)\}$, and the *$k$-ball l-path branch weight centroid* of T, denoted $B_l(T,k)$, is the set of all vertices of T with smallest k-ball l-path branch weight, i.e., $B_l(T,k) = \{x \in V(T) : \beta_l(x,k) \leq \beta_l(y,k), y \in V(T)\}$.

For special values of the parameters k and l no new central sets are obtained. When $k=0$ and $l=0$, the 0-ball 0-path branch weight centroid is $V(T)$. The 1-ball 0-path branch weight centroid consists of all vertices of maximum degree. When $l=1$, the k-ball 1-path branch weight centroid of T is just the k-branch weight centroid which is also the k-nucleus. In particular, when $k=0$ and $l=1$, the 0-ball 1-path branch weight centroid of T is the (ordinary) branch weight centroid which is also the (ordinary) median. When $l = k+1$, the k-ball $(k+1)$-path branch weight centroid of T is just the k-ball branch weight centroid. When either $k = r(T)$ and $0 \leq l \leq r(T)+1$, or $0 \leq k \leq r(T)$ and $l = r(T)+1$, the k-ball l-path branch weight centroid of T is the (ordinary) center. For appropriate values of the parameters k and l, Reid [46] showed that these central sets, once again, consist of a single vertex or two adjacent vertices. That is, he proved

Theorem 8.20. *For $0 \leq k \leq r(T)$ and $0 < l \leq r(T) + 1$, the k-ball l-path branch weight centroid of T consists of a single vertex or two adjacent vertices.*

The next family of central sets is an extension of the processing center encountered in Section 8.2.

Definition 8.19. Let T be a tree and let k be a positive integer. A subset S of leaves of T is a *k-processable set* in T provided $|S| \leq k$. A sequence $(S_1, S_2, \ldots, S_m)$ is a *k-processable sequence* of T if S_1 is a k-processable set in T, and, for each j, $2 \leq j \leq m$, S_j is a k-processable set in $T - \cup\{S_i : 1 \leq i \leq j-1\}$. Let $x \in V(T)$. The *k-processing number* of x, denoted $p_k(x)$, is the smallest possible index m so that there exists a k-processable sequence $(S_1, S_2, \ldots, S_m)$ of T with $x \in S_m$, i.e., $p_k(x) = \min\{m :$ there exists a k-processable sequence$(S_1, S_2, \ldots, S_m), x \in S_m\}$. The *k-processing center* of T, denoted $P_k(T)$, consists of all vertices of T with largest k-processing number, i.e., $P_k(T) = \{v \in V(T) : p_k(x) \leq p_k(v), x \in V(T)\}$.

Cikanek and Slater [10] introduced this family of central sets as part of a multiprocessor job scheduling problem. If $k = 1$, $P_1(T) = Proc(T)$, the processing center from Section 8.2, which is the median. If k is at least the number of leaves of T, then $P_k(T) = C(T)$. They showed

Theorem 8.21.

(a) For all trees T and for all positive integers k, $P_k(T)$ consists of either a single vertex or two adjacent vertices.
(b) There exists a tree T and a positive integer k so that $P_k(T) \cap P_{k+1}(T) = \emptyset$.
(c) There exists a tree T of order n so that $T[\cup\{P_k(T) : 1 \leq k \leq n\}]$ is not connected.
(d) There exists a tree T of order n so that $\cup\{P_k(T) : 1 \leq k \leq n\}$ is not contained in any path of T.

Condition (c) in Theorem 8.21 is sufficient to show that the collection of k-processing centers is not the same as the collection of k-nuclei or the collection of k-branch weight centroids or the collection of k-centra or the collection of k-ball branch weight centroids.

In some applications both the minimax response distance and the minimax total travel distance need to be considered. One way to do this is to consider combinations of the eccentricity function $e(\cdot)$ and the total distance function $D(\cdot)$.

Definition 8.20. Let T be a tree and let λ be a real number, $0 \leq \lambda \leq 1$. For $x \in V(T)$ define $g_\lambda(x) = \lambda D(x) + (1-\lambda)e(x)$. The *cent-dian* of T, denoted $C_\lambda(T)$, consists of all vertices of T that minimize the function $g_\lambda(\cdot)$, i.e., $C_\lambda(T) = \{x \in V(T) : g_\lambda(x) \leq g_\lambda(y), y \in V(T)\}$. A vertex in $C_\lambda(T)$ is called a *cent-dian*.

Note that $C_0(T) = C(T)$ and $C_1(T) = M(T)$. Consequently, if $C(T) = M(T)$, then $C_\lambda(T) = C(T) = M(T)$, for all λ, $0 \leq \lambda \leq 1$. Halpern [21] introduced this concept for tree networks in 1976, where $g_\lambda(\cdot)$ is defined for all points of the tree network, not just for vertices of T, and $C\lambda(T)$ consists of all points minimizing $g_\lambda(\cdot)$. So, a centdian in a tree network T might be located at a point that is not a vertex. It will be convenient to describe Halpern's result for tree networks, then interpret it for ordinary trees. Recall that the absolute center of T is unique, but either there is a unique absolute median located at a vertex or the set of absolute medians consists of all points located on an edge, including the two end vertices of that edge. So, either there is a unique absolute vertex median that is closest to the absolute center or the absolute center is contained on the edge on which all absolute medians are located and it is equidistant from the two end vertex medians of that edge. Halpern proved

Theorem 8.22. *Let c denote the absolute center of a tree network T and let λ be a real number, $0 \leq \lambda \leq 1$.*

(a) If there is a unique absolute median (located at a vertex) m that is closest to c, then $C_\lambda(T)$ contains a vertex, is contained in the path $P(m, c)$ of T from m to c, and consists of either a single vertex, or one complete edge, or (if c is not a vertex) a partial edge containing a vertex and c. Moreover, each of these possibilities occurs for suitable λ.
(b) If the vertex weights are all integers, then for every λ in the range $\frac{1}{2} \leq \lambda \leq 1$, $C_\lambda(T)$ is equal to the set of absolute medians of T.

Halpern described which possibility in part (*a*) occurs based on the value of $\lambda^{-1} - 1$. Moreover, when the absolute center is on an edge xy, but different from x and y, he showed that by subdividing the edge xy by making the point c a vertex with weight $\lambda^{-1} - 1$, then the absolute median of the new tree network and the centdian of the original tree network are the same points. This reduces the search for a centdian to the search for an absolute median.

Once again we see that when these comments are restricted to ordinary trees, rather than tree networks, this central set $C_\lambda(T)$ consists of a single vertex or two adjacent vertices. In 2002 Win and Myint [69] independently discovered most of the tree network version of Theorem 8.22 by using the definitions explicitly given in Definition 8.20 (instead of using continuity arguments as in [21]), but they named $C_\lambda(T)$ the λ-cendian of T. They proved, among other things, that for any vertex u on the path $P(m, c)$ (given in part (*a*) of Theorem 8.22) there exist infinitely many λ, where $0 < \lambda < 1$, so that $C_\lambda(T) = \{u\}$, and for any two adjacent vertices u and v on $P(m, c)$ there exists a λ, where $0 < \lambda < 1$, so that $C_\lambda(T) = \{u, v\}$.

8.5. Central subtrees

We now turn our attention to central sets that might induce a subtree different from K_1 or K_2. We present these concepts in increasing complexity of the induced subtree. As was the case for p-centers, the motivating idea for p-medians was to place p facilities to "best" serve customers at all of the vertices. As was done in Section 8.4 for R-centers, we now consider other subsets of the vertices at which customers are located and define the corresponding concepts of median and branch weight centroid.

Definition 8.21. Suppose $R = \{R_1, R_2, \ldots, R_m\}$ is a collection of m subtrees of T and $x \in V(T)$. The *R-distance* of x, denoted $D_R(x)$, is the sum of the shortest distances to all R_i, i.e., $D_R(x) = \sum\{d(x, R_i) : 1 \leq i \leq m\}$. The *R-branch weight* of x, denoted $bw_R(x)$, is the largest number of members of R entirely contained in one connected component of $T - x$, i.e., $bw_R(x) = \max\{k :$ some k of the subtrees in $R$are completely contained in one component of$T - x\}$. The *R-median* of T, denoted $M_R(T)$, consists of all vertices of T of smallest R-distance, i.e., $M_R(T) = \{x \in V(T) : D_R(x) \leq D_R(v), v \in V(T)\}$, and the *R-branch weight centroid*, denoted $Bw_R(T)$, consists of all vertices of T with smallest R-branch weight, i.e., $Bw_R(T) = \{x \in V(T) : bw_R(x) \leq bw_R(v), v \in V(T)\}$.

We consider some special cases. If $m = 1$, then $D_R(x) = d(x, R_1)$, and, for all $x \in R_1$, $bw_R(x) = 0$, while for all $x \notin R_1$, $bw_R(x) > 0$. That is, for $m = 1$, $M_R(T) = Bw_R(T) = R_1$. When $|V(R_i)| = 1$, for all i, $1 \leq i \leq m$, define S to be the union of all of the R_i's, i.e., $S = \cup\{V(R_i) : 1 \leq i \leq m\}$. Then the R-distance of x is just the sum of the distances between x and all vertices of S, i.e., $D_R(x) = \sum\{d(x, s) : s \in S\}$. And, the R-branch weight of x is the maximum number of vertices of S in any connected component of $T - x$, i.e., $bw_R(x) = \max\{|S \cap V(C)| : C$ a connected component of $T - x\}$. This implies that if $|V(R_i)| = 1$, for all i, $1 \leq i \leq m$, and $S = V(T)$, then the R-median of T is just the ordinary median of T and the R-branch weight centroid of T is just the ordinary branch weight centroid of T. And, if $S = Lf(T)$, the set of leaves of T, then we obtain the leaf median and the leaf branch weight centroid of Section 8.2. For other subsets S, new central sets are possible. In 1978 [55] Slater proved

Theorem 8.23. *Suppose that T is a tree and $R = \{R_1, R_2, \ldots, R_m\}$ is a collection of m subtrees of T such that $|V(R_i)| = 1$, for all i, $1 \leq i \leq m$,*

(a) $M_R(T)$ consists of the vertices of a path.
(b) If $|M_R(T)| \geq 2$, then $|\cup\{V(R_i) : 1 \leq i \leq m\}|$ is even; if $|\cup\{V(R_i) : 1 \leq i \leq m\}|$ is odd, then $M_R(T)$ consists of a single vertex.
(c) $Bw_R(T) = M_R(T)$.

Again, part (c) is a Zelinka type result for R-centrality. A bit more is possible when R does not consist of a collection of singletons, as was shown by Slater [59] who proved the following theorem.

Theorem 8.24. *Suppose that T is a tree and $R = \{R_1, R_2, \ldots, R_m\}$ is a collection of m subtrees of T.*

(a) If $\cap\{V(R_i) : 1 \leq i \leq m\} = \emptyset$*, then* $M_R(T)$ *consists of the vertices of a path.*
(b) If $I = \cap\{V(Ri) : 1 \leq i \leq m\} \neq \emptyset$*, then* $M_R(T) = I$ *(which is also equal to* $C_R(T)$ *by Theorem 8.16(b)).*
(c) $M_R(T) \subseteq Bw_R(T)$*, and (as in Theorem 8.23) if* $|V(R_i)| = 1$*,* $1 \leq i \leq m$*, or if* $m = 1$*, then* $M_R(T) = Bw_R(T)$*.*

Broadcasting in a communications network is a process of message dissemination whereby a message, originated by one member (or site), is spread to all members of the network. This gives rise to the concept of a broadcasting center.

Definition 8.22. If k is a positive integer, then *k-broadcasting* in a tree T is accomplished by placing a series of calls (or transmissions) over the communication lines of the network, i.e., the edges. This is to be done as quickly as possible subject to two conditions:

1) each call involves one caller at a vertex who sends one message to each of k (or fewer) of its vertex neighbors in one unit of time,
2) each vertex receiving a call can participate in at most one call per unit of time.

The *k-broadcast time* of a vertex x in a tree T, denoted $bc_k(x)$, is the minimum possible number of units of time required to transmit a message from x to all vertices of T via k-broadcasting. The *k-broadcast center* of T, denoted $BC_k(T)$, consists all vertices of T with smallest k-broadcast number, i.e., $BC_k(T) = \{x \in V(T) : bc_k(x) \leq bc_k(y), y \in V(T)\}$.

Of course, if the tree T has order n, then $bc_k(v) \geq \lceil \log_{k+1} n \rceil$, since during each time unit the number of informed vertices can at most be multiplied by $k+1$. In 2003 Harutyunyan and Shao [25] proved

Theorem 8.25. *Let* T *be a tree, and let* k *be a positive integer.*

(a) $BC_k(T)$ *induces either* K_1 *or a star* $K_{1,m}$ *with* $m \geq 1$*.*
(b) If $v \in V(T) - BC_k(T)$*, then* $b_k(v) = b_k(u) + d(v,u)$*, where* u *is the vertex of* $BC_k(T)$ *nearest to* v*.*

Subsequently, in 2009 Harutyunyan, Liestman, and Shao [24] published a linear algorithm for finding $BC_k(T)$ for a tree T. Much earlier, Slater, Cockayne, and Hedetniemi [63] obtained Theorem 8.25 in the case $k = 1$ and gave a linear algorithm for finding $BC_1(T)$. Work has been reported on determining those trees (and graphs) of order n in which there is a root v with $b_k(v) = \lceil \log_{k+1} n \rceil$ (see the references in [24]). These sources contain other references for work on another fascinating topic on information dissemination in trees and graphs, the topic of gossiping, a topic spawned by the "gossip problem". The gossip problem can be stated as follows. At each vertex of K_n there is a caller who knows a piece of information that is unknown to any of the other callers. They communicate by telephone, and whenever a call is made between two callers, they pass on to each other as much information as they know at the time. What is the minimum possible number of calls required so that all callers know all n pieces of information?

Next we consider what might be called edge centrality. One way to do this is to consider an edge xy as two singleton sets and employ R-centrality with $R = \{\{x\}, \{y\}\}$ as in Section 8.4 for the R-center and as in this section above for the R-median and the R-branch weight centroid. Another approach is to generalize Steiner k-centrality of this section below to sets S of size k that contain both x and y. Or, find a central 2-tree as described below in this section. The approach that we describe here is to look for an edge that is most "balanced."

Definition 8.23. If xy is an edge of a tree T, the *distance-edge difference* of xy, denoted $s(xy)$, is the absolute value of the difference of the sum of the distances between x and all vertices that are closer to x than to y and the sum of the distances between y and all vertices that are closer to y than to x, i.e.,

$$s(xy) = |(\sum\{d(x,z) : z \in V(x,y)\}) - (\sum\{d(y,w) : w \in V(y,x)\})|.$$

Of course, $s(xy) = s(yx)$. The *distance balanced edge center* of T, denoted $DBE(T)$, consists of all edges of T which have smallest distance-edge difference, i.e., $DBE(T) = \{xy \in E(T) : s(xy) \leq s(uv), uv \in E(T)\}$. The *weight-edge difference* of the edge xy, denoted $r(xy)$, is the absolute value of the difference between the number of vertices that are closer to x than to y and the number of vertices that are closer to y than to x, i.e., $r(xy) = ||V(x,y)| - |V(y,x)||$. The *weight balanced edge center* of T, denoted $WBE(T)$, consists of all edges of T which have the smallest weight-edge difference, i.e., $WBE(T) = \{xy \in E(T) : r(xy) \leq r(uv), uv \in E(T)\}$.

These definitions were introduced and studied by Reid and DePalma [49]. Analogous to the motivation for the distance balance of a vertex, consider attempting to balance a tree T at an edge. Given an edge xy, a vertex $z \in V(T) - \{x,y\}$ is considered as a unit weight on the lever arm whose length is given by the distance $d(z, \{x,y\})$. The *total moment* of xy with respect to $V(x,y)$ is $\sum\{d(x,z) : z \in V(x,y)\}$, and the *moment* of xy with respect to $V(y,x)$ is $\sum\{d(y,w) : w \in V(y,x)\}$. So, $s(xy)$ is a measure of how balanced the edge xy is in T. The edges in $DBE(T)$ are the most balanced edges in T. In considering the weight-edge difference of an edge xy, we disregard the distance of the vertices from xy and consider only the total weights of the two subtrees, $T[V(x,y)]$ and $T[V(y,x)]$ on "either side" of xy. This type of balance will be best for edges in $WBE(T)$. Reid and DePalma [49] described $DBE(T)$ in terms of the vertices in the distance balance center $DB(T)$ of Section 8.3, and similarly described $WBE(T)$ in terms of the vertices in the weight balance center $Wb(T)$ of Section 8.2.

Theorem 8.26. *Let T be a tree.*

(a) *If $DBE(T)$ consists of a single edge, say $DBE(T) = \{xy\}$, then $DB(T)$ is either $\{x\}$, $\{y\}$, or $\{x,y\}$.*

(b) If $DBE(T)$ consists of more than one edge, then $DB(T)$ consists of a single vertex x incident with each edge in $DBE(T)$, i.e., $DBE(T)$ induces a star $K_{1,m}$, for some $m \geq 2$, with x the vertex of degree m in the star.

(c) If $Wb(T)$ consists of two adjacent vertices, say $Wb(T) = \{x, y\}$, then $WBE(T) = \{xy\}$.

(d) If $Wb(T)$ consists of a single vertex, say $Wb(T) = \{x\}$, then every edge in $WBE(T)$ is incident with x, i.e., $WBE(T)$ induces a star $K_{1,m}$, for some $m \geq 1$, with x the vertex of degree m in the star.

Next, we describe a generalization of distance between two vertices that has applications to multiprocessor communications.

Definition 8.24. Suppose that S is a set of k vertices in a tree T. The *Steiner distance* of S in T, denoted $sd(S)$, is the number of edges in $T\langle S\rangle$, the smallest subtree of T that contains every vertex of S. For $2 \leq k \leq |V(T)|$, and $x \in V(T)$, the *Steiner k-eccentricity* of x in T, denoted $se_k(x, T)$ (or simply $se_k(x)$) is the largest possible Steiner distance of all k-sets of vertices in T that contain x, i.e., $se_k(x) = \max\{sd(S) : S \subseteq V(T), |S| = k, x \in S\}$. The *Steiner k-center* of T, denoted $SC_k(T)$, consists of all vertices with smallest Steiner k-eccentricity, i.e., $SC_k(T) = \{x \in V(T) : se_k(x, T) \leq se_k(y, T), y \in V(T)\}$. The *Steiner k-distance* of x, denoted $sD_k(x)$, is the sum of all Steiner distances of sets of k vertices that include x, i. e., $sD_k(x) = \sum\{sd(S) : S \subseteq V(T), |S| = k, x \in S\}$. The *Steiner k-median* of T, denoted $SM_k(T)$, consists of all vertices of T with smallest Steiner k-distance, i.e., $SM_k(T) = \{x \in V(T) : sD_k(x) \in sD_k(y), y \in V(T)\}$.

If $|S| = 2$, then $sd(S)$ is the distance between the two vertices of S. This implies that the Steiner 2-eccentricity of a vertex x is simply the ordinary eccentricity of x, and the Steiner 2-distance of x is the ordinary distance of x. So, the Steiner 2-center of T is the center of T, and the Steiner 2-median of T is the median of T. Consequently, each of the Steiner 2-center and Steiner 2-median is either a single vertex or two adjacent vertices. For $k \geq 3$, the structure of the subgraph induced by the Steiner k-center can be quite different from K_1 or K_2. Oellermann and Tian [41] gave a linear algorithm for finding the Steiner k-center of a tree and proved

Theorem 8.27. *Suppose that T is a tree and $3 \leq k \leq |V(T)|$.*

(a) The vertices of $SC_k(T)$ are exactly the vertices of a subtree of T.

(b) T is the Steiner k-center of some tree if and only if the number of leaves of T is at most $k - 1$.

Later, Beineke, Oellermann, and Pippert [2] gave a linear algorithm for finding the Steiner k-median of a tree and proved

Theorem 8.28.

(a) The k-median of any tree is connected, for all $k \geq 2$.

(b) *The vertex set of a tree H of order n is the Steiner k-median of some tree if and only if one of the following holds:*

(i) $H = K_1$,
(ii) $H = K_2$,
(iii) $n = k$, *or*
(iv) *H has at most $k - n + 1$ leaves.*

Next we consider subtrees of a tree that are central.

Definition 8.25. A subtree W of order k of a tree T of order n, $1 \leq k \leq n$, is called a central k-tree if W has minimum eccentricity among all subtrees of T of order k, i.e., $e(W) = \min\{e(W') : W' \text{ a subtree of } T \text{ of order } k\}$.

Of course, when either $k = 1$ and $|C(T)| = 1$ or $k = 2$ and $|C(T)| = 2$, then $C(T)$ induces the unique central k-tree. On a tree network, a central k-tree is a subtree of the network of total length k and minimum eccentricity among all subtrees of total length k, where a subtree is allowed to have its leaves at points of the network that are not vertices. In 1985 Minieka [34] gave a $O(n^2)$ algorithm for obtaining such a subtree in a tree network; it starts at the absolute center and grows outward. In 1993, Hakimi, Schmeichel, and Labbé [20] adapted, without proof, Minieka's approach in order to find a central k-tree that has all of its leaves at vertices. Subsequently, Shioura and Shigeno [52] gave an $O(n)$ algorithm for this by using relations to the bottleneck knapsack problem. McMorris and Reid [32] took a slightly different approach for ordinary trees. They proved

Theorem 8.29.

(a) *There is a linear time, tree-pruning procedure that yields a central k-tree of a tree, but generally such a subtree is not unique.*
(b) *Fix integers n and k, $1 \leq k \leq n$. Every tree of order k is a central k-tree for some tree of order n.*

There have been many algorithmic developments of this topic for networks, particularly for tree networks, including recent instances in which new path or tree shaped facilities are added to networks in which perhaps some path or tree shaped facilities are already located (see, for example, Bhattacharya, Hu, Shi, and Tamir [3] as well as Tamir, Puerto, Mesa, and Rodríguez-Chía [67] and the references therein).

Next we discuss a centrality concept in a tree T due to Nieminen and Peltola [40] that involves a subsidiary algebraic structure in determining a central set in a tree T. In this discussion care must be taken to distinguish distance in the tree T and distance in the associated algebraic structure.

Definition 8.26. For two subtrees S_1 and S_2 of a tree T, the *meet* of S_1 and S_2, denoted $S_1 \wedge S_2$, is the intersection of the two subtrees whenever the intersection is nonempty, and the *join* of S_1 and S_2, denoted $S_1 \vee S_2$, denotes the least subtree of T that contains both S_1 and S_2. Here we do not consider the empty graph to be a subtree of T. So, the subtrees of T form a join-semilattice $\mathcal{R}(T)$ of subtrees of T in which, in general, there is no least element. The *distance* in $\mathcal{R}(T)$ between S_1 and S_2 through $S_1 \wedge S_2$ (when defined) is the distance through $S_1 \vee S_2$. So, distance in $\mathcal{R}(T)$ is the same as the distance in the (undirected) graph distance in the (undirected) Hasse diagram graph $G_{\mathcal{R}}$ of $\mathcal{R}(T)$. The distance between S_1 and S_2 in $\mathcal{R}(T)$, or the $\mathcal{R}$-distance between S_1 and S_2, denoted $d_{\mathcal{R}}(S_1, S_2)$, is defined to mean the usual graphical distance in $G_{\mathcal{R}}$. The $\mathcal{R}$-eccentricity of a subtree S of T, denoted $e_{\mathcal{R}}(S)$, is the largest $\mathcal{R}$-distance from S to the other subtrees of T, i.e., $e_{\mathcal{R}}(S) = \max\{d_{\mathcal{R}}(S, S') : S' \in \mathcal{R}(T)\}$. A least central subtree of T is a smallest subtree among all those subtrees of T with least $\mathcal{R}$-eccentricity, i.e., a subtree of smallest order in the set $\{S \in \mathcal{R}(T) : e_{\mathcal{R}}(S) \leq e_{\mathcal{R}}(S'), S' \in \mathcal{R}(T)\}$. The subtree center of T is the union of all of the least central subtrees of T.

Note that the $\mathcal{R}$-eccentricity of a subtree S is not the same as $e(S)$, the eccentricity of S, since $e(S)$ involves distance in the tree T, while the $\mathcal{R}$-eccentricity involves distance in the Hasse diagram graph $G_{\mathcal{R}}$. A least central subtree of T need not be unique, but the subtree center of T is unique. Of course, if the subtree center is a single vertex, then that is the unique least central tree. In 1999 Nieminen and Peltola [40] introduced these concepts and described properties of least central subtrees as a way to describe the subtree center. They proved

Theorem 8.30. *Suppose that T is a tree.*

(a) Any two least central subtrees of T have a nonempty intersection. Thus, the subtree center of T is a subtree of T.
(b) The least central subtree of T is a single vertex if and only if T is either a path with an odd number of vertices or a star (i.e., $K_{1,m}$, for some $m \geq 2$).
(c) There is at least one least central subtree of T containing at least one vertex of the branch weight centroid of T.
(d) The (ordinary) center of T intersects at least one least central subtree of T.

A recent addition to the library of central sets in trees is due to Dahl [11].

Definition 8.27. Let x be a vertex in a tree T of order $n \geq 2$. Let $< x >$ denote the n-vector whose terms are the n distances between x and all vertices of T, arranged in non-increasing order (so, the n-th entry in $< x >$ has the value $d(x, x) = 0$, and the first entry in $< x >$ has the value $e(x)$). If v_1 and v_2 are vertices in T with $< v_1 >= (r_1, r_2, \ldots, r_n)$ and $< v_2 >= (s_1, s_2, \ldots, s_n)$, write $v_1 \prec v_2$ to mean that the sum of the first k terms in $< v_1 >$ is less than or equal to the sum of the first k terms in $< v_2 >$ for all k, $1 \leq k \leq n$, i.e., $\sum\{r_i : 1 \leq i \leq k\} \leq \sum\{s_i : 1 \leq i \leq k\}$, for all k, $1 \leq k \leq n$. Vertices $v_1, v_2 \in V(T)$ are *majorization-equivalent* if $v_1 \prec v_2$ and $v_2 \prec v_1$.

Lemma 8.2. *A tree contains at most one pair of distinct majorization-equivalent vertices.*

Definition 8.28. The majorization center of a tree T, denoted M_T, is the subset of $V(T)$ defined as follows: $M_T = \{v_1, v_2\}$ if v_1 and v_2 are majorization-equivalent, and $M_T = \cap\{V(v_1, v_2) : v_1v_2 \in E(T), v_1 \prec v_2\}$ otherwise.

Dahl [11] proved, among other things, that the subtree induced by the majorization center could have any tree structure.

Theorem 8.31. *Suppose that T is a tree of order n.*

(a) M_T induces a subtree of T that contains no leaf of T, except when $n = 2$.
(b) The center $C(T)$ of T, the median $M(T)$ of T, and the distance balance center $DB(T)$ of T are all contained in M_T.
(c) There is a tree W so that $T = M_W$.

Dahl also described a way to produce other tree centers. The set of vectors in $\mathcal{R}^n$ with terms in $\mathcal{R}^+$ is denoted $\mathcal{R}^n_+$. Start with a function $f : \mathcal{R}^n_+ :\rightarrow \mathcal{R}$ that is Shur-convex (i.e., $f(x) \leq f(y)$ whenever $x, y \in \mathcal{R}^n_+$ and $x \prec y$) and nondecreasing (i.e., $f(x) \leq f(y)$ whenever $x, y \in \mathcal{R}^n_+$ and $x \leq y$, componentwise). Here, $x \prec y$ is defined on n-tuples and means that the sum of the first k terms in x is less than or equal to the sum of the first k terms in y for all k, $1 \leq k \leq n$. Define $F : V(T) \rightarrow \mathcal{R}$ by $F(v) = f(< v >)$ for $v \in V(T)$. Declare that an F-center consists of all vertices that minimize the function $F(\cdot)$. Dahl showed that each vertex in this F-center is in M_T, and he gave specific examples of appropriate functions $F(\cdot)$.

8.6. Disconnected central sets

In this section we discuss central sets that induce disconnected subgraphs of a tree T.

Definition 8.29. The *cutting number* of a vertex in a tree T, denoted $c(x)$, is the number of (unordered) pairs of distinct vertices $\{u, v\}$ so that u and v are in distinct components of $T - x$. The *cutting center* of T, denoted $Cut(T)$, consists of all vertices of T with largest cutting number, i.e., $Cut(T) = \{x \in V(T) : c(x) \geq c(y), y \in V(T)\}$.

This measure was introduced and studied by Harary and Ostrand in 1971 [23]. In some respects this set is reminiscent of the previously discussed telephone center of Section 8.2 that turned out to be the median. In each middle set discussed in Sections 8.1-8.5, the vertices of the middle set induce a connected subtree. That is not the case for the cutting center. Harary and Ostrand proved

Theorem 8.32.

(a) *For each tree T there is a path that contains all of the vertices in $Cut(T)$.*
(b) *Given a path and any non-empty subset C of its vertices, the path can be extended to a tree T with $Cut(T) = C$.*

In particular, $Cut(T)$ need not induce a subtree of T. In fact, connected components induced by $Cut(T)$ can be arbitrarily far apart, and, indeed, the subgraph induced by $Cut(T)$ might have no edges at all. In 2004 Chaudhuri and Thompson [8] described a quadratic algorithm for finding the cutting center of a tree. In 1978 Chinn [9] independently rediscovered this concept. She defined the path number of a vertex x in a tree T, denoted $p(x)$, as the number of paths of T that contain x as an interior vertex, and defined the *path centrix* of T, denoted $Cx(T)$, to be all vertices in T that maximize the function $p(.)$, i.e., $Cx(T) = \{x \in V(T) : p(x) \geq p(y), y \in V(T)\}$. Of course, $Cut(T) = Cx(T)$.

In 1975 Slater [53] introduced another middle set of a tree T that might induce a structure more akin to the possibilities encountered by the cutting center than the possibilities of many previously discussed middle sets. At first glance, the terms involved in the definition suggest that this is a "median" version of the security center discussed in Section 8.2.

Definition 8.30. The *security index* of a vertex x in a tree T, denoted $si(x)$, is the integer $si(x) = \sum\{(|V(x,v)| - |V(v,x)|) : v \in V(T) - \{x\}\}$. The *security centroid* of T, denoted $Si(T)$, consists of all vertices of with largest security index, i.e., $Si(T) = \{x \in V(T) : si(x) \geq si(y), y \in V(T)\}$.

Slater [53] proved that the subgraph induced by this central set might be far from being connected, and in fact, might contain no edges at all. He proved the following theorem

Theorem 8.33. *For any positive integer n there is a tree T such that $Si(T)$ consists of n independent vertices (i.e., no two adjacent), and if $n \geq 3$, T can be chosen so that no three vertices in $Si(T)$ lie on the same path.*

Another central set for trees that need not induce a connected subgraph is the harmonic center. Suppose, because of the weight of fuel, for example, that a delivery vehicle can only carry a partial load of goods to a customer. That is, the greater the distance to a customer, the smaller the load that can be delivered. Suppose that only $(\frac{1}{d})$-th of a load can be delivered from vertex site x to a customer at vertex y, where $d = d(x,y)$. A desirable site at which to place a warehouse of the goods might be a vertex site from which the greatest total sum of (full or partial) loads can be delivered to customers at all the other vertices.

Definition 8.31. The *harmonic weight* of a vertex x in a tree T, denoted $h(x)$, is the sum of the reciprocals of the distances from x to the other vertices in T, i.e.,

$h(x) = \sum\{\frac{1}{d(x,y)} : y \in V(T) - \{x\}\}$. The *harmonic center* of T, denoted $H(T)$, consists of all vertices of T of maximum harmonic weight, i.e., $H(T) = \{x \in V(T) : h(y) \leq h(x), y \in V(G)\}$.

This concept was suggested by Reid and studied by Laskar and McAdoo [30] who proved that the harmonic center has some similarities to the cutting center and the security centroid.

Theorem 8.34.

(a) *The harmonic center of a tree T contains no leaf of T, unless $|V(T)| \leq 2$.*
(b) *The tree that is a path has harmonic center consisting of either a single vertex or two adjacent vertices.*
(c) *For any positive integer p, there is a tree T so that, for some integer c, the vertex set of $H(T)$ can be partitioned into p sets, each containing c vertices, and T contains no edge between any two of the sets.*

8.7. Connected central structures

In some of the central sets described above, we could just as well have couched the discussion in terms of central substructures induced by those central sets. In fact, many of the authors referenced above did just that in their original articles. In this section we will shift the focus to specific types of central substructures of a tree T. If S is a subforest of T or a subset of $V(T)$ and if there is no other restriction on the set S, then, proceeding as in the definitions of the ordinary center, median, and branch weight centroid, we see that the S that minimizes $e(S)$, $D(S)$ and $bw(S)$ is $S = T$ or $S = V(T)$. For $|V(T)| \geq 3$, the whole tree is usually not considered very central in itself. However, specific structural restrictions on S make for some interesting central structures. In particular, by restricting S to be a path we obtain path-centrality.

Definition 8.32. Suppose that T is a tree. A *path center* of T, denoted $PC(T)$, is a path P in T of shortest length so that $e(P) \leq e(P')$ for every path P' in T. A *core* (or *path median*) of T, denoted $Co(T)$, is a path P in T so that $D(P) \leq D(P')$ for every path P' in T. A *spine* (or *path branch weight centroid*) of T, denoted $Sp(T)$, is a path P of shortest length so that $bw(P) \leq bw(P')$, for every path P' in T.

The path center concept seems to have originated independently in 1977 by S. M. Hedetniemi, Cockayne, and S. T. Hedetniemi [26] and by Slater [60]. The former authors gave linear algorithms for finding the path center of a tree and proved the following theorem

Theorem 8.35. *If T is a tree, then $PC(T)$ is a unique subpath of T and its vertex set contains the center of T.*

Subsequently, Slater [60] showed that there are trees in which a core (path median) is not necessarily unique, and there are trees in which no core contains any median vertex. That is, there are trees in which the vertices of minimum distance may not be contained in any path of minimum distance. Morgan and Slater [1980] presented a linear algorithm for finding a core of a tree that also outputs a list of all vertices that are in some core. See also Slater [56]. On the other hand, analogous to the situation with path centers, Slater [60] proved

Theorem 8.36. *For a tree T, $Sp(T)$ is a unique subpath of T, and its vertex set contains both the median (and branch weight centroid) of T and the cutting center of T.*

As mentioned in Section 8.5, in the context of networks, there have been many algorithmic developments of this topic, particularly for cores, as well as for other median-like central structures (see the references in [3], [20], [38], [52], [67]).

A generalization of some of the concepts above is what might be called "degree constraint" centrality introduced recently by Pelsmajer and Reid [43].

Definition 8.33. Suppose that T is a tree and I is a non-negative integer. Let $\mathcal{T}_I$ denote the set of all subtrees of T with maximum degree at most I. A *$\mathcal{T}_I$-center* of T is a tree of least order in the set $\{W \in \mathcal{T}_I : e(W) \leq e(W'), W' \in \mathcal{T}_I\}$. A *$\mathcal{T}_I$-centroid* of T is a tree of least order in the set $\{W \in \mathcal{T}_I : bw(W) \leq bw(W'), W' \in \mathcal{T}_I\}$. A *$\mathcal{T}_I$-median* of T is a tree in the set $\{W \in \mathcal{T}_I : D(W) \leq D(W'), W' \in \mathcal{T}_I\}$.

When $I = 0$ we obtain singleton sets that are center vertices, branch-weight centroid vertices, and median vertices, respectively, so may not be unique when the center, centroid, or median contain two vertices. The $\mathcal{T}_1$-center and the $\mathcal{T}_1$-centroid are the subtrees induced by the center and median respectively. When there are two vertices in the median (equal to the centroid) or if the mother tree is a single vertex, then the $\mathcal{T}_1$-median is the subtree induced by the median. Otherwise, the $\mathcal{T}_1$-median is the subtree induced by the unique median x and one neighbor of x in a connected component of largest order of $T - x$. And, when $I = 2$, these generalizations are the path-center, path-centroid (core), and path-median (spine), respectively. As a core need not be unique, a $\mathcal{T}_2$-median of T need not be unique. So, this concept generalizes path centers, cores, and spines. And, when I is at least the maximum degree of T, each of the central subtrees in Definition 8.33 yields T.

We present a further generalization, also introduced by Pelsmajer and Reid [43], that includes the concepts in Definition 8.33. The results and proofs in [43] describe a theoretical framework into which several previous results fit. The basis for this depends on the concept in the next definition.

Definition 8.34. The set $\mathcal{T}$ is a hereditary class of trees if $\mathcal{T}$ is a non-empty set of trees such that for each $T \in \mathcal{T}$, every subtree of T is also in $\mathcal{T}$.

Examples of hereditary class of trees include trees of maximum degree at most I, trees of order at most k, trees with at most k leaves, trees of diameter at most d (for $d = 2$, these are stars, i.e., $K_{1,n}$, $n \geq 1$), caterpillars (including all paths and stars), lobsters (trees with a path of eccentricity at most 2), subdivisions of stars (including lobsters), all subtrees of a fixed set of trees, and the union and intersection of two hereditary classes of trees.

Definition 8.35. Suppose that $\mathcal{T}$ is a hereditary class of trees and T is a fixed tree. Denote by $\mathcal{T}'$ the set of subtrees of T that are in $\mathcal{T}$, i.e., $\mathcal{T}' = \{W \in \mathcal{T} : W \text{ is a subtree of } T\}$. A *$\mathcal{T}_I$-center* of T is a subtree of smallest order in the set $\{W \in \mathcal{T}' : e(W) \leq e(W'), W' \in \mathcal{T}'\}$. A *$\mathcal{T}_I$-centroid* of T is a subtree of smallest order in the set $\{W \in \mathcal{T}' : bw(W) \leq bw(W'), W' \in \mathcal{T}'\}$. A *$\mathcal{T}$-median* of T is a subtree in the set $\{W \in \mathcal{T}' : D(W) \leq D(W'), W' \in \mathcal{T}'\}$.

Pelsmajer and Reid [43] discussed linear algorithms for finding a $\mathcal{T}$-center, a $\mathcal{T}$-centroid and a $\mathcal{T}$-median.

Theorem 8.37. *Suppose that T is a tree and that $\mathcal{T}$ is a hereditary class of trees.*

(a) The $\mathcal{T}$-center of T is unique, unless $\mathcal{T} = \{K_1\}$ and $|C(T)| = 2$.
(b) If I is a positive integer, then the $\mathcal{T}_I$-center of T is unique and contains the $\mathcal{T}_{I'}$-center for all integers I', $1 \leq I' \leq I$.
(c) The $\mathcal{T}$-centroid of T is unique unless $\mathcal{T} = \{K_1\}$ and $|Bw(T)| = 2$.
(d) If I is a positive integer, then the $\mathcal{T}_I$-centroid of T is unique and contains the $\mathcal{T}_{I'}$-centroid for all integers I', $1 \leq I' \leq I$.
(e) If I is a positive integer, then a $\mathcal{T}_I$-median need not be unique and need not contain the median.

In conclusion, many of the definitions of central sets mentioned in this survey involved a minimax procedure, but there were instances of maximax, maximin, and minimin procedures (e.g., the telephone center, the security center, and the weight balance center, respectively). Other interesting sets arise when more than one of these procedures is used in the same context, e.g., peripheral vertices, vertices of maximum eccentricity. But they will have to await another survey. And, surely there are more concepts of central sets in trees waiting to be discovered.

References

1. A. Ádám, The centrality of vertices in trees, *Studia Sci. Math. Hungar.* **9** (1974), 285–303.
2. L.W. Beineke, O.R. Oellermann, R.E. Pippert, On the Steiner median of a tree, *Discrete Appl. Math.* **68** (1996) 249–258.
3. B. Bhattacharya, Y. Hu, Q. Shi, A. Tamir, Optimal algorithms for the path/tree shaped facility location problems in trees, *Proc. Intl. Symp. on Algorithms and Comp.* 2006, Lecture Notes in Computer Science 4288, Springer, Berlin/Heidelberg (2006), pp.379–388.

4. J.A. Bondy, U.S.R. Murty, *Graph Theory*, Graduate Texts in Mathematics 244, Springer, New York (2008).
5. F. Buckley, Facility location problems, *College Math. J.* **18** (1987) 24–32.
6. A. Cayley, Solution of problem 5208, *Mathematical Questions with Solutions from the Educational Times* **27** (1877) 81–83 (also in Mathematical Papers, Cambridge 1889-1889, Vol. X, 598-600).
7. G. Chartrand, O.R. Oellermann, *Applied and Algorithmic Graph Theory*, McGraw-Hill, Inc., International Series in Pure and Applied Mathematics, New York, NY (1993).
8. P. Chaudhuri, H. Thompson, A self-stabilizing graph algorithm: Finding the cutting center of a tree, *Int. J. Comput. Math.* **81** (2004) 183–190.
9. P.Z. Chinn, The path centrix of a tree, *Congress. Numer.* **21** (1978) 195–202.
10. D.G. Cikanek, P.J. Slater, The k-processing center of a graph, *Congress. Numer.* **60** (1987) 199–210.
11. G. Dahl, Majorization and distances in trees, *Networks* **50** (2007) 251–257.
12. M.S. Daskin, *Network and Discrete Location: Models, Algorithms, and Applications*, Wiley-Interscience Series in Discrete Mathematics and Optimization, John Wiley & Sons, Inc., New York (1995).
13. Z. Drezner, H.W. Hamacher (Editors), *Facility Location: Applications and Theory*, Springer-Verlag, Berlin (2002).
14. R.C. Entringer, D.E. Jackson, D.A. Snyder, Distance in graphs, *Czech. Math. J.* **26** (1976) 283–296.
15. D.P. Foster, R. Vohra, An axiomatic characterization of a class of locations on trees, *Oper. Res.* **46** (1998) 347- 354.
16. O. Gerstel, S. Zaks, A new characterization of tree medians with applications to distributed sorting, *Networks* **24** (1994) 23–29.
17. N. Graham, R.C. Entringer, L.A. Székely, New tricks for old trees: Maps and the pigeonhole principle, *Amer. Math. Monthly* **101** (1994) 664–667.
18. S.L. Hakimi, Optimal locations of switching centers and the absolute centers and medians of a graph, *Oper. Res.* **12** (1964) 450–459.
19. S.L. Hakimi, Optimum distribution of switching centers in a communications network and some related graph theoretical problems, *Oper. Res.* **13** (1965) 462–475.
20. S.L. Hakimi, E.F. Schmeichel, M. Labbé, On locating path-or-tree-shaped facilities on networks, *Networks* **23** (1993) 543–555.
21. J. Halpern, The location of a center-median convex combination on an undirected tree, *J. Regional Sci.* **16** (1976) 235–245.
22. G.Y. Handler, P.B. Mirchandani, *Location on Networks*, MIT Press, Cambridge, MA (1979).
23. F. Harary, P. Ostrand, The cutting center theorem for trees, *Discrete Math.* **1** (1971) 7–18.
24. H.A. Harutyunyan, A. Liestman, B. Shao, A linear algorithm for finding the k-broadcast center of a tree, *Networks* **53** (2009) 287–292.
25. H.A. Harutyunyan, B. Shao, Optimal k-broadcasting in trees, *Congress. Numer.* **160** (2003) 117–127.
26. S.M. Hedetniemi, E.J. Cockayne, S.T. Hedetniemi, Linear algorithms for finding the Jordan center and path center of a tree, *Transportation Sci.* **15** (1981) 98–114.
27. S.T. Hedetniemi, private communication, 2002.
28. R. Holzman, An axiomatic approach to location on networks, *Math. Oper. Res.* **15** (1990) 553–563.
29. C. Jordan, Sur les assemblages des lignes, *J. Reine Angew. Math.* **70** (1869) 185–190.

30. R.C. Laskar, B. McAdoo, Varieties of centers in trees, *Congress. Numer.* 183 (2006) 51–63.
31. F.R. McMorris, H.M. Mulder, and F.S. Roberts, The median procedure on median graphs, *Discrete Appl. Math.* **84** (1998) 165–181.
32. F.R. McMorris, K.B. Reid, Central k-trees of trees, *Congress. Numer.* **124** (1997) 139–143.
33. F.R. McMorris, F.S. Roberts, C. Wang, The center function on trees, *Networks* **38** (2001) 84–87.
34. E. Minieka, The optimal location of a path or tree in a tree network, *Networks* **15** (1985) 309–321.
35. P.B. Mirchandani, R.L. Francis, (editors), *Discrete Location Theory*, Wiley-Interscience Series in Discrete Mathematics and Optimization, John Wiley & Sons Inc., New York (1990).
36. S. Mitchell, Another characterization of the centroid of a tree, *Discrete Math.* **24** (1978) 277–280.
37. H. Monsuur, T. Storcken, Centers in connected undirected graphs: An axiomatic approach, *Oper. Res.* **52** (2004) 54–64.
38. C.A. Morgan, P.J. Slater, A linear algorithm for a core of a tree, *J. Algorithms* **1** (1980) 247–258.
39. H.M. Mulder, M.J. Pelsmajer, K.B. Reid, Axiomization of the center function on trees, *Australian J. Combinatorics* **41** (2008) 223–226.
40. J. Nieminen, M. Peltola, The subtree center of a tree, Centrality concepts in network location, *Networks* **34** (1999) 272–278.
41. O.R. Oellermann, S. Tian, Steiner centers in graphs, *J. Graph Theory* **14** (1990) 585–597.
42. O.Ore, *Theory of Graphs*, American Mathematical Society, Colloquium Publications, XXXVIII, Providence, R.I. (1962).
43. M.J. Pelsmajer, K.B. Reid, Generalized centrality in trees, in this volume (2010).
44. W. Piotrowski, A generalization of branch weight centroids, *Applicationes Mathematicae* **XIX** (1987) 541–545.
45. K.B. Reid, Centroids to centers in trees, *Networks* **21** (1991) 11–17.
46. K.B. Reid, k-ball l-path branch-weight centroids of trees, *Discrete Appl. Math.* **80** (1997) 243–250.
47. K.B. Reid, Balance vertices in trees, *Networks* **34** (1999) 264–271.
48. K.B. Reid, The latency center of a tree, unpublished manuscript (2001).
49. K.B. Reid, E. DePalma, Balance in trees, *Discrete Math.* **304** (2005) 34–44.
50. G. Sabidussi, The centrality index of a graph, *Psychometrika* **31** (1966) 581–603.
51. E. Shan, L. Kang, A note on balanced vertices in trees, *Discrete Math.* **280** (2004) 265–269.
52. A. Shioura, M. Shigeno, The tree center problems and the relationship with the bottleneck knapsack problems, *Networks* **29** (1997) 107–110.
53. P.J. Slater, Maximin facility location, *J. Research of the National Bureau of Standards Sect. B* **79B** (1975) 107–115.
54. P.J. Slater, Appraising the centrality of vertices in trees, *Studia Sci. Math. Hungar.* **12** (1977) 229–231.
55. P.J. Slater, Centers to centroids in graphs, *J. Graph Theory* **2** (1978) 209–222.
56. P.J. Slater, Centrality of paths and vertices in a graph: Cores and pits, in *The Theory and Applications of Graphs* (Kalamazoo, Mich., 1980), pp. 529–542, Wiley, New York (1981).
57. P.J. Slater, The k-nucleus of a graph, *Networks* **11** (1981) 233–242.

58. P.J. Slater, Accretion centers: A generalization of branch weight centroids, *Discrete Appl. Math.* **3** (1981) 187–192.
59. P.J. Slater, On locating a facility to service areas within a network, *Oper. Res.* **29** (1981) 523–531.
60. P.J. Slater, Locating central paths in a graph, *Transportation Sci.* **16** (1982) 1–18.
61. P.J. Slater, Some definitions of central structures, *Graph Theory Singapore* 1983, Lecture Notes in Math. 1073, pp. 169–178, Springer, Berlin (1984).
62. P.J. Slater, A survey of sequences of central subgraphs, *Networks* **34** (1999) 244–249.
63. P.J. Slater, E.J. Cockayne, and S. T. Hedetniemi, Information dissemination in trees, *SIAM J. Comput.* **10** (1981) 692–701.
64. P.J. Slater, C. Smart, Center, median, and centroid subgraphs, *Networks* **34** (1999) 303–311.
65. J.J. Sylvester, On recent discoveries in mechanical conversion of motion, *Proceedings of the Royal Institution of Great Britain* **7** (1872-1875) 179–198 (also in Mathematical Papers, Cambridge 1904-1912, Vol. III, 7-25).
66. J.J. Sylvester, On the geometrical forms called trees, *Johns Hopkins University Circulars* I (1882) 202–203 (also in Mathematical Papers, Cambridge 1904-1912, Vol. III, 640–641).
67. A. Tamir, J. Puerto, J.A. Mesa, A.M. Rodríguez-Chía, Conditional location of path and tree shaped facilities on trees, *J. of Algorithms* **56** (2005) 50–75.
68. B.C. Tansel, R.L. Francis, T.J. Lowe, Location on networks: A survey. I. The p-center and p-median problems, *Management Sci.* **29** (1983) 482–497.
69. B.C. Tansel, R.L. Francis, T.J. Lowe, Location on networks: A survey. II. Exploiting tree network structure, *Management Sci.* **29** (1983) 498–511.
70. R.V. Vohra, An axiomatic characterization of some locations on trees, *Eur. J. Oper. Res.* **90** (1996) 78–84.
71. Z. Win, A note on the medians and the branch weight centroids of a tree, unpublished manuscript (1993).
72. Z. Win, Y. Myint, The cendian of a tree, *Southeast Asian Bull. of Math.* **25** (2002) 757–767.
73. B. Zelinka, Medians and peripherians of trees. *Arch. Math. (Brno)* **4** (1968) 87–95.
74. B. Zelinka, Generalized centers of gravity of trees, *Mathematics (Geometry and Graph Theory)* (Czech), Univ. Karlova, Prague, (1970) pp. 127–136.

Chapter 9

The Port Reopening Scheduling Problem

Fred S. Roberts*

DIMACS Center, Rutgers University
Piscataway, NJ 08854 USA
froberts@dimacs.rutgers.edu

When a port needs to be reopened after closure due to a natural disaster or terrorist event or domestic dispute, certain goods on incoming ships might be in short supply. This paper formulates the problem of scheduling the unloading of waiting ships to take into account the desired arrival times, quantities, and designated priorities of goods on those ships. Several objective functions are defined, special cases are discussed, and the relevant literature is surveyed.

Introduction

Global trade is critical to economic well-being. Over 90% of international trade is by sea. An efficient, effective, secure system of ports is crucial to international trade. Ports are also crucial to the national supply chain of critical products: fuel, food, medical supplies, etc. This paper is concerned with ports being shut down in part or entirely, by natural disasters like hurricanes or ice storms, terrorist attacks, strikes or other domestic disputes. How do we aid port operators and government officials to reschedule port operations in case of a shutdown? Shutting down ports due to hurricanes is not unusual and so provides insight into current methods for reopening them. Unfortunately, scheduling and prioritizing in reopening the port after a hurricane is typically done very informally. This paper seeks to formalize an approach to reopening a port in an efficient way that responds to a variety of priorities. If a port is damaged, we envision a number of vessels waiting to dock and be unloaded. In what order should the unloading take place? The decision of how to reopen a damaged port is rather complex. There are multiple goals, including minimizing economic and security impacts of delays in delivery of critical supplies.

*This paper is dedicated to Buck McMorris. Our collaborations over the years have been a source of pleasure and inspiration to me. Not only is he a colleague, but I am pleased to call him a friend. This work was supported by the US Department of Homeland Security under a grant to Rutgers University. Many of the ideas in this paper come from joint work with N.V.R. Mahadev and Aleksandar Pekeč, in joint papers [37; 38]. These ideas are modified here to apply to the port reopening scheduling problem.

These problems can be subtle. For example, if an ice storm shuts down a port, perhaps the priority is to unload salt to de-ice. It wasn't a priority before. These subtleties call for the kind of formal approach that is outlined here.

9.1. Formalizing the Problem

Let us suppose that we are interested in a set G of n goods. We can think of G as consisting of items labeled $i = 1, 2, \ldots, n$. Each ship will have various quantities of these goods, usually packed in containers. For simplicity, let us disregard the fact that different goods are in different containers – that complication can be added later – and then think of a ship as corresponding to a vector $x = (x_1, x_2, \ldots, x_n)$, where x_i is the quantity of good i on the ship. Again for simplicity, we assume that x_i is an integer for each i. Real data describing the contents of all containers on a ship is on the ship's manifest and is collected by (in the US) Customs and Border Protection before the ship's arrival. Unfortunately, there is a large amount of such data, there is inconsistency in the units used to describe the contents (e.g., 1000 bottles of water vs. 1000 cases of bottles of water), the descriptions of contents are often imprecise or vague (e.g., "fruit" or "household goods"), and so it is not easy to get a vector like x.

Let us also assume that once a ship docks, we proceed to unload all of its cargo so it can leave the port and open a spot for another ship. For simplicity, the unloading time is assumed to be the same for each ship, though a realistic problem will consider unloading times. (In the literature, we distinguish between common and noncommon *processing times.*) Thus, we can consider integer *time slots* for unloading. We can think of a schedule Σ that assigns to each ship a time slot, the first, second, third, etc., during which it will dock and be unloaded. Thus, Σ can be thought of as a vector $(\sigma_1, \sigma_2, \ldots, \sigma_k)$ where σ_i gives the time slot during which ship i will be unloaded. Let us also assume that as we are aiming to reopen a port, we have enough berths for c ships at once. A schedule is *capacity-acceptable* if no timeslot gets more than c ships. Next, assume that ships are ready to dock and there is no delay after they are chosen. We just have to choose which ships to unload in which order, taking account of the capacity c of the port. The problem becomes more complicated (and more realistic) if we assume that some ships are not yet nearby and so each ship has a delay time before it could arrive and be unloaded.

Because some goods can rapidly become in short supply, we assume that there is a desired quantity d_i of good i (which we assume to be an integer) and that it is required no later than time t_i, i.e., the t_i^{th} unloading time slot. Let $d = (d_1, d_2, \ldots, d_n)$ and $t = (t_1, t_2, \ldots, t_n)$. The problem gets more realistic if we have vectors giving the desired quantity at time 1, the desired quantity at time 2, etc., but we shall disregard this complication here. Different goods also have different priorities p_i, with $p = (p_1, p_2, \ldots, p_n)$. For example, not having enough fuel or medicine or food may be much more critical than not having enough cookware, which is reflected in a higher p_i. We will have more to say about the priorities in Section 9.5.

In what follows, we will discuss a special case, namely where each d_i is 1. That allows us to concentrate on whether or not all the desired goods arrive in time, i.e., before desired arrival times. It also allows us to think of a schedule Σ for ship arrivals as corresponding to a schedule S that gives the arrival time S_i of the first item i for each i.

What makes one (capacity-acceptable) schedule better than another? We can assume that there is some objective function that takes into account demands for goods that are not met by the schedule either in terms of quantity or in terms of arrival time. Let $F(S, t, p)$ be the penalty assigned to schedule S given the desired arrival time and priority vectors t, p. (Recall that we are disregarding the vector d since we assume each component is 1. Also, the port capacity c is part of the problem, but not part of the penalty function.) Suppose we seek to minimize F. We shall discuss different potential penalty functions below.

The problem is similar to a number of problems that have been considered in the literature. For example, Mahadev, Pekeč, and Roberts [37; 38; 49] considered a problem posed by the Air Mobility Command (AMC) of the US Air Force. Suppose that we wish to move a number of items such as equipment or people by vehicles such as planes, trucks, etc. from an origin to a destination. Each item has a desired arrival time, we are penalized for missing the desired time, and penalty is applied not only to late (*tardy*) arrival but also early arrival. In our port problem, while the emphasis is on tardy arrivals, early arrivals of goods could be a problem if we add a complication of port capacity for storing goods until they are picked up. We will consider early arrival penalties, though we will not consider the details of port capacity for storing goods. In the AMC problem, it is assumed that each trip from origin to destination takes the same amount of time (though this assumption can be weakened) and there are only a limited number of spots for people or goods on the vehicles. The items have different priorities. For example, transporting a VIP may be more important than transporting an ordinary person and transporting fuel may be more important than transporting cookware. The penalty for tardy or early arrival depends on the priority.

Another similar problem arises in the workplace if we have a number of tasks to perform, a number of processors on which to perform them, each task has a desired completion time and we are penalized for missing that time. We assume that once started, a task cannot be interrupted (*non-preemption*) and that, for simplicity, each task takes the same amount of time. We have only a limited number of processors, so are only able to schedule a number of tasks each time period. Assume that the tasks can have different priorities. We seek to assign tasks to processors each time period so that the total penalty is minimized. This problem without the priorities is a well-studied problem in the machine scheduling literature. Some examples of papers on machine scheduling with earliness and tardiness penalties are [11; 15; 44;

8; 6; 3; 25; 26; 42; 4; 9; 39; 21; 16; 14; 55; 31]. Some survey papers on scheduling with objective functions are [2; 7; 29]. A general reference on scheduling, which includes a considerable amount of material on earliness/tardiness penalties, is [43]. Also, [22; 32; 58] are extensive surveys on scheduling with earliness and tardiness penalties.

9.2. Penalty Functions

We shall consider *summable* penalty functions, those where

$$F(S,t,p) = \sum_{i=1}^{n} g(S_i, t_i, p_i). \tag{9.1}$$

Simple cases of summable penalty functions are those that are *separable* in the sense that

$$g(S_i, t_i, p_i) = \begin{cases} h_T(p_i) f(S_i, t_i) & \text{if } S_i > t_i \\ h_E(p_i) f(S_i, t_i) & \text{if } S_i \le t_i \end{cases} \tag{9.2}$$

where h_T and h_E are functions reflecting the tardiness and earliness contributions to the penalties of the priorities of the goods. A simple example of a summable, separable penalty function is given by

$$F(S,t,p) = \sum_{i=1}^{n} p_i |S_i - t_i|.$$

This penalty function arises in the literature of single machine scheduling with earliness and tardiness penalties and *noncommon weights.* (See [11; 15; 44; 8; 6; 3; 25; 26].) Here, $h_T(p_i) = h_E(p_i) = p_i$. If $f(S_i, t_i) = |S_i - t_i|$, it is sometimes convenient to rewrite the penalty function (9.1) resulting from (9.2) as follows. Let $T_i(S) = T_i = max\{0, S_i - t_i\}$, $E_i(S) = E_i = max\{0, t_i - S_i\}$. Then (9.1) is equivalent to

$$F(S,t,p) = F_{sumE/T}(S,t,p) = \sum_{i=1}^{n} h_T(p_i) T_i + \sum_{i=1}^{n} h_E(p_i) E_i. \tag{9.3}$$

If we replace $h_T(p_i)$ and $h_E(p_i)$ by constants α_i and β_i respectively, then we simply have a weighted sum of earliness and tardiness.

A variant of the function (9.3) arises if we change to $h_E(p_i) = 0$, so we only penalize tardiness, i.e.,

$$F(S,t,p) = \sum_{i=1}^{n} h_T(p_i) |S_i - t_i| \delta(S_i, t_i),$$

where

$$\delta(S_i, t_i) = \begin{cases} 1 \text{ if } & S_i > t_i \\ 0 \text{ if } & S_i \leq t_i \end{cases}$$

or, equivalently,

$$F(S,t,p) = F_{sumT}(S,t,p) = \sum_{i=1}^{n} h_T(p_i)T_i.$$

Another case that disregards the priorities or, alternatively, has constant but different h_T and h_E is:

$$F(S,t,p) = \sum_{i=1}^{n} \alpha|S_i - t_i|\delta(S_i,t_i) + \sum_{i=1}^{n} \beta|S_i - t_i|\gamma(S_i,t_i),$$

where

$$\gamma(S_i, t_i) = \begin{cases} 1 \text{ if } & S_i \leq t_i \\ 0 \text{ if } & S_i > t_i \end{cases}$$

In the case where we disregard priorities and the α and β correspond to different weighting factors for tardiness and earliness, respectively, this penalty function arises in single machine schedule with *nonsymmetric* earliness and tardiness penalties and *common weights* (see [42; 4; 15]. The case $\alpha = \beta$ is equivalent to the penalty function studied by [27; 54; 5; 23; 15; 57; 24].

The function $F_{sumE/T}(S,t,p)$ is considered for the case where all t_i are the same in [3; 11; 15; 44; 8; 6; 25; 26]. The more general function allowing differing t_i is considered in [12; 18; 19; 20; 10; 1; 40; 41].

Still other penalty functions are only concerned with minimizing the maximum deviation from desired arrival time, rather than a weighted sum of deviations. Of interest, for example, is the function

$$F(S,t,p) = F_{maxE/T}(S,t,p)$$

$$= max\{h_T(p_1)T_1, h_T(p_2)T_2, \ldots, h_T(p_n)T_n, h_E(p_1)E_1, h_E(p_2)E_2, \ldots, h_E(p_n)E_n\}$$

where we maximize the weighted maximum deviation, including consideration of earliness deviations. If we are only interested in tardiness, we would consider instead

$$F(S,t,p) = F_{maxT}(S,t,p) = max\{h_T(p_1)T_1, h_T(p_2)T_2, \ldots, h_T(p_n)T_n\}.$$

The objective function $F_{maxE/T}(S,t,p)$ is considered in [52; 17], for example, while $F_{maxT}(S,t,p)$ is considered by many authors. Two survey papers describing work with these functions are [29; 2], with the former emphasizing constant weighted tardiness.

Since there are so many potential criteria for a good solution to the port reopening scheduling problem, it is surely useful to look at it as a multi-criteria problem. For a survey from this point of view, see [58].

9.3. The Case of Common Desired Arrival Times

The most trivial case of the problem we have formulated is where all the desired arrival times t_i are the same time τ (this is sometimes called in the scheduling literature the case of *common due dates*). Let us assume that the penalty function is summable and separable, that $f > 0$, that $h_E(p_i) = 0$ (there are no earliness penalties), and that $h_T(p_i)$ is an increasing function of p_i. Let us also assume that $c = 1$, i.e., we can only unload one ship at a time and, moreover, assume that there is only one kind of good on each ship. In this case, a simple greedy algorithm suffices to find a schedule that minimizes the penalty.

To explain this, suppose we rank the goods in order of decreasing priority, choosing arbitrarily in case of ties. The greedy algorithm proceeds as follows. Schedule the first good on the list at time 1, the next on the list at time 2, and so on until the last is scheduled at time n. To see that the resulting greedy schedule S_G minimizes the penalty, suppose that we have a schedule and the set of goods scheduled up to time τ is A and the set scheduled after time τ is B. Switching any two goods within A does not change the penalty. The order of elements in B that will minimize the penalty given the split of goods into A and B is clearly to put those of higher priority close to τ. Now consider the possibility of switching a good i in the set A_G associated with greedy schedule S_G with a good j in the set B_G associated with S_G, to obtain a schedule $S_G(i,j)$. Then if j is given time slot $\tau + r$ in S_G, we have

$$F(S_G,t,p) - F(S_G(i,j),t,p) = h_T(p_j)f(S_{G_j},t_j) - h_T(p_i)f(S_{G(i,j)_i},t_i)$$

$$= h_T(p_j)f(\tau + r,\tau) - h_T(p_i)f(\tau + r,\tau)$$

$$= [h_T(p_j) - h_T(p_i)]f(\tau + r,\tau) \leq 0,$$

since $p_j \leq p_i$, $h_T(u)$ is increasing, and $f > 0$. We conclude that a switch cannot decrease the penalty.

Things get more complicated if $c > 1$ or ships can have more than one good. Consider for example the case where we use the penalty function $F_{sumT}(S,t,p)$, we take $c = 1$, and there are two ships, ship 1 with goods vector $(1,0,0,1)$ and ship 2 with goods vector $(0,1,1,0)$, and assume that $p_1 > p_2 > p_3 > p_4$. What is a greedy algorithm? One natural idea here is to choose first the ship that has the item of highest priority and schedule that as close to desired arrival time as possible, and first if all t_i are the same as we are assuming in this section. Let us say the desired arrival times are given by $t = (1, 1, 1, 1)$. Then we would schedule ship 1 first, at time 1, then ship 2 at time 2, obtaining a goods arrival schedule $S = (1, 2, 2, 1)$ and penalty $F(S, t, p) = h_T(p_2) + h_T(p_3)$. However, if we schedule ship 2 first, we get a goods arrival schedule $S^* = (2, 1, 1, 2)$ with penalty $F(S^*, t, p) = h_T(p_1) + h_T(p_4)$, which might be lower than $F(S, t, p)$, depending on the function h_T and the values of the p_i.

To show how complicated things get very quickly, consider another simple situation where we use the penalty function $F_{sumT}(S, t, p)$, we take $c = 1$, and there are three ships, ships 1, 2, 3 with goods vectors (1,0,0,0), (0,1,0,0), and (0,0,1,1), respectively, with $t = (2, 2, 2, 2), p_1 > p_2 > p_3 > p_4$. A natural greedy algorithm would say choose ship 1 first since it has the highest priority good, and put it at time 1, then ship 2 next since it has the second highest priority good, and put it at time 2, and finally ship 3 at time 3. This gives rise to a penalty of $h_T(p_3) + h_T(p_4)$. Scheduling ship 3 at time 2, ship 1 at time 1, ship 2 at time 3 gives a penalty of $h_T(p_2)$, which might be smaller. Thus, even with constant desired arrival times and only one ship per time slot, finding an algorithm that would minimize penalty presents intriguing challenges.

9.4. Nonconstant Desired Arrival Times

It is interesting to observe that in the case of nonconstant arrival times, our intuition about the problem is not always very good. It seems reasonable to expect that if the priorities change, but the ratios of priorities p_i/p_j do not change, then an optimal schedule won't change. For example, consider the penalty function $F_{sumE/T}(S, t, p)$. One example of an increasing function h_T is given by $h_T(u) = 2^{u-1}$. Now consider the case where each ship has only one kind of good, $c = 1$, and we have $t = (1, 2, 2, 2), p = (1, 2, 2, 2)$. An optimal schedule is given by $S = (1, 2, 3, 4)$. Yet, if we multiply each priority by 2, getting $p^* = (2, 4, 4, 4)$, then the schedule S is no longer optimal, since it has a penalty of 24 while the schedule $S^* = (2, 3, 4, 1)$ has penalty 22. (This example is taken from [37].)

9.5. Meaningful Conclusions

The truth or falsity of a conclusion about optimality of a schedule can sometimes depend on properties of the scales used to measure the variables. Discussions of this point in the literature of scheduling have concentrated on the scales used to measure the priority of a good. Mahadev, Pekeč, and Roberts [37; 38] point out that the conclusion that a particular schedule is optimal in one of the senses defined in Section 9.2 can be meaningless in a very precise sense of the theory of measurement.

Thus, one needs to be very careful in drawing the conclusion of optimality of a schedule. To explain what this means, we note that in using scales of measurement, we often make arbitrary choices such as choosing a unit or a zero point. In measuring mass, for example, we can use, grams, kilograms, pounds, etc. In measuring temperature, we can use, for example, Fahrenheit or Centigrade. An *admissible transformation* of scale transforms one acceptable scale into another. For example, in changing from kilograms to pounds, it multiplies all values by 2.2, and in changing from degrees Centigrade to degrees Fahrenheit, it multiplies by 9/5 and then adds 32. In measurement theory, a statement involving scales is called *meaningful* if its truth or falsity is unchanged after applying admissible transformations to all of the scales in question. For an introduction to the theory of measurement, see [30; 36; 56; 45]. For further information about the theory of meaningfulness, see [36; 45; 46; 48; 50; 51]. For applications of the concept of meaningfulness to combinatorial optimization, see [47; 48; 51; 13].

What properties does the priority scale have? Specifically, what transformations of the priority scale are reasonable to allow? Quaddus [44] thinks of the priorities as "costs" but suggests that techniques of preference and value theory as in the classic work of Keeney and Raiffa [28] might be relevant, suggesting that priorities are more like utility measures. In the literature of utility theory, various kinds of admissible transformations of utility values are considered, including those where we change just the unit and those where we change both unit and zero point. Let us first consider the case where the priorities p_i are unique up to choice of unit. In this case, we talk about a *ratio scale* and the admissible transformations are functions of the form $\phi(u) = \alpha u$. Consider the claim that schedule S is optimal under penalty function $F_{sumE/T}(S, t, p)$. This means that for any other schedule S^*,

$$F_{sumE/T}(S, t, p) \leq F_{sumE/T}(S^*, t, p). \tag{9.4}$$

Consider the case where $h_T(u) = u, h_E(v) = v$ for all u, v. Then we consider Equation (9.4) meaningful if its truth is unchanged if we replace any p_i by $\phi(p_i) = \alpha p_i$. Clearly (9.4) is meaningful in this sense. It is even meaningful with nonconstant functions h_T, h_E if these functions satisfy the equations $h_T(\alpha u) = \alpha h_T(u), h_E(\alpha v) = \alpha h_E(v)$. Thus, under these conditions, the statement (9.4) is meaningful in the sense of measurement theory. Similar conclusions hold if we replace the penalty function with $F_{sumT}(S, t, p), F_{maxE/T}(S, t, p)$, or $F_{maxT}(S, t, p)$.

However, consider the situation where priorities are only determined up to change of both unit and zero point. Here we say that priorities are measured on an *interval scale* and admissible transformations take the form $\phi(u) = \alpha u + \beta$. In this case, the truth of the statement (9.4) can depend on the choice of unit and zero point. Consider for example the case of $n = 4$ goods, with $t = (2, 2, 2, 1), p = (9, 9, 9, 1)$. Consider the penalty function $F_{sumE/T}(S, t, p)$ with $h_T(u) = u, h_E(v) = v$ for all u, v. It is easy to see that the schedule $S = (1, 2, 3, 4)$ is optimal. However, consider the admissible transformation $\phi(u) = \alpha u + \beta$, where $\alpha = 1/8$ and $\beta = 7/8$. After this admissible transformation, we change p to $p^* = (2, 2, 2, 1)$. Then $F(S, p^*, t) = 7$ while $F(S^*, p^*, t) = 6$ for $S^* = (4, 1, 2, 3)$. This shows that the conclusion that S is optimal in this case is meaningless. (This example is due to [37].)

An extensive analysis of the meaningfulness of conclusions for scheduling problems under a variety of penalty functions and a variety of assumptions about admissible transformations of priorities is given in [37; 38].

9.6. Closing Comments

This paper has set out the port reopening scheduling problem. We have seen that even a very simplified version leads to rather subtle issues. Among the special assumptions we have considered are:

- all desired amounts are one unit, i.e., $d_i = 1$ for all i;
- in reopening, the port has limited capacity of one ship at a time, i.e., $c = 1$;
- all goods have the same desired arrival time t_i;
- all goods have only one desired arrival time t_i, rather than specifying a minimum amount desired per time;
- all ships have the same unloading time;
- all ships are ready to dock without delay;
- there is no problem storing unloaded but undemanded goods at the port;
- each ship has only one kind of good.

Even making all or most of these special assumptions leaves a complex scheduling problem. Removing these special assumptions leads to a wide variety of challenging problems, as we noted when we removed the last one.

We have also considered a variety of penalty functions. Certainly there are others that should be considered. Moreover, we have not tried to formulate a multi-criteria optimization problem that might also be appropriate.

We have also not discussed the problem of how one determines priorities and desired arrival times of the goods in question. We can envision a number of approaches to this, for example having each stakeholder (government, port operators, shippers) providing these priorities and times and then using some sort of consensus procedure. We could also create a bidding procedure for obtaining them†. The measurement-theoretic properties of the priorities (the kinds of scales they define or admissible transformations they allow) also need to be understood better.

†Thanks to Paul Kantor for suggesting this idea; the details present another interesting research challenge.

References

1. T.S. Abdul-Razaq, C.N. Potts, Dynamic Programming State-Space Relaxation for Single-Machine Scheduling, *Opnl. Res. Soc.* **39** (1988) 141–152.
2. T.S. Abdul-Razaq, C.N. Potts, L.N. van Wassenhove, A Survey of Algorithms for the Single Machine Total Weighted Tardiness Scheduling Problem, *Discrete Appl. Math.* **26** (1990) 235–253.
3. M.U. Ahmed, P.S. Sundararaghavan, Minimizing the Weighted Sum of Late and Early Completion Penalties in a Single Machine, *IIE Trans.* **22** (1990) 288–290.
4. U. Bagchi, Y. Chang, R. Sullivan, Minimizing Absolute and Squared Deviations of Completion Times with Different Earliness and Tardiness Penalties and a Common Due Date, *Naval Res. Logist.* **34** (1987) 739–751.
5. U. Bagchi, R. Sullivan, Y. Chang, Minimizing Mean Absolute Deviation of Completion Times About a Common Due Date, *Naval Res. Logist.* **33** (1986) 227–240.
6. K.R. Baker, G.W. Scudder, On the Assignment of Optimal Due Dates. *J. Opnl. Res. Soc.* **40** (1989) 93–95.
7. K.R. Baker, G.W. Scudder, Sequencing with Earliness and Tardiness Penalties: A Review, *Opns. Res.* **38** (1990) 22–36.
8. C. Bector, Y. Gupta, M. Gupta, Determination of an Optimal Due Date and Optimal Sequence in a Single Machine Job Shop, *Int. J. Prod. Res.* **26** (1988) 613–628.
9. M. Birman, G. Mosheiov, A Note on a Due-Date Assignment on a Two-machine Flow-shop, *Computers and Opns. Res.* **31** (2004) 473–480.
10. S. Chand, H. Schneeberger, Single Machine Scheduling to Minimize Weighted Earliness Subject to No Tardy Jobs, *European J. Opnl. Res.* **34** (1988) 221–230.
11. T. Cheng, An Algorithm for the CON Due Date Determination and Sequencing Problem, *Comp. Opns. Res.* **14** (1987) 537–542.
12. T.C.E. Cheng, Dynamic Programming Approach to the Single-Machine Sequencing Problem with Different Due-Dates, *Computers Math. Applic.* **19**, (1990) 1–7.
13. M. Cozzens, F.S. Roberts, Greedy Algorithms for T-Colorings of Graphs and the Meaningfulness of Conclusions about Them, *Comb. Inf. and Syst. Sci.* **16** (1991) 286–299.
14. F. Della Croce, J.N.D. Gupta, R. Tadei, Minimizing Tardy Jobs in a Flowshop with Common Due-date, *European Journal of Operational Research* **120** (2000) 375–381.
15. H. Emmons, Scheduling to a Common Due Date on Parallel Common Processors, *Naval Res. Logist.* **34** (1987) 803–810.
16. A. Federgruen, G. Mosheiov, Heuristics for Multi-machine Minmax Scheduling Problems with General Earliness and Tardiness Costs, *Naval Research Logistics* **44** (1996) 287–299.
17. M.C. Ferris, M. Vlach, Scheduling with Earliness and Tardiness Penalties, *Naval Res. Logist.* **39** (1992) 229–245.
18. T.D. Fry, R.D. Armstrong, J.H. Blackstone, Minimizing Weighted Absolute Deviation in Single Machine Scheduling, *IIE Trans.* **19** (1987) 445–450.
19. T.D. Fry, G. Leong, A Bi-Criterion Approach to Minimizing Inventory Costs on a Single Machine When Early Shipments are Forbidden, *Comp. Opns. Res.* **14** (1987) 363–368.
20. T.D. Fry, G. Leong, R. Rakes, Single Machine Scheduling: A Comparison of Two Solution Procedures, *Omega*, **15** (1987) 277–282.
21. M. Garey, R. Tarjan, G. Wilfong, One Processor Scheduling with Symmetric Earliness and Tardiness Penalties, *Mathematics of Operations Research* **13** (1988) 330–348.

22. V. Gordon, J.-M. Proth, C. Chu, A Survey of the State-of-the-Art of Common Due Date Assignment and Scheduling Research, *European J. Oper. Res.* **139** (2002) 1–25.
23. N.G. Hall, Single and Multi-Processor Models for Minimizing Completion Time Variance, *Naval Res. Logist.* **33** (1986) 49–54.
24. N.G. Hall, W. Kubiak, S.P. Sethi, Earliness-Tardiness Scheduling Problems, II: Deviation of Completion Times About a Restrictive Common Due Date, *Opns. Res.* **39** (1991) 847–856.
25. N.G. Hall, M.E. Posner, Earliness-Tardiness Scheduling Problems I: Weighted Deviation of Completion Times About a Common Due Date, *Opns. Res.* **39** (1991) 836–846.
26. J.A. Hoogeveen, S.L. van de Velde, Scheduling Around a Small Common Due Date, *European J. Opnl. Res.* **55** (1991) 237–242.
27. J.J. Kanet, Minimizing the Average Deviation of Job Completion Times about a Common Due Date, *Naval Res. Logist.* **28** (1981) 643–651.
28. R.L. Keeney, H. Raiffa, *Decisions with Multiple Objectives: Preferences and Value Tradeoffs*, Wiley, New York, 1976.
29. C. Koulamas, The Total Tardiness Problem: Review and Extensions, *Opns. Res.* **42** (1994) 1025–1041.
30. D.H. Krantz, R.D. Luce, P. Suppes, A. Tversky, *Foundations of Measurement*, Volume I. Academic Press, New York, 1971.
31. V. Lauff, F. Werner, On the Complexity and Some Properties of Multi-Stage Scheduling Problems with Earliness and Tardiness Penalties, *Computers and Opns Res.* **31** (2004) 317–345.
32. V. Lauff, F. Werner, Scheduling with Common Due Date, Earliness and Tardiness Penalties for Multimachine Problems: A Survey, *Mathematical and Computer Modelling* **40** (2004) 637–655.
33. W.-P. Liu, J.B. Sidney, Bin Packing Using Semiordinal Data, *O. R. Lett.* **19** (1996) 101–104.
34. W.-P. Liu, J.B. Sidney, Ordinal Algorithms for Packing with Target Center of Gravity, *Order* **13** (1996) 17–31.
35. W.-P. Liu, J.B. Sidney, A. van Vliet, Ordinal Algorithms for Parallel Machine Scheduling, *O. R. Lett.* **18** (1996) 223–232.
36. R.D. Luce, D.H. Krantz, P. Suppes, A. Tversky, *Foundations of Measurement*, Volume III. Academic Press, New York, 1990.
37. N.V.R. Mahadev, A. Pekeč, F.S. Roberts, Effect of Change of Scale on Optimality in a Scheduling Model with Priorities and Earliness/Tardiness Penalties, *Mathematical and Computer Modelling* **25** (1997) 9–22.
38. N.V.R. Mahadev, A. Pekeč, F.S. Roberts, On the Meaningfulness of Optimal Solutions to Scheduling Problems: Can an Optimal Solution be Non-optimal?, *Opns Res.* **46 Suppl** (1998) S120–S134.
39. G. Mosheiov, A Common Due-date Assignment Problem on Parallel Identical Machines, *Computers and Opns Res.* **28** (2001) 719–732.
40. P.S. Ow, T.E. Morton, Filtered Beam Search in Scheduling, *Int. J. Prod. Res.* **26** (1988) 35–62.
41. P.S. Ow, T.E. Morton, The Single Machine Early/Tardy Problem, *Mgmt. Sci.* **35** (1989) 177–191.
42. S. Panwalkar, M. Smith, A. Seidmann, Common Due Date Assignment to Minimize Total Penalty for the One Machine Scheduling Problem, *Opns. Res.* **30** (1982) 391–399.
43. M. Pinedo, *Scheduling: Theory, Algorithms, and Systems*, Third Ed., Springer, 2008.

44. M. Quaddus, A Generalized Model of Optimal Due-Date Assignment by Linear Programming, *Opnl. Res. Soc.* **38** (1987) 353–359.
45. F.S. Roberts, *Measurement Theory with Applications to Decisionmaking, Utility and the Social Sciences*, Addison-Wesley, Reading, MA. 1979. Digital printing, Cambridge University Press, 2009.
46. F.S. Roberts, Applications of the Theory of Meaningfulness to Psychology, *J. Math. Psychol.* **29** (1985) 311–332.
47. F.S. Roberts, Meaningfulness of Conclusions from Combinatorial Optimization, *Discr. Appl. Math.* **29** (1990) 221–241.
48. F.S. Roberts, Limitations on Conclusions Using Scales of Measurement, in: A. Barnett, S.M. Pollock, M.H. Rothkopf (eds.), *Operations Research and the Public Sector*, Elsevier, Amsterdam, 1994, pp. 621–671.
49. F.S. Roberts, A Functional Equation that Arises in Problems of Scheduling with Priorities and Lateness/earliness Penalties, *Mathematical Computer Modelling* **21** (1995) 77–83.
50. F.S. Roberts, Meaningless Statements, *Contemporary Trends in Discrete Mathematics*, DIMACS Series, Vol 49. American Mathematical Society, Providence, RI. 1999, pp. 257–274.
51. F.S. Roberts, Meaningful and Meaningless Statements in Epidemiology and Public Health, in: B. Berglund, G.B. Rossi, J. Townsend, L. Pendrill (eds.), *Measurements with Persons*, Taylor and Francis, to appear.
52. J. Sidney, Optimal Single-Machine Scheduling with Earliness and Tardiness Penalties, *Opns. Res.* **25** (1977) 62–69.
53. W.E. Smith, Various Optimizers for Single-stage Production, *Naval Res. Logist.* **3** (1956) 59–66.
54. P.S. Sundararaghavan, M.U. Ahmed, Minimizing the Sum of Absolute Lateness in Single-Machine and Multimachine Scheduling, *Naval Res. Logist.* **31** (1984) 325–333.
55. C.S. Sung, J.I. Min. Scheduling in a Two-machine Flowshop with Batch Processing Machine(s) for Earliness/tardiness Measure under a Common Due date, *European Journal of Operational Research*, **131** (2001) 95–106.
56. P. Suppes, D.H. Krantz, R.D. Luce, A. Tversky, *Foundations of Measurement*, Volume II, Academic Press, New York, 1989.
57. W. Szwarc, Single Machine Scheduling to Minimize Absolute Deviation of Completion Times from a Common Due Date, *Naval Res. Logist.* **36** (1989) 663–673.
58. V. t'Kindt, J.-C. Billaut, Multicriteria Scheduling Problems: A Survey, *RAIRO Opns Res.* **35** (2001) 143–163.

Chapter 10

Reexamining the Complexities in Consensus Theory

Donald G. Saari
Institute for Mathematical Behavioral Sciences, University of California, Irvine, CA 92697-5100 USA
dsaari@uci.edu

Several results in consensus theory indicate that it is impossible to obtain conclusions should the rules satisfy certain seemingly innocuous but desired properties. As these conclusions create a barrier for progress, it is important to find ways to circumvent the difficulties. In this paper, a source of these complications is identified, and ways to combat the problem are indicated.

10.1. The problem

Fred McMorris with coauthors such as Barthélémy [2], Day [3], and Powers [2; 4; 5] have proved that all sorts of roadblocks and obstacles clutter the landscape of consensus theory. As an illustration and motivated by the seminal Arrow's Theorem [1], McMorris and Powers [5] proved it is impossible to use rules based on seemingly innocuous and reasonable conditions to carry out the seemingly simple task of assembling into a consensus the divergent views about the structure of a simple tree such as in Fig. 10.1.

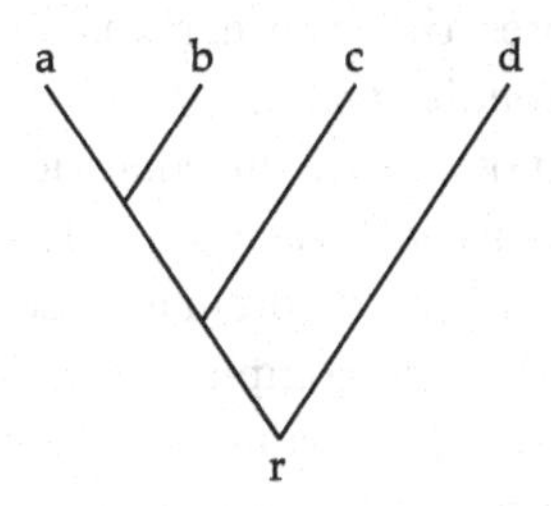

Fig. 10.1. Simple tree

What makes these results troubling is that they erect barriers; barriers that frustrate attempts toward making progress. Something must be done, and it is clear what it should be: the requirements imposed on the aggregation rules must be modified. The mystery, of course, is to understand in what ways this change can and should be accomplished.

A way to obtain insight into this puzzle is to carefully reexamine Arrow's Theorem and the negative assertions in consensus theory. In doing so, my goal is to go beyond the formal proofs to determine "why" the various conclusions hold. By answering the "why" question, information is obtained about how to circumvent the negative conclusions. Basic ideas are developed and discussed in this paper; a more extensive report will appear elsewhere.

10.2. Arrow's Theorem and surprising extensions

Arrow's result is among the most influential theorems in the areas of social choice and consensus theory. The statement of his theorem is simple; each voter has a complete transitive ranking of the alternatives; there are no other restrictions. The societal outcome also must be a complete transitive ranking.

Only two conditions are imposed upon the rule:

(1) *Pareto*: If everyone ranks some pair in the same manner, then that unanimous choice is the pair's societal ranking.
(2) *Independence*: The ranking of each pair is strictly determined by the voters' relative rankings of that pair. Specifically, if $\mathbf{p}_1$ and $\mathbf{p}_2$ are any two profiles where each voter ranks a specified pair in the same manner, then the societal ranking for the pair is the same for both profiles.

Arrow's striking conclusion is that with three or more alternatives, only one rule always satisfies these condition: a *dictatorship*. In other words, the rule is equivalent to identifying a particular person (or, in the context of a decision problem where voters are replaced with criteria, a specific criterion) so that for all possible profiles the rule's ranking always agrees with the preferences of the identified person. The dictatorial assertion underscores the true conclusion that no reasonable decision rule can always satisfy these conditions.

To analyze Arrow's result, notice how the conditions imposed on the rule require it to concentrate on what happens with each pair. Thus, differing from standard interpretations, I prefer to treat Arrow's theorem as describing a "divide-and-conquer" methodology. Namely, to handle the complexity of finding a societal ranking, an Arrovian rule divides the problem into pairs. After finding appropriate conclusions for each pair, the whole (the societal ranking) is an assembly of these pairwise rankings.

To carry this line of thought another step, it is reasonable to expect for each profile that an appropriate societal ranking does exist. Arrow's result asserts that:

Suppose a complete, transitive societal ranking exists for each profile. There exist settings where this ranking, which is based on the complete transitive inputs from two or more voters (or, for a decision problem with complete, transitive inputs from two or more criteria) cannot be found by independently determining the outcomes for each of the pairs.

The whole need not resemble the assembly of the parts; there always exist settings where this divide-and-conquer technique fails.

A way to better understand this theorem is to explore the mathematical role played by each of Arrow's conditions. It is well understood that "independence" is the crucial condition. Indeed, independence is the provision that, de facto, creates a divide-and-conquer method by requiring the approach to separately find an outcome for each pair independent of what is being done for any other pair. This condition is extensively discussed throughout this article.

The second condition is Pareto; it is a particular case of independence in that it specifies the societal outcome for the special setting where there is unanimity about the ranking of a pair. Mathematically, this condition serves two roles. The first function is to guarantee that each pair has at least two different societal outcomes; e.g., for $n \geq 3$ candidates, the Pareto condition ensures there are at least $n!$ attainable societal outcomes. (An unanimous selection of each of the $n!$ transitive rankings requires it to be a societal ranking.)

The full power of this aspect of the Pareto condition — the full set of $n!$ societal rankings — is not needed to prove Arrow's result. As shown in [6; 7], the negative aspect of Arrow's theorem holds even after replacing Pareto with what I called "involvement." This is where for each of at least two pairs that share an alternative, there are attainable societal outcomes (i.e., with supporting profiles) that rank this pair in at least two of the three possible ways.

Involvement includes Pareto and negative Pareto (where the ranking is the opposite of what voters want (Wilson [12])) as special cases. But "involvement" is significantly more flexible because it can be satisfied even with only two attainable societal rankings! With four alternatives $\{A, B, C, D\}$, for instance, suppose that a rule allows only the two societal rankings $A \succ B \succ C \succ D$ and $B \succ C \succ A \succ D$. This rule fails to satisfy Pareto, but it does satisfy the involvement condition for the pairs $\{A, B\}$ and $\{A, C\}$. For another example, replace $B \succ C \succ A \succ D$ with $B \sim C \sim A \succ D$.

The second role of Pareto is to impose a "direction" on the societal outcome; e.g., should all voters rank a pair in the same way, that is the pair's societal ranking. But for our purposes, the directional component of the Pareto condition can be ignored; rather than Pareto, the more inclusive "involvement" suffices. By replacing Pareto with "involvement" (Sect. 4) a negative conclusion follows (i.e., either the rule fixes an outcome for all profiles, or the outcome is determined by a single individual – a dictator or a negative dictator where the societal ranking is the opposite of what he wants).

The connection of Arrow's Theorem with divide-and-conquer methodologies suggests that it may be possible to extend these ideas in all sorts of new directions that involve various disciplines. This I have done in a manner general enough to have applications in economics, organizational design, engineering design, nano-technology, problems of dark matter in astronomy, etc. The approach is to find appropriate compatibility conditions that include issues coming from these different host areas. Ways in which this has been done are described in my papers Saari [9; 10]; other papers are in preparation. The point is that in all of these areas, problems are encountered. Stated loosely, my assertion is that

> *with any divide-and-conquer methodology, where the complexity is addressed by dividing the problem into component parts from which answers are independently determined, where information comes from at least two sources, and where there are compatibility conditions on both the inputs and the outcome, there exist setting where this approach cannot succeed.*

In other words, inefficiencies and errors must be expected even if division-of-labor approaches are carried out with the best of intent.

These results raise the stakes because divide-and-conquer approaches are central to almost everything we do. It is what we do in universities in the pursuit of knowledge where the division-of-labor defines various schools and then different departments. It is the approach used in organizational design where the structure specifies who does what and who sends what to whom. It is what a company does to handle products by dividing the labor into units of design, manufacturing, and sales. Without question, there is a need to find how to obtain positive assertions. That is the topic of the next section.

10.3. Still another vote

An argument that captures the source of the problem comes from my recent book [8]. To introduce the ideas, consider the following common experience of still another ballot. To pose a specific instance, suppose a committee of three is charged with assembling a three-person scientific board that must have a member from the physical sciences, the social sciences, and mathematics. The candidates are

Phys. Sciences	Soc. Sciences	Mathematics
Antti	Katrina	Erkki
Bob	Dave	Fred

(10.1)

Each committee member votes for one person from each of the three lists. Suppose each of Antti, Katrina, and Erkki wins with a 2:1 vote. Will the selection committee be happy with the outcome?

A way to analyze this conclusion is to adopt a reverse engineering perspective by listing all possible supporting profiles. Then, check each profile to determine whether the outcome is appropriate. The five different profiles are:

Voter 1	**Voter 2**	**Voter 3**
Antti, Katrina, Erkki	Bob, Dave, Fred	Antti, Katrina, Erkki
Bob, Katrina, Erkki	Antti, Dave, Fred	Antti, Katrina, Erkki
Antti, Dave, Erkki	Bob, Katrina, Fred	Antti, Katrina, Erkki
Antti, Katrina, Fred	Bob, Dave, Erkki	Antti, Katrina, Erkki
Antti, Katrina, Fred	Bob, Katrina, Erkki	Antti, Dave, Erkki

(10.2)

The outcome appears to be reasonable for the first four profiles. After all, for each of these profiles, the first two voters have directly opposing views. Thus their complete tie is broken by the last voter. The last profile is more problematic; about the best we can state is that each winning candidate won by 2:1.

Whatever conclusion is reached about the last profile, a comforting observation is that the outcome is appropriate for at least 80% of the supporting profiles; i.e., in general the conclusion is reasonable. Even stronger, as Katri Sieberg and I proved [11], the outcome for paired comparison voting rules never reflects the actual, specified profile; it reflects an outcome that is appropriate for the largest set of the supporting profiles. If the actual profile happens to be in this set, the outcome is appropriate; if the actual profile is an outlier with respect to this set, there may be difficulties.

This comment about the outlier suggests examining more carefully the fifth profile. To do so, suppose that this profile, the outlier, is the actual profile. Suppose that instead of having all three winners coming from Finland, and all three losers coming from the US, each voter wanted to have a mixed-cultural board. This comment accurately reflects how they voted; each voter voted for at least one representative from each country.

A natural objection to this comment is that nothing is built into the rule that would allow it to capture added and hidden conditions such as this mixed-cultural intent, or maybe a mixed-gender objective, or a mixed-race goal, or ... Without including in the rule the intent to respect these conditions, it is understandable why the rule ignores, even severs, the voters' objective to coordinate among the component parts. To ensure an outcome that respects these conditions, use more appropriate rules.

To reach the main message of this section, first convert the Eq. 10.1 example into an example that is equivalent because only the names are changed. Change Antti and Bob to, respectively, $A \succ B$ and $B \succ A$; Katrina and Dave to, respectively, $B \succ C$ and $C \succ B$, and Erkki and Fred to, respectively $C \succ A$ and $A \succ C$. Doing so converts Eq. 10.2 into

Voter 1	Voter 2	Voter 3
$A \succ B, B \succ C, C \succ A$	$B \succ A, C \succ B, A \succ C$	$A \succ B, B \succ C, C \succ A$
$B \succ C \succ A$	$A \succ C \succ B$	$A \succ B, B \succ C, C \succ A$
$C \succ A \succ B$	$B \succ A \succ C$	$A \succ B, B \succ C, C \succ A$
$A \succ B \succ C$	$C \succ B \succ A$	$A \succ B, B \succ C, C \succ A$
$A \succ B \succ C$	$B \succ C \succ A$	$C \succ A \succ B$

(10.3)

A comparison of Eqs. 10.2 and 10.3 shows that a "same-culture" ranking in Eq. 10.2 becomes a cyclic ranking in Eq. 10.3. The societal outcome of "Antti, Katrina, Erkki", for instance, is converted into the societal cyclic ranking of $A \succ B, B \succ C$, $C \succ A$. Each of the six multi-cultural rankings, however, is equated with a transitive ranking.

Of importance is that the outlier choice of Eq. 10.2 (the last profile) is the Condorcet triplet

$$A \succ B \succ C, \quad B \succ C \succ A, \quad C \succ A \succ B. \tag{10.4}$$

This profile creates the so-called "paradox of voting" where even though each voter has a complete, transitive ranking of the three alternatives, the societal outcome is a cycle.

The reason for the "paradox of voting" now is clear. The majority vote over pairs severs the voters' intent to have a multi-cultural board; it severs the voters' connecting conditions of transitivity in the translated problem. Because the voting rule cannot capture the intent of the multi-cultural voters, or the individual rationality condition that the voters have transitive preferences, the actual domain for the pairwise vote differs from what is intended. Our intended domain consists of the six transitive rankings while the actual domain includes the two cyclic rankings of the three pairs. With the actual domain, the Eq. 10.3 cyclic conclusion for the paradox of voting is highly appropriate because it accurately represents the natural conclusion for a full 80% of the supporting profiles. Even though we want to avoid certain profiles by requiring the voters to have transitive preferences, the rule is incapable of capturing this intent.

This severing, this dropping of crucial information, extends from the majority vote over pairs to become a property of Arrow's independence condition, and this is the theme of the next section. The important take-away message from this section is that for the kinds of negative conclusions that are discussed here, *compatibility conditions on inputs are imposed to reflect our intent. The conditions may involve transitivity, or ternary relations (from [5]), or the compatibility conditions in my divide-and-conquer results. But unless these conditions are built into the rule, the rule will ignore them.*

As described in the next section, these problems are consequences of the independence conditions. Indeed, all of the independence conditions that I have investigated, the independence condition causes the rule to ignore the carefully constructed compatibility conditions. Namely, the independence conditions force the desired and the actual domains for a rule to differ significantly. Whenever this occurs, the only way information from the parts can always be assembled into a whole is by restricting which agents are allowed to play an active role. This leads to conclusions asserting the rule is equivalent to a dictator, or an oligopoly, or some other undesired setting.

10.4. Finding the actual space of inputs

The main message of Sect. 3 is that independence conditions can introduce unexpected domains and ranges. The desired domain and range, which are defined by compatibility conditions, tend to be proper subsets of the actual domain and range over which the consensus rule acts. It is this difference between the desired and the actual spaces that creates puzzling problems in consensus theory.

To introduce the approach used in this section, let me confess that, as a mathematician, I am attracted to the standard approach of using filters and ultrafilters to prove Arrovian type theorems. Unfortunately, this elegant technique disguises what actually is happening. So, to answer my "why" concerns, my approach is based on geometry.

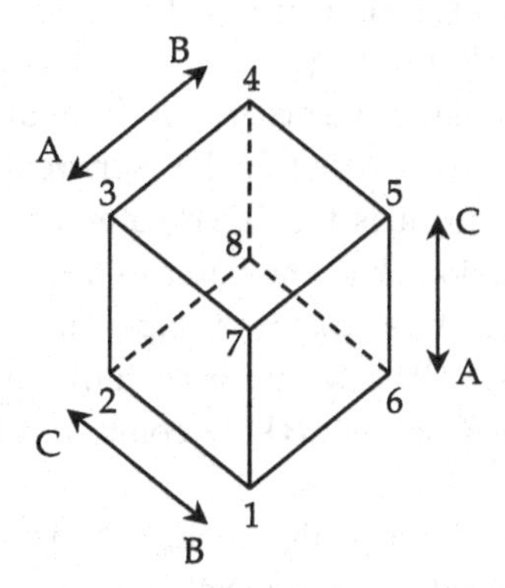

Fig. 10.2. Representation cube

To show how this happens, consider the three alternative case where decisions are made via Arrow's independence conditions. The three pairs are depicted in Fig. 10.2 where the x, y, and z directions indicate, respectively, the A, B, the B, C and the C, A rankings. For each axis, the arrows indicate the preferred alternative. For instance, all vertices on the vertical face on the left represent rankings where $A \succ B$, while the vertices on the back opposite face represent rankings where $B \succ A$. The numbering of the vertices define the rankings according to Eq. 10.5; the hidden vertex 8 is diametrically opposite vertex 7.

Number	Ranking	Number	Ranking
1	$A \succ B \succ C$	4	$C \succ B \succ A$
2	$A \succ C \succ B$	5	$B \succ D \succ A$
3	$C \succ A \succ B$	6	$B \succ A \succ C$
7	$A \succ B, B \succ C, C \succ A$	8	$A \succ C, C \succ B, B \succ A$

(10.5)

Arrow's compatibility condition of transitivity asserts that the domain and range consist of vertices 1 to 6. As we will see, the independence condition forces all eight vertices to be in the actual domain and range.

10.4.1. *Outline of Arrow's result where "involvement" replaces Pareto*

"Involvement" ensures there are at least two pairs where each has at least two different societal outcomes. For simplicity, assume that all rankings are strict. For each pair with different societal outcomes, each ranking is supported by a profile. In some order, change each voter's ranking from what it is the first profile to what it is in the second. Clearly, at some point, a particular voter's change in preferences changes the pair's societal outcome. Collect the information accompanying this setting; i.e., note the name of the agent who caused the societal change and the preferences of all other voters. (Because of the independence conditions, we only need to register each voter's ranking of the pair being considered.) Do so for all possible paths from one profile to the other, and from all possible supporting profiles for each ranking. Do so for each pair that has different outcomes.

If for all possible scenarios it is the same voter who changes the outcome, then this captures Arrow's assertion that one voter dictates the outcome. (By not using Pareto, this voter need not be a dictator; e.g., the societal outcome can be the opposite of his preferences.) Otherwise, for each of two pairs, there exist scenarios where, say, Ann determines the $\{A, B\}$ outcome and Bob determines the $\{B, C\}$ outcome.

Notice that for this to occur, certain scenarios must occur. Thus, all voters other than Ann and Bob might need to have specified preferences over the two pairs. But there always exists a transitive ranking that has a specified ranking for each of two pairs, so for each voter other than Ann and Bob, assign the voter a ranking that is consistent with these conditions, and then fix this ranking for the rest of the discussion.

Bob might also have to have a specific $\{A, B\}$ ranking to allow Ann to change $\{A, B\}$ outcomes. If it is $A \succ B$, then restrict Bob to the preferences on the front left face of the Fig. 2 cube — the face defined by the vertices $\{1, 2, 3, 7\}$. If Bob must have a $B \succ A$ ranking, then restrict Bob to the opposite back face consisting of vertices $\{4, 5, 6, 8\}$. In this construction, further require Bob's preferences to a fixed $\{A, C\}$ ranking; e.g., select either the bottom or top edge for each of the two faces. That is, Bob's moves are between the $\{1,2\}$ and the $\{6,8\}$ edges to keep the $A \succ C$ ranking, or between the $\{3,7\}$ and $\{4,5\}$ edges to keep the $C \succ A$ ranking. According to the independence condition, the rule only knows which edge is adopted; it cannot differentiate between the two defining vertices.

Similarly, to unleash Bob's power, Ann may need to have a particular $\{B,C\}$ ranking, so capture her preferences either with those on the front right face (vertices $\{1,5,6,7\}$) or the back face (vertices $\{2,3,4,8\}$). Again, so she always has the same $\{A,C\}$ ranking, select her preference rankings to both come either from the bottom edge or the top edge.

With these choices, Ann and Bob are free to independently change the societal rankings of these two pairs. In particular, Ann can force the outcome to be on the front left face representing an $A \succ B$ ranking (with vertices $\{1,2,3,7\}$) while Bob can force the outcome to be on the front right face of a $B \succ C$ ranking (with vertices $\{1,5,6,7\}$). Thus, on the Fig. 10.2 cube, the outcome is defined by the intersection of these two faces; it is the front vertical edge defined by vertices $\{1,7\}$. Similarly, Ann can force the outcome to be on the back face with a $B \succ A$ ranking defined by vertices $\{4,5,6,8\}$ while Bob can force the outcome to be on the face with a $C \succ B$ ranking defined by $\{2,3,4,8\}$. In this setting, the societal outcome is on the back vertical edge $\{4,8\}$.

Herein lies the contradiction. No voter changes his or her $\{AC\}$ ranking, so, according to the independence condition, the $\{A,C\}$ outcome remains fixed. Thus, all outcomes are either on the top or on the bottom face. If the outcomes must be on the top face, they vary between 4 and 7; if on the bottom face, the societal outcomes vary between 1 and 8. But with either face, one of the outcomes must be a cyclic ranking given by either vertex 7 or 8. As these outcomes do not satisfy the compatibility condition, it follows that the scenario involving two voters cannot occur; the societal ranking must be determined by only one voter.

10.4.2. *Finding the problem*

The problem, of course, is that the actual set of outcomes defined by the rule differs from what we want; the rule's de facto set includes cyclic rankings. Namely, the rule nicely satisfies the independence conditions by introducing a cyclic ranking. The problem is not that of the rule, it is that we reject this choice.

A way to address my "why" concerns is to establish a connection with Eq. 10.3. So suppose the rule also satisfies some monotonicity condition whereby voting for a candidate helps her. Also assume (without loss of generality) that Ann and Bob select their preferences from the top face. This means that the vertex 7 cyclic ranking occurs if Ann selects a ranking given by either vertex 3 or vertex 7, while Bob selects either vertex 5 or vertex 7. In the spirit of Eqs. 10.2, 10.3, the full space of indistinguishable (by the independence condition) supporting profiles for this cyclic outcome is given by the four pairs

$$(3,5), \quad (3,7), \quad (7,5), \quad (7,7)$$

where the first listed preference in a pair is Ann's.

As true with Eq. 10.3 and consistent with my result with Sieberg [11], the largest number (here 75%) of supporting profiles involve cyclic preferences! The cyclic outcome, in other words, is an appropriate conclusion for most of the supporting profiles in the *actual domain* – the domain that is defined and used by the rule. Stated in another manner, while we require the voters to have transitive preferences, the independence condition forces the rule to ignore this intent by creating its own domain to determine conclusions. (Without a monotonicity condition, these comments apply to another outcome from the top face.)

10.4.3. *Other compatibility and independence conditions*

A similar effect occurs with other independence conditions. An illustration comes from the earlier mentioned nice paper by McMorris and Powers [5]. Precise definitions can be found in their paper, but, by using Fig. 1, a ternary relationship over four items $S = \{a, b, c, d\}$ can be loosely viewed as a listing of all triplets (u, v, w) where the path from u to v does not intersect the path from w to r. Thus different labelings of the Fig. 1 vertices define different sets of paths, i.e., different ternary relationships. The set of all possible ternary relationships for a set S is denoted by $T(S)$, so, with N agents, a profile is $\pi \in (T(S))^N$. A consensus function is given by

$$C : (T(S))^N \to T(S). \tag{10.6}$$

The ternary independence condition is the obvious one. Namely, if each agent's listing for a triplet is the same for two profiles π and π', then outcome for that triplet is the same for both profiles. The result is that when the independence condition is accompanied with a condition such as Pareto, or involvement, and a finite number of agents, the consensus function is equivalent to a dictatorship.

To connect this result with the above discussion, notice that the desired domain consists of graphs of the Fig. 1 type with all possible different permutations of $\{a, b, c, d\}$ in the vertices. (For simplicity, consider only triplets representing distinct end vertices.) As such, an example of an inadmissible quadruple is given by the four triplets (a, b, c), (b, c, d) (these two triplets are consistent with Fig. 1), (d, a, b) (which is not consistent as the path from d to a intersects the path from b to r), and (d, a, c). The independence condition, however, ensures that this set of four triplets *belongs to the actual domain of any consensus function that satisfies the ternary independence condition.*

A question is whether the undesired can accompany the approved; namely, for a specified profile, will the compatibility requirements keep out the unwashed? To see that the undesired can join (and distort) the analysis, consider the four-agent profile representing the four ways to label Fig. 1, going from left to right, as

$$[a, b, c, d], \quad [b, c, d, a], \quad [c, d, a, b], \quad [d, a, b, c]. \tag{10.7}$$

Accompanying the ternary relationships of these four choices[a] are indistinguishable settings (because of the independence condition) where an agent has the four triplets $(a, b, c), (b, c, d), (c, d, a), (d, a, b)$, which clearly are inconsistent. But because of the independence condition, the rule cannot recognize this inconsistency, so it becomes part of the rule's actual, de facto domain. This actual set of indistinguishable triplets accompanying Eq. 10.7, then, has features similar to that of Eq. 10.3 where most of the choices are undesired.

In other words, the intended domain of $T(S)$ has $4! = 24$ ternary relationships. (This can be reduced to 12.) The actual domain that is being used by the consensus function includes products of all of the parts, so it has $(3!)^4 = 1296$ relationships! The consensus function acts over the actual domain rather than over the intended one. An argument similar to that used above with Arrow's theorem shows that this explains the dictatorship conclusion.[b]

10.4.4. *Positive conclusions?*

The way to find positive conclusions is to redesign the independence conditions (the divide-and-conquer methodologies) to permit an appropriate level of agreement between the actual and the desired domains. In terms of the above comments and recognizing that the independence conditions partly define the acceptable rules, the goal is to design appropriate independence conditions that allow the rules to more accurately reflect the desired compatibility conditions. As an example using transitive rankings, I show in my book [8] how Arrow's theorem, and other problematic conclusions, can be converted into positive statements just by forcing the rule to respect the specified compatibility conditions.

The way this is done is to replace the usual information about the relative ranking of a pair to one that also specifies how many alternatives separate the two specified alternatives. With the $A \succ B \succ C \succ D$ ranking, for instance, instead of just using the $A \succ D$ relative ranking information, I use the $(A \succ D, 2)$ "intensity information" where the "2" indicates that A and D are separated by two other alternatives.

Arrow's independence condition uses the relative ranking of a pair; my IIIA condition for a pair is that if each voter in the two profiles has the same intensity information for this pair, then the pair has the same relative societal ranking for both profiles.

The actual domain defined by my condition remains larger than the desired one, but it suffices to obtain positive conclusions. The mathematical reason is that

[a]The central role played by this kind of configuration will be discussed elsewhere.

[b]Differences are minor and expected; e.g., in the transition from (a, b, c) to (c, b, a), the first change may be (a, c, b).

the restriction to transitive preferences now becomes a meaningful condition by imposing a useful restriction on the actual domain. To illustrate with the Eq. 10.4 profile, the associated set of supporting profiles for Arrow's independence condition is given by Eq. 10.3; with my independence condition, all Eq. 10.3 rankings that include a cycle are expelled to leave only the actual profile.

This change significantly alters the conclusion: My independence condition, which imposes a minimal coordination among what happens with each of the parts, replaces Arrow's dictator with the Borda Count (where an N-candidate ballot is tallied by assigning $N-j$ points to the jth positioned candidate). Preliminary results (to be reported elsewhere) indicate that a positive conclusion also is obtained for the Fig. 1 tree problem by modifying the ternary conditions so that they respect a similar intensity condition.

10.5. Summary

Problems similar to that described above accompany other independence conditions. The reason is that standard independence conditions define the "parts" of a structure. By separating the analysis into the parts, which in effect divorces any intended connection among the parts, this independence condition requires the actual domain for the rule to be the product of all of the defined parts. In contrast, the compatibility conditions define the desired domain that is only a proper subset of this product space. The rule respects and is making its computations based on the actual domain, not on the desired one. As such, problems must be anticipated. While the problems need not be dictatorial kinds of conclusions, to deal with the differences between the actual and desired domains, the consensus rule must have distorted properties.

Solutions include altering the independence conditions to provide some level of coordination about what is happening in other parts when determining the outcome for a specific part. More about this will be discussed elsewhere.

References

1. K.J. Arrow, *Social Choice and Individual Values* 2nd. ed., Wiley, New York, [1952] 1963.
2. J.P. Barthélémy, F.R. McMorris, R.C. Powers, Dictatorial consensus functions on n-trees, *Mathematics of the Social Sciences* **25** (1992), 59–64.
3. W.H.E. Day and F.R. McMorris, Axiomatics in group choice and bioconsensus, pp. 3–35. In: *Bioconsensus*, Vol 61, *DIMACS Series in Discrete Mathematics and Theoretical Computer Science*, ed., M.F. Janowitz, F.-J. Lapointe, F.R. McMorris, B. Mirkin, F.S. Roberts, American Math Society, Providence, R.I., 2003.
4. F.R. McMorris, R.C. Powers, Consensus functions on trees that satisfy an independence axiom, *Discrete Applied Mathematics* **47** (1993), 47–55.
5. F.R. McMorris, R.C. Powers, The Arrovian Program from weak orders to hierarchical and tree-like relations, pp 37-45 in *Bioconsensus*, Vol 61, *DIMACS Series in Discrete*

Mathematics and Theoretical Computer Science, ed., M.F. Janowitz, F.-J. Lapointe, F.R. McMorris, B. Mirkin, F.S. Roberts, American Math Society, Providence, R.I., 2003.
6. D.G. Saari, *Geometry of Voting*, Springer-Verlag, New York, 1994.
7. D.G. Saari, *Basic Geometry of Voting*, Springer-Verlag, New York, 1995.
8. D.G. Saari, *Disposing Dictators, Demystifying Voting Paradoxes*, Cambridge University Press, New York, 2008.
9. D.G. Saari, Source of complexity in the social and managerial sciences: An extended Sen's Theorem. *Social Choice & Welfare*, 2010.
10. D.G. Saari, Aggregation and multilevel design for systems: finding guidelines, UCI Preprint, 2009.
11. D.G. Saari, K.K. Sieberg, The sum of the parts can violate the whole, *Amer. Pol. Science Review* **95**(2001) 415–433.
12. R. Wilson, Social choice theory without the Pareto principle, *Jour. Economic Theory* **5** (1972) 478–486.

Chapter 11

The Contributions of F.R. McMorris to Discrete Mathematics and its Applications

George F. Estabrook, Terry A. McKee, Henry Martyn Mulder, Robert C. Powers, Fred S. Roberts

Department of Ecology and Evolutionary Biology, University of Michigan
Ann Arbor MI 4809-1048 USA
gfe@umich.edu

Department of Mathematics & Statistics, Wright State University
Dayton, Ohio 45435 USA
terry.mckee@wright.edu

Econometrisch Instituut, Erasmus Universiteit
P.O. Box 1738, 3000DR Rotterdam, Netherlands
hmmulder@ese.eur.nl

Department of Mathematics, University of Louisville
Louisville, KY 40292 USA
rcpowe01@louisville.edu

DIMACS Center, Rutgers University
Piscataway, NJ 08854 USA
froberts@dimacs.rutgers.edu

In this chapter we discuss the contributions of F.R. McMorris to discrete mathematics and its applications on the occasion of his retirement in 2008.

Introduction

End of August 2008 F.R. McMorris, Buck for his friends and colleagues, retired as Dean of the College of Science and Letters at IIT, Illinois Institute of Technology, Chicago. He also celebrated his 65-th birthday. To commemorate these events a two-day conference was held early May 2008 at IIT. In addition this volume is written in honor of his contributions to mathematics and its applications. The focus of the volume is on areas to which he contributed most. The chapters show the broadness of his interests and his influence on many co-authors and other mathematicians. Here we survey his work.

First some basic facts. At the moment of finishing this volume Math. Reviews lists 2 books, the editorship of 3 conference proceedings, and 108 published papers of Buck McMorris, and there are still many to come. There are 53 co-authors listed, and again there are more to come. Of course, the papers and co-authors listed in Math. Reviews are not all. The areas to which he has contributed, by number of publications, are: Combinatorics; Group theory and generalizations; Biology and other natural sciences; Game theory, economics, social and behavioral sciences; Operations research and mathematical programming; Order, lattices and ordered algebraic structures; Statistics; Computer science.

Below we will highlight many of his contributions. Some characteristics of his way of working are: an open mind, keen on fundamental mathematics with a relevance for applications, always taking a broad view: try to formulate a 'master plan' that may be a guide for creating many specific questions and open problems. What seems to be equally important is that Buck has become a dear friend for many of his co-authors. That fact was fundamental for the success, mathematically and socially, of the above mentioned celebration conference.

Many of his co-authors are mentioned below, some are contributors of this volume. Much to our regret, because of the focus chosen, not all of his important co-authors are represented as author in this volume. Therefore we mention them here by number of their collaborations with Buck McMorris, according to MathSciNet: C.S. Johnson Jr., Bill Day, Frank Harary, Mike Jacobson, John Luedeman, Ewa Kubicka, Grzegorz Kubicki, Jean-Pierre Barthélémy, Robert Brigham, Hans-Hermann Bock, Jerald Kabell, and Chi Wang.

McMorris started his mathematical career in semi-groups. But even in his early work we can already discern some of his future interests in discrete mathematics and its applications: mathematical biology, intersection graphs, voting theory, consensus theory. In the early eighties his focus shifted from semigroups to discrete mathematics, with an emphasis on graph theory, while his early interests remained. In the sections below we highlight these. Of course, not all of his publications can be discussed. We hope that the choices made provide a clear picture of his work and interests.

11.1. Mathematics of Evolutionary Biology

McMorris published 10 papers motivated by a concept from evolutionary systematic biology called character compatibility. Seven were between 1975 and 1981, the early days of his involvement with this concept: four of these included Estabrook as a coauthor [16; 17; 18; 19], two were entirely his own [41; 42] and one more with Zaslavsky [56]. McMorris et al. [55] addresses an abstract issue in graph theory related to an unresolved question from character compatibility, and Day et al. [13] apply McMorris' potential compatibility test to look for randomness in about 100 published data sets. For his last publication on character compatibility [20], he worked again with his original coauthor to examine the relationship between geologic stratigraphic data and compatibility.

To understand the relevance of McMorris' contributions to character compatibility analysis, it is useful to understand some of the concepts of evolutionary systematic biology. This subfield of biology seeks to estimate the tree of ancestor-descendant relationships among species, consequent of their evolution, and then use these evolutionary relationships to recognize higher taxa (groups of species in genera, families, etc). In the late 19th century, systematic biologists realized that similarities and differences with respect to a basis for comparison among a group of related species under study could be the basis for an hypothesis about the relationships among species and the ancestors from which they evolved, their so-called ancestor relation. Such hypotheses are expressed as characters, which group species together into the same character state if they are considered to be the same with respect to a basis for comparison, and then arrange these character states into a character state tree to indicate where speciation events associated with the observed changes are hypothesized to have occurred. By mid 20th century, some natural scientists also realized that some pairs of such hypotheses based on different bases for comparison could be logically incompatible, i.e., they could not both be true. At that time, scientists began to develop tests for, and ways to resolve, incompatibilities to estimate the ancestor relation from these hypotheses. Wilson (1965) [66] is among the earliest published works to present an explicit test for the compatibility of (two state) characters. Estabrook (2008) [21] provides an in-depth discussion of biological concepts of character state change, and the nature of compatibilities and incompatibilities among characters that arise from them. Estabrook (this volume) provides explicit explanations of McMorris' contributions motivated by this concept. Here we will summarize briefly what we consider to be his most significant contributions.

McMorris recognized a bi-unique correspondence between character state trees for a collection of related species and trees of subsets (ordered by inclusion) of that collection, which enabled a simple test for compatibility that identified the states involved with contradictions when the test failed. He realized that character state trees themselves enjoy a lower semi lattice order under the relation "is a refinement of", and described a simple test to recognize when a pair of character state trees were in that relation. Qualitative taxonomic characters are characters with their character states, but no explicitly hypothesized character state tree. Two qualitative taxonomic characters are potentially compatible if there exists character state trees for each that are compatible with each other. Estabrook had conjectured a simple test for potential compatibility, see [16; 6], which McMorris proved to be correct. Potential compatibility raises an unresolved issue: Several qualitative taxonomic characters can be pairwise potentially compatible but in some cases character state trees for each do not exist so that they remain pairwise compatible as character state trees. Simple criteria to recognize such cases have not yet been discovered.

This is related to chordal graphs [24; 55]. McMorris' last publication addresses stratigraphic compatibility [20] and raises questions related to functional graphs. For an in-depth treatment of the papers discussed here, the reader is referred to Estabrook (this volume).

11.2. Contributions to Intersection Graph Theory

As in many parts of discrete mathematics, McMorris introduced or popularized significant new ideas in intersection graph theory, sometimes with a conference talk proclaiming a "master plan" for developing the idea. Six of these contributions are described below, with further discussion of each available in the 1999 SIAM monograph *Topics in Intersection Graph Theory* [40].

Upper bound graphs [40, §4.4]. The 1982 McMorris & Zaslavsky paper [57] combines McMorris's interests in partially ordered sets and graph representations. The *upper bound graph* G of a partial ordering $(P, <)$ has vertex set P, with distinct vertices adjacent in G if and only if the corresponding elements of P have a common upper bound in P. Reference [57] characterizes upper bound graphs by the existence of complete subgraphs $Q_1, \ldots, Q_k$ that cover $E(G)$ such that, for each $j \leq k$, there exists a vertex v_j in G where $v_j \in Q_j$ but $v_j \notin Q_i$ for $i \neq j$; moreover, each Q_i can be assumed to be an inclusion-maximal complete subgraph of G. This characterization has spawned many related results, often coupled to applicable topics such as competition graphs.

Bipartite intersection graphs [40, §7.2]. The 1982 Harary, Kabell & McMorris paper [26] generalizes classical intersection graphs, and interval graphs in particular. A *bipartite intersection graph* G has $V(G)$ partitioned into sets X and Y, with each $x \in X$ and $y \in Y$ assigned sets S_x and T_y such that vertices x and y are adjacent in G if and only if $S_x \cap T_y \neq \emptyset$; furthermore, G is a *bipartite interval graph* if each S_x and T_y is an interval of the real line. Others have subsequently introduced concepts of directed intersection and interval graphs that are structurally interconnected with their bipartite counterparts.

***p*-Intersection graphs [40, §6.1].** The 1991 Jacobson, McMorris & Scheinerman paper [33] generalizes standard (1-)intersection graphs, significantly generalizing traditional intersection graph theory. The *p-intersection graph* G of a multiset $\{S_1, \ldots, S_k\}$ of subsets of an underlying finite set $\mathcal{X}$ has vertices $v_1, \ldots, v_k$, with distinct vertices v_i and v_j adjacent in G if and only if $|S_i \cap S_j| \geq p$. In particular, this extends the well-studied (but notoriously hard) concept of the *intersection number* of a graph—the minimum cardinality of $\mathcal{X}$ such that G is an intersection graph of subsets of $\mathcal{X}$—to *p-intersection numbers*.

Tolerance intersection graphs [40, §6.3]. The 1991 papers by Jacobson, McMorris & Mulder [32] and Jacobson, McMorris & Scheinerman [33] introduce this very general concept. The *ϕ-tolerance intersection graph* G of a family $\{S_1, \ldots, S_k\}$ of subsets of an underlying finite set $\mathcal{X}$ has each subset of S assigned a *measure* $\mu(S)$, has each S_i assigned a *tolerance* t_i, and has a binary function $\phi(x, y)$ that is often $\min\{x, y\}$, with distinct vertices v_i and v_j adjacent in G if and only if $\mu(S_i \cap S_j) \geq \phi(t_i, t_j)$. Tolerance intersection graphs generalize both p-intersection graphs and the previously-studied "tolerance graphs," which can now be described as interval graphs with μ the length of an interval and $\phi(x, y) = \min\{x, y\}$.

Sphere-of-influence graphs [40, §7.11]. The 1993 Harary, Jacobson, Lipman & McMorris paper [25] promotes ideas that were motivated by pattern recognition and computer vision problems. Suppose $\mathcal{X}$ is any finite set of points in the plane and each $x \in \mathcal{X}$ is associated with the open disc centered at x with radius equal to the minimum distance from x to the other points of $\mathcal{X}$. A *sphere-of-influence graph* G has vertices that correspond to such open discs, with distinct vertices adjacent in G if and only if the corresponding open discs have nonempty intersection. One basic question from [25] is which complete graphs are sphere-of-influence graphs—K_8 is; K_9 is conjectured to be; K_{12} is not. Closed sphere-of-influence graphs and ϕ-tolerance sphere-of-influence graphs have also been studied.

Probe interval graphs [40, §3.4.1] The 1998 McMorris, Wang & Zhang paper [54] developed tools that were directly motivated by work in physical mapping of DNA. A graph G is a *probe interval graph* if $V(G)$ contains a subset P and each vertex corresponds to an interval of the real line, with distinct vertices adjacent in G if and only if at least one of them is in P and their corresponding intervals have nonempty intersection. Reference [54] contributes structural information about probe interval graphs and has led to considerable recent work in this active research area.

11.3. Competition Graphs and their Generalizations

Buck McMorris has made some very interesting contributions to the study of competition graphs and the related phylogeny graphs. These topics combine his interests in graph theory with his interests in biology. This work was done in collaboration with Roberts (3 papers) and others.

11.3.1. *Competition Graph Definitions and Applications*

The study of competition graphs has given rise to a very large literature, some of which is surveyed in the articles [34; 37; 63]. Suppose $D = (V, A)$ is a digraph. Its *competition graph* $C(D)$ is the graph $G = (V, E)$ with the same vertex set and an edge $\{x, y\}$ in E for $x \neq y$ if and only if there is a vertex a in V so that arcs (x, a) and (y, a) are in D. Competition graphs were introduced by Cohen [9] in connection with a problem of ecology. The vertices of D represent species in an ecosystem and there is an arc from u to v if u preys on v. We call such a digraph a *food web*. There is an edge between species x and y in $C(D)$ if and only if x and y have a *common prey* a in D, *i.e.*, if and only if x and y compete for a. In the literature of competition graphs, it is very common to study the special case where D is an acyclic digraph without loops, as is commonly the case for food webs.

A variant of the competition graph idea is called the phylogeny graph because it was motivated by a problem in phylogenetic tree reconstruction. We say that G is the *phylogeny graph* $P(D)$ of $D = (V, A)$ if $G = (V, E)$ and there is an edge between $x \neq y$ in E if and only if either (x, a) and (y, a) are in A for some a in V, or (x, y) is in A or (y, x) is in A. If D is a digraph without loops and D' is the corresponding digraph with a loop added to each vertex, then it is easy to see that the phylogeny graph of D is the competition graph of D'. This observation was first made by Buck McMorris in a personal communication to Roberts. Roberts and Sheng [64] introduced the term phylogeny graph because of a possible connection of this concept to the problem of phylogenetic tree reconstruction. It is appropriate that Buck should have played a role in this notion of phylogeny graph because of his longstanding interest and many contributions to the theory and practice of phylogenetic tree reconstruction.

The notion of competition graph also arises in a variety of other non-biological contexts. (See [61].) Suppose the vertex set of D can be divided into two classes, A and B, and all arcs are from vertices of A to vertices of B. (We do not assume that A and B are disjoint.) Then we sometimes seek the restriction of the competition graph to the set A. This idea arises for instance in communications where A is a set of transmitters, B is a set of receivers, and there is an arc from u in A to v in B if a message sent at u can be received at v. We then note that x and y in A *interfere* if signals sent at x and y can be received at the same place, *i.e.*, if and only if x and y are adjacent in the competition graph (restricted to A). The problem of channel assignment in communications can be looked at as the problem of coloring the *interference graph.*

The idea also arises in coding. Suppose A is a transmission alphabet, B is a receiving alphabet, and there is an arc from u in A to v in B if when symbol u is sent, symbol v can be received. Then symbols x and y in the transmission alphabet are *confusable* if they can be received as the same letter, *i.e.*, if and only if x and y are adjacent in the competition graph (restricted to A). We often seek a minimum set of mutually non-confusable symbols in a transmission alphabet – this is the problem of finding a maximum independent set in the competition graph (restricted to A).

Competition graphs arise in scheduling in situations where we have conflicting requests. Suppose that A is the set of users of a facility and B the set of facilities, and an arc from u in A to v in B means that user u wishes to use facility v. Then users x and y *conflict* if they both wish to use the same facility. In another scheduling application, A is a set of users of a fixed facility and B the set of times that facility might be used, and an arc from u in A to v in B means that user u

wishes to use the facility at time v. Users x and y conflict if they both wish to use the facility at the same time. The competition graph is sometimes called the *conflict graph.*

Competition graphs arise in studies of the structure of models of complex systems arising in modeling of energy and economic systems. In such models, we often use matrices and set up linear programs. Let A be the set of rows of a matrix M and B the set of columns, and take an arc from u to v if the u, v entry of M is nonzero. Then in a corresponding linear program, the constraints corresponding to rows x and y involve a common variable with nonzero coefficients if and only if x and y are adjacent in the competition graph. In the literature, the competition graph is called the *row graph* of matrix M. The row graph is useful in understanding the structure of linear programs.

11.3.2. *Competition Numbers and Phylogeny Numbers*

As noted earlier, Buck McMorris has made extensive contributions to the theory and applications of interval graphs. Interval graphs have played a central role at the interface between mathematics and biology, and the connection between interval graphs and competition graphs has been a primary force in leading to the great interest in competition graphs. In ecology, a species' normal healthy environment is characterized by allowable ranges of different important factors such as temperature, humidity, pH, etc. If there are p factors and each is taken to be a dimension in Euclidean p-space, then if the ranges on the different factors are independent (a simplifying assumption), the species can be represented by a box in p-space. This box is called the species' *ecological niche.* An old ecological principle says that two species compete if and only if their ecological niches overlap. (That is why the competition graph is sometimes called the *niche overlap graph.*) Cohen [9; 10; 11] asked if, given an independent notion of competition, we could assign each species in an ecosystem to an ecological niche in such a way that competition between species corresponds to overlap of niches. In particular, he started with a food web or digraph with an arc from u to v if u preys on v, defined the corresponding competition graph, and asked if the competition graph could be represented as the intersection graph of boxes in p-space. More specifically, he asked for the smallest such p, which is known as the *boxicity* of the competition graph. Cohen [9] made the remarkable observation that in a large number of examples of food webs, the boxicity of the competition graph always turned out to be 1, i.e., that the competition graph was always an interval graph. In other words, only one ecological dimension sufficed to account for competition. The interpretation of this dimension was (and is) unclear. Although later examples were found by Cohen and others to show that not every competition graph had boxicity 1, Cohen's original observation and the continued preponderance of examples with boxicity 1 led to a large literature devoted to attempts to explain the observation and to study the properties of competition graphs.

In attempting to explain the observation that most real world food webs have competition graphs that are interval graphs, Roberts [62] asked whether this was just a property of the construction, *i.e.*, whether most acyclic digraphs have competition graphs that are interval graphs. He noted that if G is any graph, then G plus sufficiently many isolated vertices is a competition graph of an acyclic digraph. Roberts then defined the *competition number* $k(G)$ of a graph G as the smallest r so that G plus r isolated vertices is a competition graph of an acyclic digraph. Thus, any algorithm for recognizing competition graphs of acyclic digraphs will also compute the competition number, and conversely.

11.3.3. *p-Competition Graphs*

A large number of variations of the notion of competition graph have given rise to interesting problems and questions. To define one such variation, suppose $D = (V, A)$ is a digraph. The *p-competition graph* of D has vertex set V and an edge between x and y in V if there are distinct vertices a_1, a_2, ..., a_p in V so that (x, a_i) and (y, a_i) are arcs in D for $i = 1, 2, ..., p$. In terms of the ecological motivation, x and y compete if and only if they have at least p common prey. This idea was studied by Buck McMorris and collaborators in a series of three papers: [30; 35; 36]. A variety of results analogous to those about ordinary competition graphs are known. Paper [36] by McMorris and coauthors gave necessary and sufficient conditions for a graph with n vertices to be the p-competition graph of some acyclic digraph.

It also provides similar results for arbitrary digraphs (loops allowed) and arbitrary digraphs (loops not allowed). Graph-theoretically, the most interesting results arise if one studies p-competition graphs of arbitrary digraphs. So far, most of the interesting results are about the case $p = 2$. Paper [36] showed that every triangulated graph is a 2-competition graph of an arbitrary digraph. So is every unicyclic graph except the 4-cycle.

The question of what complete bipartite graphs $K(m, x)$ are 2-competition graphs of arbitrary digraphs leads to some very interesting (and difficult) combinatorial questions. In paper [30], McMorris and colleagues showed that $K(2, x)$ is a 2-competition graph of an arbitrary digraph if and only if $x = 1$ or $x \geq 9$ and that $K(3, x)$ is not a 2-competition graph of an arbitrary digraph if x = 3, 4, 5, 7, 8, 11. Then, Jacobson [31] showed that $K(3, x)$ is a 2-competition graph of an arbitrary digraph for $x \geq 38$. The situation for $K(3, 6)$ and $K(3, 37)$ remains open, to our knowledge.

Also of interest is a concept analogous to competition number. The *p−competition number* $k_p(G)$ is the smallest r so that G together with r isolated vertices is a p-competition graph of some acyclic digraph. McMorris and his colleagues [35] showed that this is well-defined. In this same paper, they showed the surprising result that for every m, there is a graph G with $k_p(G) \leq k(G) - m$.

11.3.4. *Tolerance Competition Graphs*

As noted above in Section 11.2, the 1991 papers by Jacobson, McMorris and Mulder [32] and Jacobson, McMorris and Scheinerman [33] introduced a very general concept called a ϕ-tolerance intersection graph. An analogous notion for competition graphs was introduced by Brigham, McMorris, and Vitray [7; 8]. Let ϕ be a symmetric function assigning to each ordered pair of natural numbers another natural number. We say that $G = (V, E)$ is a *ϕ-tolerance competition graph* if there is a directed graph $D = (V, A)$ and an assignment of a nonnegative integer t_i to each vertex v_i in V such that, for $i \neq j$,

$$\{v_i, v_j\} \in E(G) \leftrightarrow |\{a : (v_i, a) \in A\} \cap \{a : (v_j, a) \in A\}| \geq \phi(t_i, t_j).$$

A *2-ϕ-tolerance competition graph* is a ϕ-tolerance competition graph in which all the t_i are selected from a 2-element set. Characterizations of such graphs, and relationships between them, are presented for ϕ equal to the minimum, maximum, and sum functions, with emphasis on the situation in which the 2-element set is $\{0, q\}$.

11.4. Location Functions on Graphs

As mentioned in Section 11.2 Buck McMorris used the idea of a "master plan" to generate all kinds of interesting questions and problems. This inspired his coauthor Mulder to use this meta-concept as well. The first instance was Mulder's "Metaconjecture" mentioned in Mulder (this volume). Trees and hypercubes share being median graphs. In a sense they are the extreme cases within this class, in the class of all median graphs with n vertices, the trees realize the minimum number of edges: $n-1$, and the n-cube Q_n realizes the maximum number of edges 2^n. The following has served as a "master plan" in the sense of McMorris.

Metaconjecture. *Let $\mathcal{P}$ be a property that makes sense, which is shared by the trees and the hypercubes. Then $\mathcal{P}$ is shared by all median graphs.*

The reader is referred to Mulder (this volume) for the incentive this Metaconjecture has given. In the spirit of this Metaconjecture one also tries to generalize results on trees to median graphs whenever possible. This was the motivation for Buck McMorris to study the axiomatic characterization of locations functions on median graphs. Location functions are are a specific instance of consensus functions. A consensus function is a model to describe a rational way to obtain consensus among a group of agents or clients. The input of the function consists of certain information about the agents, and the output concerns the issue about which consensus should be reached. The rationality of the process is guaranteed by the fact that the consensus function satisfies certain "rational" rules or "consensus axioms". For a location function on a network the input is the position of the clients in the network, and the output is the set of preferred locations. For a full discussion of the axiomatic characterizations of three important location functions see McMorris, Mulder, and Vohra (this volume), where the details of the results discussed below can be found.

A central problem in location theory and consensus theory is to find those points in a set X that are "closest" to any given profile $\pi = (x_1, x_2, \ldots, x_k)$. Most of the work done in this area focuses on developing algorithms to find these points [12; 58]. In recent years, there have been axiomatic studies of the procedures themselves and these have resulted in a much better understanding of the process of location and consensus [4; 5; 23; 27; 28]. Without any conditions imposed, a *location function* (*consensus function*) on X is simply a mapping $L : X^* \to 2^X - \emptyset$, where X^* is the set of all profiles of all finite lengths and $2^X - \emptyset$ denotes the set of all nonempty subsets of X.

Let $\delta : X \times X^* \to \mathbb{R}$ be a function such that $\delta(x, \pi)$ represents a measure of "remoteness" of x to the profile π. An attractive class of location functions on (X, δ) is defined by letting $L(\pi) = \{x \in X : \delta(x, \pi) \text{ is minimum}\}$. Two important location functions in this class are the *median function Med*, defined by letting $\delta(x, \pi) = \sum_{i=1}^{k} \delta(x, x_i)$, where $\pi = x_1, x_2, \ldots, x_k$, and the *center function Cen*, defined by letting $\delta(x, \pi) = \max\{\delta(x, x_1), \delta(x, x_2), \ldots, \delta(x, x_k)\}$.

In the continuous case we consider connected networks $N = (V, A)$ with vertex set V and set of arcs A. Think of N as being embedded in n-space. The arcs are curves with a length. The set X is the set of all vertices *and* all interiors points on the arcs. In the discrete case we consider connected graphs $G = (V, E)$ with vertex set V and edge set E. Now the set X is the set of vertices V. Note that there might be big differences between the continuous and the discrete case. For instance, the center function Cen is single-valued in the continuous case but not in the discrete case. Also proof techniques may be quite different.

In 1996 Vohra [65] characterized the median function on tree networks axiomatically, where only three simple axioms were needed. In a tree network the set X is the set of all vertices *and* interior points on the arcs, where arcs can have any length. This is the "continuous case". Rephrased Vohra's axioms are as follows. the *segment* $S(x, y)$ between x and y in a tree network is the set of all point on the path between points x and y. For two profiles π and ρ we denote the concatenation of these by π, ρ.

(A) Anonymity: for any profile $\pi = x_1, x_2, \ldots, x_k$ on X and any permutation σ of $\{1, 2, \ldots, k\}$, we have $L(\pi) = L(\pi^\sigma)$, where $\pi^\sigma = x_{\sigma(1)}, x_{\sigma(2)}, \ldots, x_{\sigma(p)}$.

(B) Betweenness: [Continuous] $L(x, y) = S(x, y)$, for all $x, y \in X$.

(C) Consistency: If $L(\pi) \cap L(\rho) \neq \emptyset$ for profiles π and ρ, then $L(\pi, \rho) = L(\pi) \cap L(\rho)$.

Note that it is easy to show that Med satisfies these three axioms. But Vohra proved the 'converse' as well: any consensus function on a tree network satisfying (A), (B), and (C) necessarily is the median function on the tree network.

When McMorris started to work on the discrete case for Med he realized that it should be done on median graphs. Now the betweenness axiom has to be adapted to the discrete case. The *interval* $I(u, v)$ between vertices x and y in a graph $G = (V, E)$ is the set of vertices lying on the shortest paths between x and y.

(B) Betweenness: [Discrete] $L(u, v) = I(u, v)$, for all $u, v \in V$.

In [46] it was proved that the median function Med on cube-free median graphs is characterized by the three obvious axioms (A), (B), and (C). A median graph G is *cube-free* if G does not contain a 3-cube Q_3. Such graphs are a nice generalization of trees. They seem to be quite esoteric, but there is a one-to-one correspondence between the class of connected triangle-free graphs and a subclass of the cube-free median graphs, see [29]. For the class of all median graphs McMorris, Mulder and Roberts [46] needed an extra 'heavy duty' axiom. These results were extended in [43; 44], where also the ordered case, viz. distributive and median semilattices, is discussed. Another interesting case initiated by McMorris is the *t-Median Function*, see [45]. We omit details.

There is not as much known for Cen on graphs. Foster and Vohra [22] studied the center function on tree networks. A breakthrough occurred when Buck McMorris and coauthors [53] succeeded in characterizing the center function on trees as we have defined it. The result is that a location function L on a tree T is the center function Cen if and only if L satisfies the following four axioms. For a profile π we denote the set of vertices in π by $\{\pi\}$. For a vertex x we denote by $\pi \setminus x$ the subprofile of π by deleting all occurrences of x from π. For a profile π on a tree T we denote by $T(\pi)$ the smallest subtree of T containing all of π.

(Mid) Middleness: [Discrete] Let u, v be two not necessarily distinct vertices of a tree T. Then $L(u, v)$ is the middle of the unique path joining u and v in T.

(QC) Quasi-consistency: If $L(\pi) = L(\rho)$ for profiles π and ρ, then $L(\pi, \rho) = L(\pi)$.

(R) Redundancy: Let L be a location function on a tree T. If $x \in T(\pi \setminus x)$ then $L(\pi \setminus x) = L(\pi)$.

(PI) Population Invariance: If $\{\pi\} = \{\rho\}$ then $L(\pi) = L(\rho)$.

A shorter proof if this result can be found in [60]. A closer look at that proof yields that an analogous result holds for the continuous case, i.e., for tree networks.

McMorris and his coworkers are still continuing research on these location functions, but also other nice instances as the *Mean Function.* A *mean vertex* of π is a vertex v minimizing

$$\sum_{1\leq i\leq k} [d(v, x_i)]^2.$$

The *mean* of π is the set of mean vertices of π. The *Mean Function Mean* on G is the function $Mean : VX^* :\rightarrow 2^V - \emptyset$ with $Mean(\pi)$ being the mean of π.

11.5. Contributions to Bioconsensus: An Axiomatic Approach

Buck McMorris has made many contributions to the area of mathematical consensus and a few of these contributions will be mentioned in this section. We first describe what is meant by a consensus function and then we introduce two well known axioms a given consensus function may or may not satisfy.

Let $\mathcal{D}$ be the set of all (finite) discrete structures of a particular type. (e.g., $\mathcal{D}$ could be a set of specialized labelled graphs, unlabelled graphs, digraphs, partially ordered sets, acyclic digraphs, hypergraphs, partitions, networks, etc.) A *consensus function* on $\mathcal{D}$ is a map $C : \mathcal{D}^k \longrightarrow \mathcal{D}$, where $k \geq 2$ is a positive integer. A major aspect of the *consensus problem* for $\mathcal{D}$ is to find "good" consensus functions that can capture the common agreement of an input *profile* $P = (D_1, \ldots, D_k)$ of members of $\mathcal{D}$, i.e. $C(P)$ should consist of the element (or elements) of $\mathcal{D}$ that best represents whatever similarity that all of the D_i's share. If possible, a good function C should not only have this "consensus" aspect, but additionally should satisfy mathematical properties that enable it to be understood in order that it can be effectively computed exactly or with approximating algorithms. The consensus problem for discrete structures has been a very active area of research with much of it stimulated by the axiomatic approach to social choice (voting theory) pioneered by K. Arrow in the 1950's. In the classical theory developed by Arrow and others, $\mathcal{D}$ is usually taken to be the set of all weak or linear orders on a given set of alternatives S. Many of the axioms are given in terms of the "units of information" (building blocks) of members of $\mathcal{D}$, which in the case for partial orders, are the ordered pairs of S making up the order relation. (Other discrete structures obviously have other types of building blocks.) For example, in generic terms, a property that is universally accepted as being desirable for data aggregation is the following: A consensus function $C : \mathcal{D}^k \longrightarrow \mathcal{D}$ is *Pareto* (P) if whenever $P = (D_1, \ldots, D_k)$ is a profile and 'unit of information' x is in every D_i, then x is in $C(P)$. The Pareto condition simply requires the preservation of the unanimous agreement portion of the input data profile. Another seemingly reasonable property is the following: A consensus function C is *independent (of irrelevant alternatives)* (I) if whenever profiles P and P' agree on a subset $X \subseteq S$, then $C(P)$ and $C(P')$ agree on X. This independence condition also seems to be a good one and captures an aspect of a "stable" consensus function. Of course, what it means to "agree" must be carefully defined. When $\mathcal{D}$ is the set of all weak orders on S (reflexive, transitive and complete relations on S), agreement of two weak orders simply means that they are equal as sets of ordered pairs when restricted to elements in X. Profiles then are said to agree on X if they agree term by term on X. The famous Impossibility Theorem of Arrow essentially says that the only consensus functions on weak orders (where $|S| \geq 3$) satisfying both (P) and (I) are the *dictatorships*, i.e., there is an index j such that for any profile P, if x is strictly preferred to y in D_j, then x is strictly preferred to y in $C(P)$ [1].

Buck McMorris , along with his coauthors, has extended Arrow's Impossibility Theorem in many different directions. For example, in 1983, McMorris and Neumann proved an analog of Arrow's Theorem for tree quasi-orders [47]. A tree quasi-order is a binary relation ρ on a finite set S such that ρ is reflexive, transitive, and $(z,x),(z,y) \in \rho$ implies that $(x,y) \in \rho$ or $(y,x) \in \rho$ for all $x,y,z \in S$. The last condition is called the *tree condition* and it is a generalization of completeness. In 2004, McMorris and Powers extended the tree quasi-order version of Arrow's Theorem by dropping the Pareto condition and replacing it with two profile conditions [49]. In this case, it was shown that the resulting class of consensus functions are quasi-oligarchic. In 1991, Barthélémy, McMorris and Powers, using a carefully constructed independence axiom, established a version of Arrow's Theorem for consensus functions defined on the set $\mathcal{H}(S)$ of all hierarchies of S [2]. A *hierarchy* on S is a collection H of subsets of S such that $S, \{x\} \in H$ for all $x \in S$; $\emptyset \notin H$; and $A \cap B \in \{A, B, \emptyset\}$ for all $A, B \in H$. In 1995, Barthelemy, McMorris and Powers investigated eight different versions of independence conditions for consensus functions on $\mathcal{H}(S)$ and the complete relationships among these eight conditions were determined [3]. In 2003, a deeper connection was made between consensus functions on weak orders and consensus functions on $\mathcal{H}(S)$ with the possibility of having an infinite number of voters. A key to this connection is to view a hierarchy as a special type of ternary relation. Using this viewpoint, it was shown how to embed and extend Arrow's Theorem for weak orders to a result involving ternary relations [48].

In 1952, Kenneth May gave an elegant characterization of simple majority decision based on a set with exactly two alternatives [39]. This work is a model of the classical voting situation where there are two candidates and the candidate with the most votes is declared the winner. May's theorem is a fundamental result in the area of social choice and it has inspired many extensions. In particular, in 2008, McMorris and Powers generalized May's Theorem to the case of three alternatives, but where the voters' preference relations are required to be trees with the alternatives at the leaves [51].

A popular consensus function on the set of hierarchies $\mathcal{H}(S)$ is *majority rule.* Majority rule outputs the set of clusters that appear in more than half of the input profile. The fact that the output is a hierarchy was first noted in 1981 by Margush and McMorris [38]. Although this result is easy to prove, it stands in stark contrast to the situation in classical voting theory where the majority outcome could produce what is called a *voting paradox.* In 2008, McMorris and Powers proved that the majority consensus rule on hierarchies is the only consensus function satisfying four natural and easily stated axioms [50]. The majority consensus rule is part of a larger class of consensus rules where the output is determined by a family of overlapping sets, often called *decisive sets.* Axiomatic characterizations of this class of consensus rules can be found in [47] and [52]. Finally, for a more thorough discussion of axiomatic consensus theory we refer the reader to the book by Day and McMorris [14].

References

1. K.J. Arrow, *Social Choice and Individual Values*, Wiley, New York 2nd ed. (1963).
2. J.P. Barthélémy, F.R. McMorris, R.C. Powers, Independence conditions for consensus n-trees revisited, *Applied Mathematics Letters* **4** (1991) 43–46.
3. J.P. Barthélémy, F.R. McMorris, R.C. Powers, Stability conditions for consensus functions defined on n-trees, *Mathematical and Computer Modelling* **22** (1995) 79–87.
4. J.P. Barthélémy, B. Monjardet, The median procedure in cluster analysis and social choice theory, *Math. Social. Sci.* **1** (1981) 235–268.
5. J.P. Barthélémy, B. Monjardet, The median procedure in data analysis: New results and open problems, in: H.H. Bock, ed., *Classification and Related Methods of Data Analysis*, Elsevier Science, Amsterdam, 1988, pp. 309–316.
6. D. Boulter, D. Peacock, A. Guise, T.J. Gleaves, G.F. Estabrook, Relationships between the partial amino acid sequences of plastocyanin from members of ten families of flowering plants, *Phytochenistry* **18** (1979) 603–608.
7. R.C. Brigham, F.R. McMorris, R.P. Vitray, Tolerance Competition Graphs, *Linear Algebra and Applications* **217** (1995) 41–52.
8. R.C. Brigham, F.R. McMorris, R.P. Vitray, Two-Φ-Tolerance Competition Graphs, *Discrete Applied Math.* **66** (1996) 101–108.
9. J.E. Cohen, Interval Graphs and Food Webs: A Finding and a Problem, *RAND Corporation Document* 17696-PR, Santa Monica, CA, 1968.
10. J.E. Cohen, Food Webs and the Dimensionality of Trophic Niche Space, *Proc. Nat. Acad. Sci.* **74** (1977) 4533–4536.
11. J.E. Cohen, *Food Webs and Niche Space*, Princeton University Press, Princeton, N.J., 1978.
12. M.S. Daskin, *Network and Discrete Location: Models, Algorithms, and Applications*, Wiley-Interscience, New York, 1995.
13. W.H.E. Day, G.F. Estabrook, F.R. McMorris, Measuring the phylogentic randomness of biological data sets, *Systematic Biology* **47** (1998) 604–616.
14. W.H.E. Day, F.R. McMorris, *Axiomatic consensus theory in group choice and biomathematics.* With a foreword by M. F. Janowitz. Frontiers in Applied Mathematics, 29. Society for Industrial and Applied Mathematics (SIAM), Philadelphia, PA, 2003.
15. G.F. Estabrook, L.R. Landrum, A simple test for the possible simultaneous evolutionary divergence of two amino acid positions, *Taxon.* **24** (1975) 609–613.
16. G.F. Estabrook, C.S. Johnson, F.R. McMorris, An idealized concept of the true cladistic character, *Mathematical BioSciences* **23** (1975) 263–272.
17. G.F. Estabrook, C.S. Johnson, F.R. McMorris, A mathematical foundation for the analysis of cladistic character compatibility. *Mathematical BioSciences* **29** (1976) 181–187.

18. G.F. Estabrook, F.R. McMorris, When are two qualitative taxonomic characters compatible? *J. Mathematical Biology* **4** (1977) 195–200.
19. G.F. Estabrook, F.R. McMorris, When is one estimate of evolutionary relationships a refinement of another? *J. Mathematical Biology* **10** (1980) 367–373.
20. G.F. Estabrook, F.R. McMorris, The compatibility of stratigraphic and comparative constraints on estimates of ancestor-descendant relations, *Systematics and Biodiversity* **4** (2006) 9–17.
21. G.F. Estabrook, Fifty years of character compatibility concepts at work, *J. of Systematics and Evolution* **46** (2008) 109–129.
22. D.P. Foster, R.V. Vohra, An axiomatic characterization of the absolute centre of a tree, unpublished manuscript, 1991.
23. D.P. Foster, R. Vohra, An axiomatic characterization of some locations intrees, *European J. Operational Res.* **90** (1996) 78–84.
24. F. Gavril, The intersection graphs of subtrees in trees are exactly the chordal graphs, *J. Combin. Theory, series B* **16** (1974) 47–56.
25. F. Harary, M.S. Jacobson, M.J. Lipman, and F.R. McMorris, Abstract sphere-of-influence graphs, *Math. Comput. Modelling* **17** (1993) 77–83.
26. F. Harary, J.A. Kabell, and F.R. McMorris, Bipartite intersection graphs, *Comment. Math. Univ. Carolin.* **23** (1982) 739–745.
27. P. Hansen, F.S. Roberts, An impossibility result in axiomatic location theory, *Math. Oper. Res.* **2** (1996) 195–208.
28. R. Holzman, An axiomatic approach to location on networks, *Math. Oper. Res.*, **15** (1990) 553–563.
29. W. Imrich, S. Klavžar, H.M. Mulder, Median graphs and triangle-free graphs, *SIAM J. Discrete Math.* **12** (1999) 111–118.
30. G. Isaak, S.-R. Kim, T. McKee, F.R. McMorris, F.S. Roberts, 2-Competition Graphs, *SIAM J. on Discrete Math.* **5** (1992) 524–538.
31. M.S. Jacobson, On the p-Edge Clique Cover Numbers of Complete Bipartite Graphs, *SIAM J. Discrete Math.* **5** (1992) 539–544.
32. M.S. Jacobson, F.R. McMorris, H.M. Mulder, An introduction to tolerance intersection graphs, in *Graph Theory, Combinatorics, and Applications* (Y. Alavi, G. Chartrand, O. R. Oellermann, and A. J. Schwenk, eds.) Vol. **2** (1991), Wiley, New York, 705–723.
33. M.S. Jacobson, F.R. McMorris, E.R. Scheinerman, General results on tolerance intersection graphs, *J. Graph Theory* **15** (1991) 573–577.
34. S.-R. Kim, The Competition Number and its Variants, in: J. Gimbel, J.W. Kennedy, L.V. Quintas, eds., *Quo Vadis Graph Theory?*, *Annals of Discrete Mathematics*, Vol. **55** 1993, pp. 313–325.
35. S.-R. Kim, T. McKee, F.R. McMorris, F.S. Roberts, F.S., p-Competition Numbers, *Discrete Applied Math.* **46** (1993) 87–92.
36. S.-R. Kim, T. McKee, F.R. McMorris, F.S. Roberts, p-Competition Graphs *Linear Algebra and Applications* **217** (1995) 167–178.
37. J.R. Lundgren,, Food Webs, Competition Graphs, Competition-Common Enemy Graphs, and Niche Graphs, in: F.S. Roberts (ed.), *Applications of Combinatorics and Graph Theory in the Biological and Social Sciences*, Vol. 17 of IMA Volumes in Mathematics and its Applications, Springer-Verlag, New York, 1989, pp. 221-243.
38. T. Margush, F.R. McMorris, Consensus n-trees, *Bulletin of Mathematical Biology* **43** (1981) 239–244.
39. K.O. May, A set of independent necessary and sufficient conditions for simple majority decision, *Econometrica* **20** (1952) 680–684.

40. T.A. McKee, F.R. McMorris, *Topics in Intersection Graph Theory*, Society for Industrial and Applied Mathematics, Philadelphia, 1999.
41. F.R. McMorris, Compatibility criteria for cladistic and qualitative taxonomic characters, in G.F. Estabrook ed. *Proceedings of the Eighth International Conference on Numerical Taxonomy*, Freeman, San Francisco, 1975, pp. 189–230.
42. F.R. McMorris, On the compatibility of binary qualitative taxonomic characters, *Bulletin of Mathematical Biology* **39** (1977) 133–138.
43. F.R. McMorris, H.M. Mulder, R.C. Powers, The median function on median graphs and semilattices, *Discrete Appl. Math.* **101** (2000) 221–230.
44. F.R. McMorris, H.M. Mulder, R.C. Powers, The median function on distributive semilattices, *Discrete Appl. Math.* **127** (2003) 319–324.
45. F.R. McMorris, H.M. Mulder, R.C. Powers, The t-median function on graphs, *Discrete Appl. Math.* **127** (2003) 319–324.
46. F.R. McMorris, H.M. Mulder and F.S. Roberts, The median procedure on median graphs, *Discrete Appl. Math.* **84** (1998) 165 –181.
47. F.R. McMorris, D.A. Neumann, Consensus functions defined on trees, *Mathematical Social Sciences* **4** (1983) 131–136.
48. F.R. McMorris, R.C. Powers, The Arrovian program from weak orders to ternary and quaternary relations, *Dimacs Series in Discrete Mathematics and Theoretical Computer Science* **61** (2003) 37–45.
49. F.R. McMorris, R.C. Powers, Consensus functions on tree quasiorders that satisfy an independence condition, *Mathematical Social Sciences* **48** (2004) 183–192.
50. F.R. McMorris, R.C. Powers, A Characterization of Majority Rule, *Journal of Classification* **25** (2008) 153–158.
51. F.R. McMorris, R.C. Powers, The majority decision function for trees with 3 leaves, *Annals of Operations Research* **163** (2008) 169–175.
52. F.R. McMorris, R.C. Powers, Consensus Rules based on decisive families: the Case of Hierarchies, *Mathematical Social Sciences* **57** (2009) 333–338.
53. F.R. McMorris, F.S. Roberts, C. Wang, The center function on trees, *Networks* **38** (2001) 84–87.
54. F.R. McMorris, C. Wang, P. Zhang, On probe interval graphs, *Discrete Appl. Math.* **88** (1998) 315–324.
55. F.R. McMorris, T.J. Warnow, T. Wimmer, Triangulating vertex colored graphs, *SIAM J. Discrete Math.* **7** (1994) 296–306.
56. F.R. McMorris, T. Zaslavsky, The number of cladistic characters, *Mathematical BioSciences* **54** (1981) 3–10.
57. F.R. McMorris, T. Zaslavsky, Bound graphs of a partially ordered set, *J. Combin. Inform. System Sci.* **7** (1982) 134–138.
58. P.B. Mirchandani, R.L. Francis eds., *Discrete Location Theory*, John Wiley and Sons, New York, 1990.
59. H.M. Mulder, *The interval function of a graph*, Math. Centre Tracts 132, *Math. Centre*, Amsterdam, Netherlands 1980.
60. H.M. Mulder, K.B. Reid, M.J. Pelsmajer, Axiomization of the center function on trees, *Australasian J. Combin.* **41** 223–226.
61. A. Raychaudhuri, F.S. Roberts, Generalized Competition Graphs and their Applications, in: P. Brucker, R. Pauly (eds.), *Methods of Operations Research* **49**, Anton Hain, Konigstein, West Germany, 1985, pp. 295-311.
62. F.S. Roberts, Food Webs, Competition Graphs, and the Boxicity of Ecological Phase Space, in: Y. Alavi and D. Lick (eds.), *Theory and Applications of Graphs*, Springer-Verlag, New York, 1978, pp. 477-490.

63. F.S. Roberts, Competition Graphs and Phylogeny Graphs, in: L. Lovász (ed.), *Graph Theory and Combinatorial Biology,* Bolyai Society Mathematical Studies **7**(1999), *J. Bolyai Mathematical Society*, Budapest, Hungary, 333–362.
64. F.S. Roberts, L. Sheng, Phylogeny Numbers, *Discrete Applied Mathematics* **87** (1998) 213–228.
65. R. Vohra, *An axiomatic characterization of some locations in trees*, European J. Operational Research **90** (1996) 78 – 84.
66. E.O. Wilson, A consistency test for phylogenies based on contemporary species, *Systematic Zoology* **14** (1965) 214–220.

Author Index

Ádám, A., 172, 175
Albatineh, A.N., 25–27, 33, 34, 44
Alpert, R., 38
Arabie, P., 28, 34, 35
Arrow, K.J., 72, 164, 212, 236
Arrows, K.J., 6
Asan, G., 155, 156
Avann, S.P., 94, 106, 108, 111

Bandelt, H.J., 114, 121
Baroni-Urbani, C., 29
Barthélémy, J.P., 28, 81, 110, 120, 211, 226, 237
Batagelj, V., 42
Baulieu, F.B., 26, 28, 37, 42
Beineke, L.W., 187
Benhadda, H., 26
Benini, R., 35, 40, 46
Bennani, M., 36
Bennett, E.M., 38, 43, 45
Bhattacharya, B., 171, 188
Bishop, J., 39
Bloch, D.A., 30, 33
Bock, H.H., 226
Bondy, J.A., 169
Boole, G., 62
Boorman, S.A., 28
Boulter, D., 13, 18
Boyce, R.L., 34
Braun-Blanquet, J., 40, 43, 46
Brešar, B., 112, 113, 122
Bren, M., 42
Brennan, R.L., 34, 39
Brigham, R.C., 226
Bron, C., 18
Bullen, P.S., 40
Buneman, P., 120
Buser, M.W., 29
Byrt, T., 39

Camacho, A.I., 18, 19
Campbell, D.E., 162, 163
Carlin, J.B., 39
Cartmill, M., 16
Castellan, N.J., 31
Cayley, A., 169
Changat, M., 116
Chartrand, G., 169
Chaudhuri, P., 191
Cheetham, A.H., 26
Chepoi, V., 113
Chinn, P.Z., 191
Chung, F.R.K., 117
Cicchetti, D.V., 33, 38
Cikanek, D.G., 174, 182
Cockayne, E.J., 185, 192
Cohen, J.A., 26, 30, 33, 38–40, 43, 45, 46
Cohen, J.E., 229, 231
Cole, L.C., 35, 40
Constantin, J., 110
Crawford, D.J., 17
Cureton, E.E., 45
Czekanowski, J., 34

Dahl, G., 189, 190

Daskin, M.S., 171
Davenport, E.C., 46
Day, W.H.E., 19, 28, 150, 211, 226, 238
De Mast, J., 39
Delson, E., 17
DeMarco, G.A., 17
DePalma, E., 173, 177, 186
Detyniecki, M., 42
Deza, E., 26
Deza, M.M., 26
Dice, L.R., 28, 33, 34, 36, 38–40, 43, 46
Digby, P.G.N., 31, 42, 44
Divgi, D.R., 31
Djokovic, D., 113
Dohmen, K., 64
Donoghue, M.J., 13
Doolittle, 32
Drezner, Z., 171
Driver, H.E., 28, 34, 35, 40, 43, 46
Duarte, J.M., 34

Edwards, A.W.F., 31
El-Sanhurry, N.A., 46
Ellison, P.C., 34
Entringer, R.C., 130, 169
Estabrook, G.F., 2, 4, 5, 8, 10–21, 120, 226, 227
Evans, E., 106

Feinstein, A.R., 33, 38
Femia-Marzo, P., 25, 26
Fichet, B., 36, 43
Field, R., 39
Fishburn, P.C., 162
Fitch, W.M., 13, 18
Fleiss, J.L., 25, 32, 44, 46
Foster, D.P., 171, 235
Fowlkes, E.B., 34, 35
Francis, R.L., 171
Fredrickson, G.N., 146

Galambos, J., 62
Gardner, R.C., 17
Gavril, F., 18
Gerstel, O., 174
Gioan, E., 59
Gleason, H.A., 34
Goddard, W., 29
Goldman, A.J., 102
Goldstein, A.C., 38
Goodman, G.D., 43, 45
Goodman, L.A., 32, 35, 38
Gower, J.C., 26, 28, 36, 42, 43
Graham, N., 130, 169
Graham, R.L., 117
Grupta, B.S., 18
Guggenmoos-Holzmann, I., 30
Guilford, J.P., 39, 45
Guimarães, K.S., 54

Hakimi, S.L., 170, 171, 188
Halpern, J., 182, 183
Hamacher, H.W., 171
Hamann, U., 39
Handler, G.Y., 171
Harary, F., 190, 226, 228, 229
Harutyunyan, H.A., 185
Hazel, 26
Hedetniemi, S.M., 192
Hedetniemi, S.T., 174, 185, 192
Heiser, W.J., 3, 36
Hennig, W., 17
Holley, J.W., 39
Holzman, R., 72, 89, 171
Hsu, M.J., 39
Hu, Y., 171, 188
Hubálek, Z., 26
Hubert, L.J., 34, 35
Hurlbert, S.H., 36

Imrich, W., 113, 122
Isbell, J.R., 106

Jaccard, P., 28, 33, 36, 42, 43
Jackson, D.E., 169
Jacobson, M.S., 226, 228, 229, 232, 233
Janes, C.L., 39
Janowitz, M.F., 11, 42, 81
Janson, S., 25, 26, 28, 30, 33, 34, 37–39
Jantosciak, J., 109
Johnson, C.S., 226
Johnson, H.M., 35
Jordan, C., 76, 77, 102, 127, 128, 130, 169
Jordan, K., 35
Jothi, R., 54

Kabell, J.A., 226, 228
Kang, L., 177
Kantor, P., 207
Keeney, R.L., 206
Kelly, J.S., 162, 163
Klavžar, S., 113, 122
Kraemer, H.C., 25, 30, 32, 33
Krippendorff, K., 26, 35, 39, 44
Kroeber, A.L., 28, 33–35, 40, 43, 46
Kruskal, W.H., 32, 35, 38, 43, 45
Kubicka, E.M., 29, 226
Kubicki, G., 29, 226
Kulczyński, S., 28, 40, 43, 46

Labbé, M., 188
LaDuke, J.C., 17
Landrum, L.R., 13, 14
Laskar, R.C., 192
Legendre, P., 26, 28, 36
Lenzen, M., 145, 146
LeQuesne, W.J., 16–18
Lerbosch, J., 18
Lesot, M.-J., 26–29, 37, 42
Li, A., 34
Liestman, A., 185
Light, R.J., 32, 34
Lipman, M.J., 229
Loevinger, J.A., 35
Lowe, I.J., 171
Luedeman, J.K., 226

Maddison, W.P., 13
Mahadev, N.V.R., 201, 205
Mallows, C.L., 34, 35
Margush, T., 28, 149
Martín Andrés, A., 25, 26
Maskin, E., 162
Maxwell, A.E., 33, 39
May, K.O., 4, 149, 157, 165, 237
Mayr, E., 6, 17
McAdoo, B., 192
McGarvey, D.C., 162
McKee, T.A., 3, 4
McMorris, F.R., 2–6, 10–12, 14–21, 28, 29, 76, 99, 100, 103, 105, 118, 122, 127, 128, 134, 146, 149, 150, 154, 171, 188, 199, 211, 220, 225
Meacham, C.A., 11, 16, 18, 19
Melo, L.C., 34
Mesa, J.A., 171, 188
Michener, C.D., 27, 32, 37, 38
Mihalko, D., 25
Minieka, E., 128, 171, 188
Mirchandani, P.B., 171
Mitchell, S., 173
Mokken, R.J., 35
Molenaar, I.W., 35
Monjardet, B., 150, 153, 154
Monsuur, H., 171
Morgan, C.A., 145, 193
Moulin, H., 164
Mulder, H.M., 3, 4, 73, 78, 79, 84, 105, 106, 108, 111, 118, 122, 146, 171, 228, 233–235
Murty, U.S.R., 169
Myint, Y., 183

Nebeský, L., 94, 106

Nei, M., 34
Neumann, D.A., 150, 154, 237
Nicolai's, F., 58
Nieminen, J., 188, 189
Niewiadomska-Bugaj, M., 25

Ochiai, A., 28, 34, 40, 43
Oellermann, O.R., 169, 187
Omhover, J.F., 42
Ore, O., 128, 169
Ostrand, P., 190

Paul's, C., 59
Pearson, K., 31, 32
Peirce, C.S., 32, 40
Pekeč, A., 201, 205
Pelsmajer, M.J., 4, 146, 171, 193, 194
Peltola, M., 188, 189
Pierce, J.J., 146
Pilliner, A.E.G., 33
Piotrowski, W., 168
Pippert, R.E., 187
Pisani, D., 19
Pisanski, T., 119
Polat, N., 113
Popping, R., 34
Post, W.J., 33, 39
Powers, R.C., 4, 72, 118, 211, 220, 237
Prediger, D.J., 39
Prenowitz, W., 109
Przytycka, T.M., 54
Puerto, J., 171, 188

Quaddus, M., 206
Qui, Y.-L., 17, 19

Raiffa, H., 206
Rand, W.M., 27, 34
Rao, T.R., 29
Ratliff, R.D., 36
Reid, K.B., 4, 128, 134, 171
Rifgi, M., 26
Rifqi, M., 42
Roberts, F.S., 4, 76, 171, 201, 205, 230, 232, 235
Rodríguez-Chía, A.M., 171, 188
Rogers, D.J., 37
Russel, P.F., 29

Saari, D.G., 4, 214
Sabidussi, G., 169, 171
Saks, M.E., 117
Santos, J.B., 34
Sanver, M.R., 155, 156
Scheinerman, E.R., 58, 228, 233
Schmeichel, E.F., 188
Schrijver, A., 94, 108, 111
Scott, W.A., 38, 39, 43, 45
Sen, A.K., 162
Sethuraman, J., 164
Shan, E., 177
Shao, B., 185
Sheng, L., 230
Shi, Q., 171, 188
Shigeno, M., 188
Shioura, A., 188
Sholander, M., 106–108, 111
Sibson, R., 42
Sieberg, K.K., 215, 220
Sijtsma, K., 35
Simonelli's, I., 62
Simpson, G.G., 6, 29, 40, 43, 46
Skelton, P.W., 6
Slater, P.J., 127, 128, 130, 140, 145, 168
Sneath, P.H.A., 13, 18, 26–29, 33, 36, 37, 43
Snijders, T.A.B., 33, 37, 39
Snyder, D.A., 169
Sokal, R.R., 17, 26–29, 32, 33, 36–38, 43
Stearns, R., 162
Stein, W.E., 17
Steinley, D., 25, 33–35

Storcken, T., 171
Strasser, E., 17
Sylvester, J.J., 76, 169
Székely, A., 130
Székely, J.J., 169
Sørenson, T., 34, 36, 38, 40, 43, 46

Tamir, A., 171, 188
Tanimoto, T.T., 37
Tansel, B.C., 171
Thompson, H., 191
Tian, S., 187
Tversky, A., 37

Van de Vel, M., 109
Vegelius, J., 25, 26, 28, 30, 33, 34, 37–39
Vohra, R.V., 3, 78, 105, 118, 122, 171, 234, 235
Voss, N.A., 17
Voss, R.S., 17

Wallace, D.L., 28, 33–35
Wang, C., 76, 171, 226, 229
Ward, B., 162
Warrens, M.J., 3, 25–28, 33–39, 41–44, 46
Wilkeit, E., 114
Wilson, E.O., 5, 227
Wilson, R., 213
Win, Z., 178, 183

Young, H.P., 75
Yule, G.U., 30, 31, 40, 44, 46

Zaks, S., 174
Zangerl, R., 17
Zaslavsky, T., 11, 226, 228
Zelinka, B., 130, 169, 176, 184
Zhang, P., 229
Zotenko, E., 54
Zwick, R., 39
Zysno, P.V., 30, 32, 40, 45

Subject Index

absolute p-center, 170
absolute p-median, 170
absolute q-majority rule, 155
accretion center, 173
admissible transformation, 206
aggregation rule, 155
almost median graph, 114
amalgamation, 101
Amsterdam, 99
ancestor relation, 5, 6, 20
anonymity, 74, 155, 234
Armchair, 100, 102, 112, 121
Arrow's independence condition, 216, 217
Arrow's theorem, 4, 211–213
association coefficient, 33
asymmetric relation, 160
atom, 82

balanced aggregation rule, 158
ball, 177
benzenoid, 119
benzene molecule, 118
benzene ring, 119
betweenness, 75, 76, 234, 235
betweenness relation, 107
betweenness structure, 107
bi-idempotent, 160
biology, 1–3, 33, 53, 59, 93, 120, 149, 226, 227, 229, 231
biometrics, 2, 25
bipartite intersection graph, 228
bipartite interval graph, 228
boat trip, 99
Bonferroni-type inequality, 62
Bonferroni-type probabilistic inequality, 3
Boolean lattice, 93
Borda Count, 222
Borda's rule, 75
bottleneck property, 85
branch weight, 130, 168
branch weight centroid, 128, 130, 168
branch weight of a subset, 130

$\mathcal{CC}$-tree, 58
cancellation, 78
capacity-acceptable, 200
Cartesian product, 96
caterpillar, 133
caterpillar center, 134
cent-dian, 182
center, 4, 76, 128, 130, 168
center function, 3, 76, 234
central path, 128, 130
central vertex, 76
centroid, 4, 128, 130, 168
character, 2, 5, 7, 18
character compatibility analysis, 2, 5
character state, 2, 5, 7
character state tree, 2, 5, 8, 20
chordal graph, 54
chordal graph sieve, 64
classification, 1
clustering, 1, 25, 34, 149
cluster analysis, 34, 44

clique cycle, 60
clique graph, 56
clique path, 60
clique tree, 56
Cochin Conference, 115
cograph, 58
combinatorial probability, 3
competition graph, 4, 229
competition number, 232
complement-reducible graph, 58
complete relation, 160
complete graph of order n, 129
computational biology, 3
concatenation, 75
conflict event structure, 110
conflict graph, 231
conflict model, 109
consensus function, 3, 74, 151, 236, 220
consensus problem, 236
consensus theory, 1, 3, 4, 95, 118, 211, 212, 217, 226, 234, 238
consistency, 75, 234
continuous betweenness, 76
continuous middleness, 75
contraction, 99
contraction map, 99
convex amalgamation, 101
convex character state, 14
convex closure, 79, 95
convex cover, 97
convex expansion, 97
convex set in a graph, 79
convex set in join geometry, 109
convex subgraph, 86, 95
convex subnetwork, 86
convex subset, 95
convex QTC, 18
convexity, 95, 108
copair hypergraph, 108
core, 128, 130, 192
coronation property, 82
counting rule, 150
covering graph, 82, 107, 151
cover, 151
cross-product, 31
cube, 79
cube-free, 79, 105
cube network, 84
cutset coloring, 100
cutting center, 190
cutting number, 190
$\frac{1}{2}$-Condorcet, 80
4-cycle property, 101

(D, G, x)-core, 142
decisive, 163
decisive monotonicity, 152
decisive neutrality, 152
decisive set, 152
degree, 129, 167
descendant, 100
diameter, 129, 168
dictatorial, 164
dictatorial consensus function, 153
dictatorship, 212, 237
discrete betweenness, 75, 235
discrete middleness, 75, 235
discrete structure, 115
distance balance, 176
distance balance center, 176
distance balanced edge center, 186
distance between two sets, 168
distance between two vertices, 129
distance in a graph, 73, 95
distance in a network, 73, 170
distance of a subset, 130
distance of a vertex, 130
Distribution Center Problem, 72, 77
distributive, 150
DNA, 13, 20
dually chordal graph, 57
dummy agents, 164
Dynamic Search Problem, 117

Eastern Parkway, 103
eccentricity, 129, 168
eccentricity of a subset, 130
economics, 1, 3, 93, 100, 121, 226
edge multiset, 55
epidemiology, 2, 25, 32
equilateral triangle, 112
even profile, 74, 95
evolutionary biology, 4, 226
evolutionary systematics, 2, 6
evolutionary theory, 120
evolutionary unit, 18, 20
expansion, 95, 113
Expansion Theorem, 99, 100, 101

$\mathcal{F}$-graph, 60
$\mathcal{F}$-tree, 54
faithfulness, 75
family, 53
federation, 151
federation rule, 152
Fire Station Problem, 72
food web, 230
frequency of compatibility attainment, 19

gate, 95
gated closure, 95
gated set, 95
genealogical tree, 120
generalized centrality, 1, 127
geodesic, 73, 95
graph, 3, 53, 73, 95, 167
grid, 3

Hamming distance, 28
Hamming graph, 96
harmonic center, 192
harmonic weight, 191
Hasse diagram, 7, 93
Helly hypergraph, 94
Helly property, 108
hereditary class of trees, 133
hierarchy, 149
history of a profile, 100
history of a subset, 100
history of a vertex, 100
homology of species structures, 16
Hoover bound, 63
horizontal edge, 83
"host" graph, 54
Hunter–Worsley bound, 61
hypercube, 3, 93

imprint, 113
inclusion-exclusion formula, 3, 61
independence, 212
independence of irrelevant alternatives, 162
induced path function, 116
induced subgraph, 129
interference graph, 230
interior point, 73
intersection class, 58
intersection graph, 3, 4, 53
intersection number, 228
intersection of two graphs, 97
intersection representation, 53
interval, 73, 95
interval scale, 206
invariance, 89
isometric subgraph, 102
isometric, 84, 113

join geometry, 109
join irreducible, 150
join semilattice, 150
join space, 109
join, 107, 150, 189
join-Helly property, 150
join-hull commutativity, 109
join-irreducible, 82

k-ball, 177

k-ball branch weight, 180
k-ball branch weight centroid, 180
k-ball l-path branch weight, 181
k-ball l-path branch weight centroid, 181
k-branch weight, 178
k-branch weight centroid, 178
k-broadcast center, 185
k-broadcast time, 185
k-broadcasting, 185
k-centrum, 180
k-nucleus, 178
k-distance, 177
k-processable, 182
k-processing center, 182
k-processing number, 182
k-tree core, 129
Kakutani separation property, 109
Kounias bound, 62
Kekulé, 119
Kruskal's algorithm, 55, 63

latency, 173
latency center, 174
latent vertices, 120
lattice, 150
leaf branch weight, 175
leaf branch weight centroid, 175
leaf median, 175
length, 73
length of a path, 73
length of a path in a graph, 129, 168
length of a path in a network, 170
length of a profile, 95
length of an edge, 170
lift map, 97
limited transitivity, 162
linear order, 160
Lipschitz axiom, 89
literary history, 3, 120
lobster, 133
location function, 1, 74, 234
location theory, 1, 3, 4
logistics, 1
Louisville, 103

m-cycle, 83
majority consensus, 151
majority rule, 1, 4, 161, 237
Majority Strategy, 103, 117
majorization-equivalent, 189
Maskin monotonicity, 155
maxclique, 54
maximal clique, 17
maximal Helly copair hypergraph, 108
maximum likelihood, 17
maximum pairing, 174
mean, 236
mean function, 3, 89, 235, 236
mean of a profile, 89
mean point, 89
median, 4, 96, 128, 130
median algebra, 106
median betweenness structure, 107
median function, 2, 3, 78, 151, 234
median graph, 3, 73, 78, 95, 151
median location function, 3
median network, 83
median of a graph, 168
median semilattice, 82, 107, 150
median set, 96
median set of a profile, 77
median strategy, 118
median transit function, 110
median-type structure, 3, 93
median vertex, 96, 130
median vertex of a profile, 77
meet, 107, 150, 188
meet semilattice, 150
Metaconjecture, 94, 99, 102, 117, 233
middleness, 75, 235
minimal subtree, 131
minimal vertex separator, 56

minimal vertex weak separator, 56
mismatch, 27
missing link, 120
modular graph, 95, 114
moment, 186

n-cube, 93
n-th power center of gravity, 176
n-th power distance, 176
negative match, 27
negatively transitive relation, 160
network, 73, 170
neutrality, 155
niche overlap graph, 231
node, 54

odd profile, 74, 95
odds ratio, 31
one-vertex graph, 97
opposide of a vertex, 99
optimization, 1
optimal position sequence, 117
order equivalence, 42
order interval, 107
order of a graph, 129

p-center, 145
p-competition graph, 232
p-competition number, 232
p-core, 146
p-intersection graph, 228
p-intersection number, 228
p-median, 145
p-path center, 146
pairing, 174
pairing center, 174
Pareto, 162–164, 212, 236
Pareto condition, 213
parsimony, 16
partial anonymity, 158
partial cube, 81, 113
partial order, 7, 81
partially ordered set, 150
path, 73
path branch weight centroid, 192
path center, 130, 192
path centroid, 130
path median, 130, 192
penalty function, 202
perfect elimination ordering, 65
peripheral expansion, 99
Peripheral Expansion Theorem, 99
peripheral subgraph, 99
phylogeny graph, 4, 230
phylogenetic tree, 2, 7, 10, 16
pit vertex, 145
Plurality Strategy, 117
point, 73
population invariance, 77, 235
positive match, 27
positive responsiveness, 157
poset, 150
potentially compatible QTCs, 14
power mean, 40
probe interval graph, 229
processing center, 174
processing number, 174
processing sequence, 174
processing time, 200
profile, 74, 151, 236
profile on a graph, 95
proper cover, 97
proper k-cover, 111
protein interaction graph, 3, 59, 60
pseudo-median, 114
pseudo-median graph, 114
pseudo-modular graph, 114
psychology, 1–3, 25, 34, 93, 120
Pythagorean means, 39
ϕ-tolerance competition graph, 233

qualitative taxonomic character,
14–21
quasi-consistency, 75, 235

quasi-median, 112
quasi-median expansion, 111
quasi-median graph, 112

R-branch weight centroid, 184
R-branch weight, 184
R-center, 179
R-distance, 184
R-eccentricity, 179
R-median, 184
radius, 168
random character, 19
ratio scale, 206
reachable, 83, 181
reduced network, 88
redundancy, 77, 235
redundant arc, 88, 73
relational statistics, 27
resonance graph, 119
row graph, 231

s-decisive, 152
security center, 172
security centroid, 191
security index, 191
security number, 172
segment, 74
semilattice, 107
semi-median graph, 114
sequential labeling, 173
sequential number, 173
series-parallel, 62
scheduling problem, 4, 173, 182, 199, 207
separable, 202
separation property, 108
side, 99
simple majority rule, 157
simple matching coefficient, 27
simply compatible QTC, 14
site, 110
social welfare function, 160
spanning subgraph, 3, 53
spanning tree, 3, 53
sphere-of-influence graph, 229
spine, 128, 130, 192
split, 99, 110
split half, 80
star, 57
statistics, 3
stratigraphic compatibility, 20
stratigraphic constraint, 20
Strong Metaconjecture, 99, 101, 103, 104
Steiner k-center, 187
Steiner k-distance, 187
Steiner k-eccentricity, 187
Steiner k-median, 187
Steiner distance, 187
strategy-proof, 121
subconsistency, 81
subdivision, 105
subquasi-consistency, 81
summable penalty function, 202
support of a profile, 76
svelte graph, 116
symmetric statistic, 27

$\mathcal{T}$-center, 133, 194
$\mathcal{T}$-median, 145, 194
$\mathcal{T}_D$-center, 128, 131
$\mathcal{T}_D$-centroid, 128, 131
$\mathcal{T}_D$-median, 128, 131
$\mathcal{T}_I$-center, 193, 194
$\mathcal{T}_I$-centroid, 138, 193, 194
$\mathcal{T}_I$-median, 193, 194
t-Condorcet, 81, 82
t-median function, 81, 82, 235
telephone center, 173
ternary algebra, 106
ternary distributive semilattice, 106
tetrachoric correlation, 31
time slot, 200
topological K_4, 62

total distance, 168
total moment, 186
transit axiom, 115
transit function, 110, 115
transitive relation, 160
transversal, 152
tree, 1, 3, 54, 73, 97, 128, 167
tree condition, 237
triangle-free, 105
triangular triple, 112
trivial cover, 97
trivial profile, 96
Type A statistics, 28
Type B statistics, 29
Type C statistics, 30
2×2 table, 1, 2, 27
2-tree, 62

unanimity, 75
underlying graph, 74, 116
underlying graph of a ternary graph, 106
union of two graphs, 97
unique ternary distance graph, 94
Universe of All Graphs, 105
upper bound graph, 228

value resticted, 162
vertex p-center, 170
vertex p-median, 170
voting procedure, 121
voting theory, 1, 3, 4, 121, 226, 236

weak order, 160
weakly median graph, 115
weakly modular graph, 115
weight balance center, 173
weight balance, 173
weight balanced edge center, 186
weighted distance, 170
weighted eccentricity, 170
weight-edge difference, 186
without Condorcet triples, 164

Zaan, 99
zero decisive neutrality, 153
zero Maskin monotonicity, 159

Symbol Index

(A) anonymity axiom, 74, 155, 234
$\mathcal{A}(X)$ set of all asymmetric binary relations on X, 160

(B) betweenness axiom, 75, 234, 235
$B(G)$ (branch weight) centroid of G, 130
$Bw(T)$ centroid of T, 168
$bw(S)$ branch weight of set S, 168
$bw(X)$ branch weight of subset X, 130
$bw(v)$ branch weight of v, 130, 168

(C) consistency axiom, 75, 234
$C(G)$ center of G, 130, 168
(Ca) cancellation axiom, 78
Cen center function, 76, 234
$Cen(\pi)$ center of π, 76
$C_{\mathcal{F}}$ federation rule, 152
Cf frequency of compatibility attainment, 19
$Con[W]$ convex closure of W, 79, 95
CST character state tree, 8
C_q consensus rule, 151
C_{q^*} majority consensus rule, 151

DCP Distribution Center Problem, 72
DM decisive monotonicity axiom, 152
DN decisive neutrality rule, 152
$D(S)$ sum of distances of vertices of G to set S, 130, 168
$D(x)$ sum of distances of x to all other vertices, 96, 130, 168
$D(x,\pi)$ distance of x to profile π in a graph, 77, 96
$D(x,\pi)$ distance of x to profile π in a network, 77
$d(X,Y)$ distance between sets X and Y, 130, 168
$dia(G)$ diameter of G, 129, 168
$d_G(u,v)$ distance between u and v in G, 73, 95, 168
$\delta(p,q)$ distance in network, 73

$d(u, v)$	distance between u and v, 73, 95, 129, 168
$d(v)$	degree of a vertex, 129, 167
$E(G)$	edge set of graph G, 167
$\mathcal{E}(T)$	edge multiset, 56
$e(X)$	eccentricity of subset X, 130
EU	evolutionary unit, 14
$e(S)$	eccentricity of S, 168
$e(v)$	eccentricity of v, 129, 168
$ext(S)$	extension of S, 95
(F)	faithfulness axiom, 75
$\mathcal{F} = \{S_1, \ldots, S_n\}$	family or multiset of subsets, 53
$\mathcal{F}$	subset of powers set of N, 151
F_{12}	set of edges between sides of split G_1, G_2, 98
FSP	Fire Station Problem, 72
G	graph, 73, 95, 129, 167
$G[W]$	subgraph of G induced by W, 129, 167
$G = (V, E)$	simple, loopless graph with vertex set V and edge set E, 73, 95, 12
G^k	Cartesian product of k copies of G, 96
G_{0i}	subgraph induced by the ends of F_{12} in G_i, 98
G_1, G_2	split, 99
$G_1 \Box G_2$	Cartesian product of graphs G_1 and G_2, 96
$G_1 \cap G_2$	intersection of graphs G_1 and G_2, 97
$G_1 \cup G_2$	union of graph, G_1 and G_2, 97
$G_1 - G_2$	subgraph induced by vertices in G_1 not in G_2, 97
$G\langle W \rangle$	smallest connected subgraph of G containing W, 129, 168
G_u^{uv}	subgraph of vertices closer to u than to v, 80
$G - X$	subgraph of G induced by vertices not in X, 130, 168
$G - x$	vertex deleted subgraph, 129
$\mathcal{H}(S)$	set of all hierarchies on S, 154
IIA	independence of irrelevant alternatives axiom, 162
$I(u, v)$	interval between u and v, 73, 95
$I(u, v, w)$	intersection of the intervals $I(u, v), I(v, w)$ and $I(w, u)$, 95
$I_G(u, v)$	interval between u and v in G, 73, 95
(Inv)	invariance axiom, 89
$J(u, v)$	induced path interval, 116

(K) convexity axiom, 80
κ contraction map, 99
$k(G)$ competition number, 232
$K_{m,n}$ complete bipartite graph, 168
K_n complete graph on n vertices, 129, 168
$k_p(G)$ p-competition number, 232
K_1 one-vertex graph, 97
$K_{1,k}$ star, 57
$K_4 - e$ complete graph on four vertices minus an edge, 112

$Lf(T)$ set of leaves in tree T, 167
LT limited transitivity axiom, 162
(Li) Lipschitz axiom, 89
$\mathcal{L}(X)$ set of all linear orders on X, 160
$\lambda(uv)$ length of arc uv, 73
λ_i lift map, 97
$l(P)$ length of path P in a graph, 129, 168
$l(P)$ lenght of path P in network, 170
$l(e)$ length of e in network, 170

M median function, 151
$Mean$ mean function, 89
$Mean(\pi)$ mean set of profile π, 89
Med median function, 78, 151, 234
$Med(\pi)$ median set of π, 78
$M(G)$ median set of graph G, 96, 130, 168
(Mid) middleness axiom, 75, 235
MM Maskin monotonicity axiom, 155
MN monotonic neutrality axiom, 152
$M(\pi)$ median set of profile π, 96
M_t t-median function, 81
$M_t(\pi)$ t-median set of profile π, 81

(N) neutrality axiom, 155
$N = (G, \lambda)$ network on graph G with length function λ, 73
$N_s(\pi)$ indices of the profile elements above s, 151
$n_+(R)$ set of indices for which $R_i = 1$, 155
$n_-(R)$ set of indices for which $R_i = -1$, 155
$n_0(R)$ set of indices for which $R_i = 0$, 155

$\Omega^w(\mathcal{F})$ complete graph on $\mathcal{F}$ with non-negative weight function w, 54

(PA)	partial anonymity axiom, 158
(PI)	population invariance axiom, 77, 235
(PR)	positive responsiveness axiom, 157
$\{\pi\}$	support of π, 76
π	profile, 95, 151
$\lvert\pi\rvert$	length of profile, 95
$\pi \setminus x$	all occurrences of x deleted from π, 77
π_i	subprofile of π in G_i, 103
π_u^{uv}	subprofile of elements closer to u than to v, 80
$\pi - x_i$	vertex-deleted profile, 79
(QC)	quasi-consistency axiom, 75, 235
(SQC)	subquasi-consistency axiom, 81
QTC	qualitative taxonomic character, 14
Q_n	n-cube, n-dimensional hypercube, 93
Q_3	cube, 3-cube, 79
(R)	redundancy axiom, 77, 235
$r(G)$	radius of G, 168
(SC)	subconsistency axiom, 81
$S(p,q)$	segment in network, 74
T	tree, 73, 129, 167
$T = (V, E)$	tree with vertex set V and edge set E, 73
$\mathcal{T}$	hereditary class of trees, 133
$\mathcal{T}_D$	set of all subtrees of maximum degree D, 131
T_v	subgraph of T induced by all nodes of T containing v, 54
(t-Co)	t-condorcet axiom, 81
(t-Co)	t-condorcet axiom, ordered, 82
(U)	unanimity axiom, 75
(V, E, λ)	network with vertex set V and edge set E and length function λ, 73
$V(G)$	vertex set of a graph G, 54, 167
$V(H)$	vertex set of subgraph H, 129
$V(x,y)$	subset of vertices closer to x than to y, 168
$\langle W\rangle_G$	subgraph of G induced by W, 83
$\langle W\rangle_N$	subnetwork of N induced by W, 83
$\mathcal{W}(X)$	set of all weak orders on X, 160

W_{uv} set of vertices closer to u than to v, 110
W_u^{uv} set of vertices closer to u than to v, 86
$w(v)$ weight of v, 170

X_k vertex set of unique minimal subtree with eccentricity at most k, 132
$x \wedge y$ meet of x and y, 150
$x \vee y$ join of x and y, 150

0DN zero decisive neutrality axiom, 153
(0MM) zero maskin monotonicity axiom, 159
($\frac{1}{2}$-Co) $\frac{1}{2}$-Condorcet axiom, 80
2^N power set of N, 151

$\leq$ partial order, 150
#smpl($\mathcal{F}$) number of unique members of $\mathcal{F}$, 63
$\bigvee A$ join of subset A, 150
$\bigwedge A$ meet of subset A, 150